Production, Refining, Fabrication and Recycling of Light Metals

Titles of Related Interest—

Ashby ENGINEERING MATERIALS 1
Ashby ENGINEERING MATERIALS 2
Brook IMPACT OF NON-DESTRUCTIVE TESTING
Koppel AUTOMATION IN MINING, MINERAL AND METAL PROCESSING 1989
Ruhle METAL-CERAMIC INTERFACES
Taya METAL MATRIX COMPOSITES

Other CIM Proceedings Published by Pergamon

Bergman FERROUS AND NON-FERROUS ALLOY PROCESSES
Bickert REDUCTION AND CASTING OF ALUMINUM
Chalkley TAILING AND EFFLUENT MANAGEMENT
Closset PRODUCTION AND ELECTROLYSIS OF LIGHT METALS
Dobby PROCESSING OF COMPLEX ORES
Embury HIGH TEMPERATURE OXIDATION AND SULPHIDATION PROCESSES
Jaeck PRIMARY AND SECONDARY LEAD PROCESSING
Jonas DIRECT ROLLING AND HOT CHARGING OF STRAND CAST BILLETS
Kachaniwsky IMPACT OF OXYGEN ON THE PRODUCTIVITY OF NON-FERROUS METALLURGICAL PROCESSES
Lait F. WEINBERG INTERNATIONAL SYMPOSIUM ON SOLIDIFICATION PROCESSING
Macmillan QUALITY AND PROCESS CONTROL IN REDUCTION AND CASTING OF ALUMINUM AND OTHER LIGHT METALS
Mostaghaci PROCESSING OF CERAMIC AND METAL MATRIX COMPOSITES
Plumpton PRODUCTION AND PROCESSING OF FINE PARTICLES
Purdy FUNDAMENTALS AND APPLICATIONS OF TERNARY DIFFUSION
Rigaud ADVANCES IN REFRACTORIES FOR THE METALLURGICAL INDUSTRIES
Ruddle ACCELERATED COOLING OF ROLLED STEEL
Salter GOLD METALLURGY
Thompson COMPUTER SOFTWARE IN CHEMICAL AND EXTRACTIVE METALLURGY
Twigge-Molecey MATERIALS HANDLING IN PYROMETALLURGY
Twigge-Molecey PROCESS GAS HANDLING AND CLEANING
Tyson FRACTURE MECHANICS
Wilkinson ADVANCED STRUCTURAL MATERIALS

Related Journals

(Free sample copies available upon request)

ACTA METALLURGICA
CANADIAN METALLURGICAL QUARTERLY
MATERIALS RESEARCH BULLETIN
MINERALS ENGINEERING
SCRIPTA METALLURGICA

PROCEEDINGS OF THE INTERNATIONAL SYMPOSIUM ON PRODUCTION, REFINING, FABRICATION AND RECYCLING OF LIGHT METALS
HAMILTON, ONTARIO, AUGUST 26-30, 1990

Production, Refining, Fabrication and Recycling of Light Metals

Editors

Michel Bouchard
Université du Québec à Chicoutimi
Chicoutimi, Québec

Pierre Tremblay
Alcan International Ltée,
Jonquière, Québec

Symposium organized by the Light Metals Section of
The Metallurgical Society of CIM

29th ANNUAL CONFERENCE OF METALLURGISTS OF CIM
29e CONFÉRENCE ANNUELLE DES MÉTALLURGISTES DE L'ICM

Pergamon Press
Member of Maxwell Macmillan Pergamon Publishing Corporation
New York Oxford Beijing Frankfurt São Paulo Sydney Tokyo Toronto

Pergamon Press Offices:

U.S.A.	Pergamon Press, Inc., Maxwell House, Fairview Park, Elmsford, New York 10523, U.S.A.
U.K.	Pergamon Press plc, Headington Hill Hall, Oxford OX3 0BW, England
PEOPLE'S REPUBLIC OF CHINA	Pergamon Press, 0909 China World Tower, No. 1 Jian Guo Men Wai Avenue, Beijing 1000004, People's Republic of China
FEDERAL REPUBLIC OF GERMANY	Pergamon Press GmbH, Hammerweg 6, D-6242 Kronberg, Federal Republic of Germany
BRAZIL	Pergamon Editora Ltda, Rua Eça de Queiros, 346 CEP 04011, Paraiso, São Paulo, Brazil
AUSTRALIA	Pergamon Press Australia Pty Ltd., P.O. Box 544, Potts Point, NSW 2011, Australia
JAPAN	Pergamon Press, 8th Floor, Matsuoka Central Building, 1-7-1 Nishishinjuku, Shinjuku-ku, Tokyo 160, Japan
CANADA	Pergamon Press Canada Ltd., Suite 271, 253 College Street, Toronto, Ontario M5T 1R5 Canada

Library of Congress Cataloging in Publication Data

ISBN 0-08-040416-2

Printing: 1 2 3 4 5 6 7 8 9 Year: 0 1 2 3 4 5 6 7 8 9

Printed in the United States of America

The paper used in this publication meets the minimum requirements of American National Standard for Information Sciences-Permanence of Paper for Printed Library Materials, ANSI Z 39.48-1984

Foreword

In conjunction with the 29th Annual Conference of Mertallurgists of CIM, The Light Metals Section of The Metallurgical Society of CIM, has organized an International Symposium of unprecedented scope, on the Production, Refining, Fabrication and Recycling of Light Metals.

The symposium features over thirty papers from engineers and scientists of seven countries, reporting recent developments in the production, refining, fabrication and recycling of aluminum and magnesium.

Technical papers are being presented by all the primary aluminum and magnesium producers in Canada. These contributions are complemented by exceptional papers from members of the Canadian and international academia and from other companies involved in the light metals industries, in Canada and abroad.

For the first time, the symposium will formally broach on subjects fast becoming hallmarks of our times. The session on composites comprises several papers on this developing high-potential sector of the aluminum industry. Also, much to the credit of the light metals industries, sesssions on environmental protection and recycling will be presented.

The success of the Symposium is due to the engineers and scientists reporting their findings. We wish to thank them for taking precious time to document their work appropriately and making the publication of these proceedings possible. It is only fitting that we dedicate this book to all of them.

Michel Bouchard
Pierre Tremblay
Editors

TABLE OF CONTENTS

LIGHT METAL MATRIX COMPOSITES

MELTING, ALLOYING AND TREATMENT OF LIGHT METALS

ALUMINUM AND MAGNESIUM RECYCLING

FABRICATION AND APPLICATIONS OF LIGHT METALS

Aluminum and magnesium smelting

Selection of monolithic castables for cathode barriers

D.V. Stewart and A.T. Tabereaux
Reynolds Metals Company, Manufacturing Technology Laboratory, Extractive Metallurgy Department, P.O. Box 1200, Sheffield, Alabama 35660, U.S.A.

Abstract

Bath penetration into the cathode begins immediately after the start-up of an alumina reduction cell and is particularly high during the first weeks of operation. Later on, bath penetration is considerably reduced but continues throughout the life of the cathode. The extent of the bath penetration and its composition will depend upon the cell design and lining materials used. In this work, commercially available, low-water castable refractories were evaluated and compared in laboratory tests as a possible monolithic barrier material in cathodes for protecting the low thermal conductivity insulating materials. A monolithic castable barrier would eliminate the inherent weakness of the mortar seams present in the commonly-used brick barriers in protecting against penetration and degradation by cryolitic salts, bath, and aluminum metal. Installation and effectiveness of low-water castable barriers in plant trials are discussed.

Keywords

Cathode refractories, castable, monolithic; cryolite resistance and physical and chemical properties.

Introduction

In the production of primary aluminum, one of the major costs is the electrical power requirement. It is the desire of aluminum producers to reduce the electrical energy consumption per unit of metal produced by various methods and to save thermal energy by improving the thermal insulation in the cathodes.

Corrosive cryolitic bath components, and even aluminum metal, penetrate and saturate the cathode lining, destroying cathode refractories and, thus, degrading the lower cathode thermal insulation during the reduction cell's lifetime.

Layers of materials are used in the cathode to control the rate that cryolite constituents penetrate the bottom cathode refractories to prevent this diffusion process from happening too rapidly. Materials which have been used as physical and chemical barriers in industrial cells include: super duty fireclay and chamotte bricks, ceramic tiles, metal plates, graphite foil, glass, and various chemical barriers. In an effort to address the problem of selecting refractories for application as the cathode barrier, an investigation was undertaken to evaluate commercially available, low-water castable refractories.

FIGURE 1.

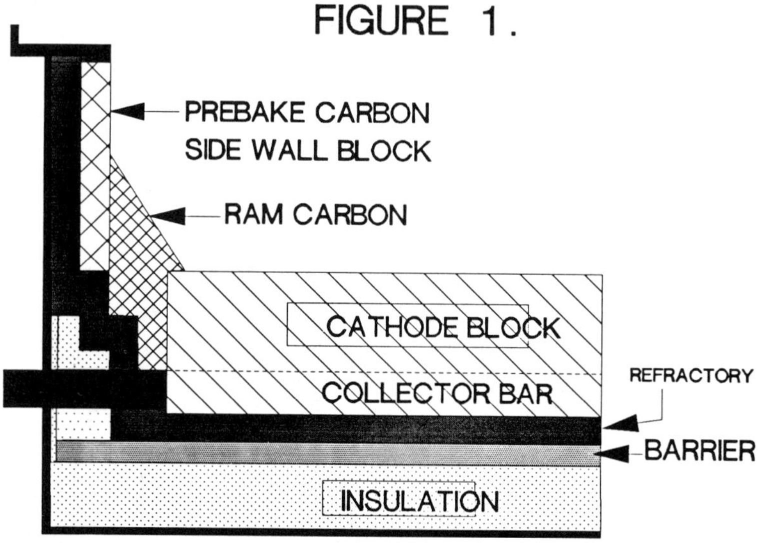

REDUCTION CELL CATHODE
CROSS SECTION VIEW

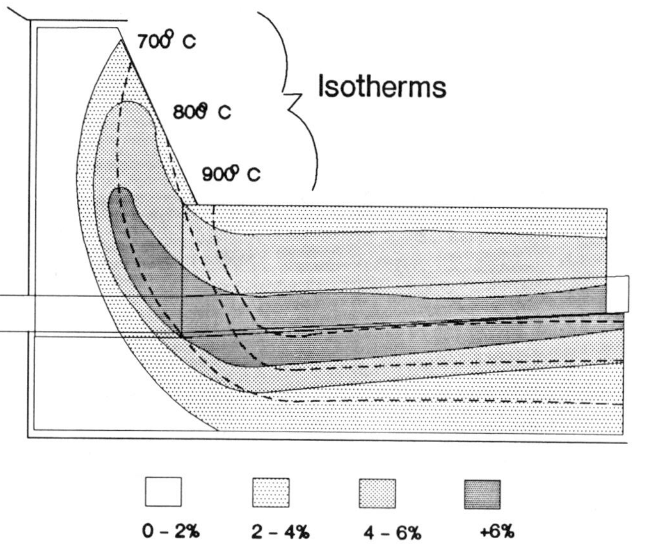

TEMPERATURE ISOTHERMS & CRYOLITIC SALT
CONTENT OF ALUMINA INSULATED CATHODE

CRITERIA FOR SELECTING REFRACTORIES
FOR CATHODE BARRIERS

1. Maximum Service Temperature
   - Cathode operating temperatures
   - Freeze plane for cryolitic bath components

2. Resistance To Attack By Cryolitic Bath Components and Aluminum at Service Temperature
   - Density and permeability (porosity & pore size distribution)
   - Chemical composition

3. Structure Strength, or Load Bearing Capability
   - Strength at operating temperatures
   - Cold crushing strength

4. Volume Stability
   - Thermal expansion
   - Chemical reation & expansion

5. Insulation
   - Thermal conductivity

6. Installation Limitations
   - Sizing tolerances
   - Equipment necessary
   - Curing and drying considerations

7. Material Availability

8. Labor Availability
   - o Training, experience, supervision

9. Cost considerations

Materials received by the user for cathode construction should be evaluated to ensure that they are within statistical control of established refractory specifications.

Maximum Refractory Service Temperature

- Refractories should have sufficient structural and chemical stability over the entire cathode temperature range during cell startup and operations.

- Refractory layer should not lose strength, disintegrate, change weight nor dimension during cell startup and operation.

- Refractory's softening or solidus temperature should not be reached, even if penetrated by cryolitic bath components.

REFRACTORY	MAXIMUM SERVICE (°F)	RECOMMENDED TEMPERATURE (°C)
Hoganus Densecast 50	3236	1780
National Refractory Kricon 32	3200	1760
Carborundum ND-84	3000	1650
Plibrico 3100 Special	3100	1700
Clayburn Kilcast 46	2800	1538
National Refractory Kricon 28	2800	1538
North American Refractories Refracrete 23	2300	1260
AP Green SM Brick	3169	1743
Clayburn Kilgard Brick	2772	1522
Pamas PMS Brick	2650	1450

COMMENTS:

- The maximum recommended service temperatures of the castables are equal to that of the three fireclay bricks used in cathodes.
- The maximum recommended service temperatures of all castables and bricks are much higher than the 900 to 1000°C temperature isotherms experienced in cathodes during startup and operations.

CATHODE OPERATING TEMPERATURES/ISOTHERMS

- It is advisable, but not always possible, to keep the main body of refractory lining below the freezing isotherms of most bath components.
- The ideal location for the castable barrier layer is between a top layer of protective refractory (bricks or vibrated powdered alumina) and bottom layer of insulation.
- This position may provide a sufficient temperature gradient, (about 750 to 850°C) to prevent contact with the low temperature eutectic corrosive cryolitic constituents for some period of time.

- The absolute worst situation for a castable barrier would be to locate it directly below the cathode blocks.
- The higher temperature gradient (900°C) and almost certain direct contact with corrosive bath constituents diffusing through the carbon cathode blocks would attack the monolithic castable refractory rapidly and increase chemical reaction kinetics.
- It should be noted that the temperature gradient on the hot face side of the castable barrier increases insulation in the layer of low density, low thermal conductivity insulation below the barrier material.

2. Density/permeability

2a. Density and apparent porosity

- Generally, as the bulk density of a refractory increases the apparent porosity decreases.
- Porosity is an indication of how permeable the refractory is to vapor.
- Less porosity means less area in the pore structure for corrosion attack to start and proceed in refractory.
- Refractory is less permeable if it has a lower porosity and finer pore size distribution.
- The more continuous pore phase generally leads to a greater permeability.
- Reduction of the permeability in refractory relates to the ability of the refractory to prevent penetration of corrosive materials (gases and/or liquids) that leads to chemical change in the material and/or redistribution of the material's chemical and physical makeup which alters the structure, and may lead to expansive growth.
- However, too high density with very low porosity in a refractory may result in poor refractory thermal shock properties.

REFRACTORY	WATER ADDED, (%)	BULK DENSITY, (g/cm^3)	APPARENT POROSITY, (%)
ND-84	5	2.68	16.5
Kricon 32	6	2.51	16.5
Kilcast 46	7	2.36	15.2
Kricon 28	5	2.35	15.8
3100 Special	9	2.29	20.4
Densecast 50	9	2.21	21.8
Refracrete 23	12	2.02	28.0
AP Green SM Brick	-	2.30	10.0
Kilgard Brick	-	2.21	14.6
Pamas PMS Brick	-	2.20	14.0

COMMENTS:

- Higher density and lower porosity are related to the amount of mix water used in the castables. By using the minimum amount of water there will less void space after curing.

- By comparison, the commercial fireclay bricks had the lowest density, 2.2 to 2.3 g/cm³, and the lowest porosity-due to their higher fired and pressed state.

2b. Pore size distribution

- Castables usually have a finer pore size than bricks, with the pore size increasing as the firing temperature increases.

- Measured porosity and pore structure in the castables are associated with the porosity connected to the surface of the material.

REFRACTORY	APPARENT POROSITY (%)	PORE SIZE DIAMETER (PORE VOLUME, % >SHOWN MICRONS) 50	40	30	20	10	5	1
Densecast 50	21.8	2	3	4	6	10	15	17
3100 Special	20.4	13	17	18	19	28	30	35
Kricon 32	16.5	16	17	19	22	26	29	35
ND-84	16.5	13	14	14	15	20	28	38
Kilgard Brick	14.6	11	11	13	15	21	43	62

COMMENTS:

- Desecast 50 had the finer porosity structure of the castables measured, and started out with a higher mix water, 9%.

- Kilgard brick had a coarser pore structure with fewer percentage of pores in the range <1 to 5 microns than the castables.

2c. Chemical composition

- Generally, a higher alumina content provides more resistance to cryolite attack in the cathode, and correspondingly a higher silica content is more prone to attack by cryolite.

REFRACTORY	MAJOR CHEMICAL COMPONENTS Al_2O_3 (%)	SiO_2 (%)	CaO (%)	Fe_2O_3 (%)	TiO_2 (%)
3100 Special	(69.3	24.1	1.5	0.8	1.3)
ND 84	61.1	33.3	0.5	1.1	0.7
Kricon 32	59.2	33.3	1.2	1.1	1.7
Densecast 50	51.4	41.2	--	1.4	1.0
Kilcast 46	(45.9	48.5	1.1	2.0	1.4)
Kricon 28	(45.6	47.7	1.5	1.3	1.8)
Refracrete 23	(32.6	49.4	14.3	0.9	1.6)

REFRACTORY	MAJOR CHEMICAL COMPONENTS				
	Al_2O_3 (%)	SiO_2 (%)	CaO (%)	Fe_2O_3 (%)	TiO_2 (%)
AP Green SM Brick	(38.5	56.5	0.4	1.8	1.5)
Kilgard Brick	25.7	62.1	0.3	8.3	0.4
Pamas PMS Brick	(31.3	60.9	0.1	2.4	1.6)

The data in parenthesis indicates manufacturer's data.

COMMENTS:

- The alumina content in the castables had a range from 32 to 69%, and the corresponding silica content had a range from 24 to 49%.
- The chemical composition of the commercial brick differed greatly with significantly lower alumina content from 26 to 38%, and higher silica content from 56 to 62%.
- A higher service temperature can be related to chemical composition; for example, Kricon 32, ND-84 and 3100 Special have a higher alumina content, (≥60%) and higher service temperatures, ≥3,000°F.
- Refracrete 23 had a high calcium content, 14%, due to a high cement content, which is usually associated with a lower refractoriness.

2d. <u>Cryolite resistance</u>

<u>Cryolitic Cup Test</u>

- High ratio bath composition (excess sodium fluoride).
  - simulates cryolitic compounds found in saturated cathodes.
  - aggressively attacks refractories.
- Test temperatures
  - established at a point where you get significant amount of attack in the majority of refractories.
- Measurement of reactivity
  - samples are cut in half.
  - area of cryolite penetration and degradation in the samples is measured and compared.
- The advantages of the cryolitic cup test are that it:
  - provides fast results with a simple test procedure.
  - provides a relative comparison of reactivity of refractories with cryolite.
- The disadvantages of the cryolite cup test are that:
  - samples are not at cathode charge potential.
  - reaction mechanism may differ from that actually experienced in cathodes.
  - different bath compositions are used by different laboratories.
- The cryolite cup test can be conducted in a carbon crucible, cathodically charged, in a molten cryolitic bath that allows

the cryolitic bath components to penetrate the refractory sample.

- The advantage of this test procedure is that refractory samples are more related to actual cathode conditions.

- The disadvantage of this procedure is that it involves a more complex experimental test setup and conditions.

COMMENTS:

- The cryolite resistance of low-thermal conductivity insulation has been shown to be related to density/porosity properties in cup tests.

- Cryolite cup tests may give misleading results if the attack mechanism in some processes is due to corrosive vapor:
    - cup test is by a liquid phase in a refractory structure, and
    - requires a test that simulates attack along a thermal gradient and pressure gradient.

REFRACTORY	CRYOLITE RESISTANCE AT 950°C (48-288 hours) REACTED AREAS, (in^2)	AVERAGE RELATIVE DIFFERENCE, (%)*
Densecast 50	0.33	-45
Refracrete 23	0.38	-37
ND 84	0.42	-30
3100 Special	0.43	-28
Kricon 32	0.49	-18
Kilcast 46	0.53	-12
Kricon 28	0.63	+ 5
Pamas SM Brick	0.57	
Kilgard Brick	0.61	
AP Green SM Brick	0.63	
Vermiculite Slab	2.73	+ 355

*Compared with the average brick reacted area, 0.60 in^2.

COMMENTS:

- Refractories which have a lower infiltration/reacted area demonstrate the better resistance to cryolitic bath attack in the cup tests.
- An important conclusion drawn from this data is that all of the castables demonstrated relatively good resistance to

cryolite components.

- All of the castables, except one, demonstrated better resistance to cryolite than the fire clay bricks commonly used in cathode construction.

- When cryolitic bath components contact low density, low thermal insulation-there is considerable attack, and degradation of the material in a short period of time, as evident by the vermiculite sample, + 355% compared with the bricks.

<u>Cryolite Resistance and Bulk Density/Porosity Relationship for Castables</u>

- Surprisingly, Densecast 50, Refracrete 23, and Plibrico 3100 castables demonstrated the least amount of attack by cryolite in the above comparison data, but had the lowest bulk density and the highest apparent porosity.

- Tests conducted with individual castables demonstrated an increased infiltration and attack by cryolite with increased mix water content, demonstrating that the cryolite resistance of castable refractories is also related to density and porosity.

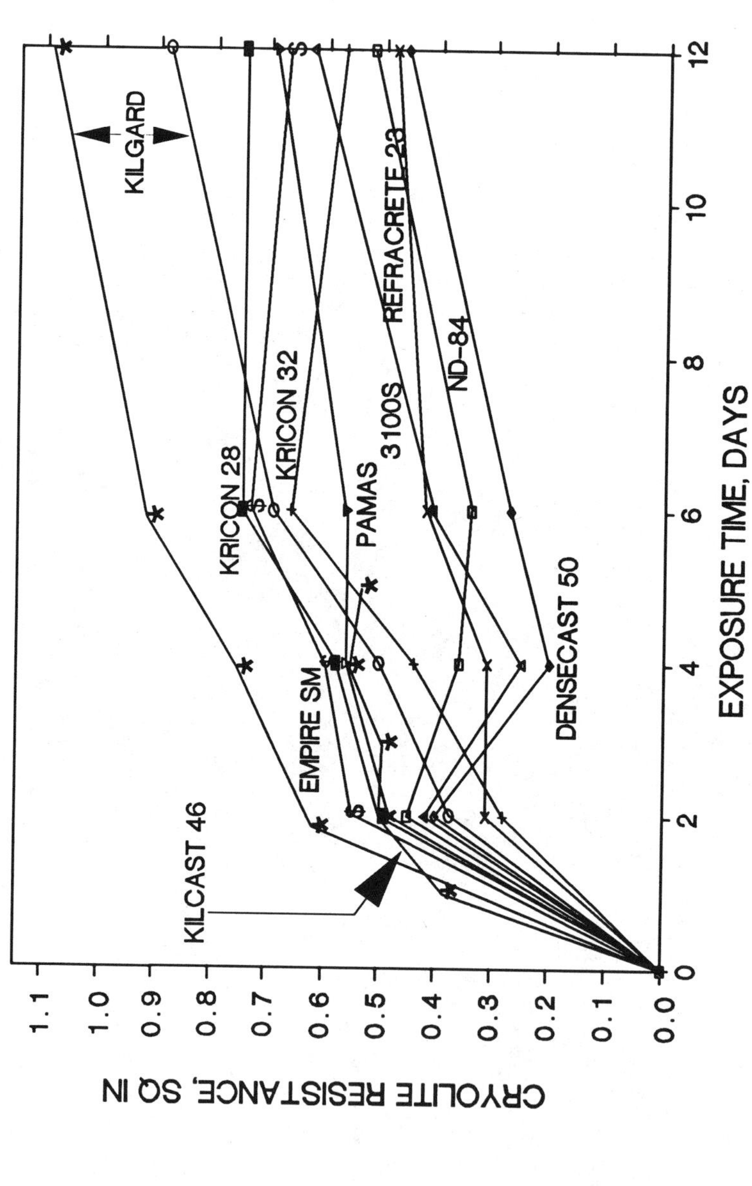

FIGURE 2. CRYOLITE RESISTANCE OF BARRIER MATERIALS

3. Structure Strength-Load Bearing Capability

MODULUS OF RUPTURE AND COLD CRUSHING STRENGTH

REFRACTORY	MODULUS OF RUPTURE @900°C (lb/in.²)	@200°C (lb/in.²)	COLD CRUSHING STRENGTH (lb/in.²)
Densecast 50	(2185)	(2200)	(11380)
3100 Special	2180	--	--
Kilcast 46	1850	(1850)	(6500)
Kricon 28	(1400)	(1900)	(7450)
ND 84	(1235)	(655)	--
Refracrete 23	(1200)	(750)	(2700)
Kricon 32	1170	(1400)	(7600)
Kilgard Brick	4490	(3250)	(11000)
AP Green SM Brick	2170	(1200)	(4000)
Pamas PMS Brick	1730	(1641)	(4922)

The data in parenthesis indicates manufacturer's data.

COMMENTS:

- Hot strengths (HMOR) of all of the castables are lower than the Kilgard brick.
- Cold crushing strengths of the castable refractories are almost equivalent for the other commercial fire bricks.
- Hot strength of the mortar joints between bricks is considerably less than the castables, for example for commercial silicon carbide and alumina mortars:

REFRACTORY	MORTAR LAP-JOINT STRENGTH @900°C (lb/in.²)
SK 80 CC SiC Mortar	670
Super Duty Mortar	500
High Alumina Mortar	350
Carbofrax No. 4 SiC Mortar	246

4. Volume Stability

COEFFICIENT OF THERMAL EXPANSION

- Consideration of thermal expansion must be made in the design and selection of cathode materials to allow for the eventual expansion of refractories at operating temperatures.

Refractory	THERMAL EXPANSION AT 900°C CTE x 10^{-6} (°C)	100 in (in)	300 in (in)	TOTAL LINEAR CHANGE, (%)
Kilcast 46				(- 0.2)
3100 Special	2.6	0.2	0.4	--
Densecast 50	4.4	0.4	1.2	(+ 0.7)
ND 84	7.9	0.7	2.1	(+ 1.0)
Kricon 32	9.5	0.9	2.6	(+ 0.9)
Kricon 28				(+ 0.5)
Refracrete 23				(- 2.0)
Kilgard Brick	5.6	0.5	1.5	
AP Green SM Brick				
Pamas PSM Brick				

The data in parenthesis indicates manufacturer's data.

COMMENTS:

- Densecast 50 and 3100 Special castables have lower thermal expansion values, 4.4 - 2.2 x 10^{-6} per °C:
    - compared to brick, would need less lateral expansion.

- ND-84 and Kricon 32 castables have the highest thermal expansion, 7.9 - 9.5 x 10^{-6} per °C compared with brick, 5.6 x 10^{-6} per °C:
    - would need an additional 0.5 to 1.0 inch of lateral expansion of material in a reduction cell cathode.

5. Insulation

THERMAL CONDUCTIVITY

- The thermal conductivity of the castable is not as critical as some of the other information as castable barrier is not used primarily for insulation purposes.

- Thermal conductivity data is collected for the castables to provide thermal information for modeling cathode heat losses.

REFRACTORY	THERMAL CONDUCTIVITY AT 600°C (W/m°C)		RELATIVE DIFFERENCE, (%)*
Densecast 50	0.9	(0.9)	-40.0
ND 84	1.4	-	- 6.7
Kricon 32	1.5	-	0.0
3100 Special	1.6	-	+ 6.7
Kricon 28		-	
Kilcast 46		-	
Refracrete 23		-	
Kilgard Brick	1.5	-	
AP Green SM Brick		-	
Pamas PSM Brick		-	

*Compared with the brick sample.
The data in parenthesis indicates manufacturer's data.

COMMENTS:

- Densecast 50 castable had the lowest thermal conductivity, 0.9 W/m°C, compared with the other castables, because it had:
  - lower density, and
  - higher porosity.

- The thermal conductivities of the other castable refractories are about the same, ±7%, compared with commercial brick commonly used in reduction cell cathodes.

6. Installation limitations

6a. A water requirement for mixing

- The amount of mix water required to prepare a castable is important due to:
  - its direct impact on the physical properties, for example: density and porosity, which may then alter the castable's chemical reactivity.

EFFECT OF CASTING WATER ON PHYSICAL PROPERTIES OF CASTABLE

CASTING WATER CONTENT:	4%	5%	6%
Casting Comments:	Flow ok	Flow ok	Too wet
Bulk Density, lb/ft.³	148.2	146.8	146.0
Modulus of Rupture, lb/in.²	1515	1453	1247
Cold Crushing, lb/in.²	10206	7782	5725

Increased Mix Water Content In The Castable Results In:

- o Decreased Bulk Density
- o Decreased Modulus of Rupture
- o Decreased Cold Crushing Strength

6b. Curing and Drying Considerations

REFRACTORY	MIXING WATER (weight %)	LOSS AT 500°C (weight %)	WEIGHT DIFFERENCE (%)
Kricon 32	6	7.1	1.1
Densecast 50	9	10.1	1.1
3100 Special	9	10.5	1.5
ND 84	5	7.1	2.1

COMMENTS:

- Each castable lost more weight than can be accounted for from the mix water only.

- Discrepancy in weight loss can be attributed to material vaporization during curing:
  - moisture in the castable powder initially,
  - use of organic lubricants, and
  - part of the castable being vaporized.

- The weight loss of each castable had stabilized at about 300 C.

- Castable layer would have to achieve approximately 300 C for six hours to be cured enough to obtain sufficient strength to allow the continuation of the construction of the cathode.

6c. Measured Cathode Heat Losses

Cathode bottom shell temperatures were measured on commercial HS Soderberg cells in order to compare the relative performance of the monolithic castable-steel barrier design and cells with no barriers. The bottom shell temperatures were taken in the same manner and locations on all cells.

CATHODE DESIGN	MAXIMUM TEMP., °C	AVERAGE TEMP., °C	HEAT LOSS, (KW)
NON-BARRIER (Alumina Insulation only)	233 ± 2	203 ± 6	28.4 ± 1.7
CASTABLE BARRIER (& Alumina Insulation)	180 ± 26	132 ± 16	14.1 ± 2.5
	53	71	14.3
Percent Change	22.8	35	50.4

CATHODE DESIGN	MAXIMUM TEMP., °C	AVERAGE TEMP., °C	HEAT LOSS, (KW)
BRICK-STEEL & VERMICULITE INSULATION	143 ± 35	109 ± 21	11.1 ± 3.7
CASTABLE-STEEL & VERMICULITE INSULATION	149 ± 2	111 ± 17	11.1 ± 1.3

COMMENTS:

- Average shell bottom temperature of 15 cathodes with castable barriers is about 70 degrees cooler than cathodes without barriers, showing that the former
  - conserve about 14.3 kw (50.4%) from reduced heat loss from the exposed cathode bottom surface, and
  - indicate stable barrier and insulation performance.

Conclusions

Low water castable refractories were evaluated, and compared with bricks, for use as a cell barrier to protect the bottom insulation. The criteria used in the selection of commercially available materials was reviewed.

The castables, which can form a monolithic refractory layer with no mortar seams, demonstrated properties equal to that of the bricks:

- High maximum service temperatures, ≥3000°F.
- High load bearing strength at operating temperatures, ≥ 2000 lb/in^2.
- Dense, low porosity and low permeability with fine pore structure.
- High resistance to cryolite (even compared with dense fired bricks).
- Low thermal expansion, or equal to brick, at operating temperatures.

In addition, the castables can provide faster installation and are less labor intensive to install than bricks. Plant results, to date, demonstrate that the castables can be an effective barrier to provide energy conservation.

References

1. Tabereaux, A.T., "Thermal Insulation Materials for Reduction Cell Cathodes," Light Metals 1982, p. 571.
2. Bratland, D., "Refractories in the Aluminum Industry," Light Metals 2, 1976, p. 247.
3. Johansson, S., "An Alternative Method for Evaluation of Resistance of Pot Insulation to Bath Attack," Light Metals 1986, p. 501.
4. Stewart, D.V. and Tabereaux, A.T., "Evaluation of Castable Barriers In Reduction Cell Cathodes," Light Metals 1989, p. 153.
5. Meyer, H.J. and Rowden, C.E., "Pot Relining Insulation and Refractory Specification Development," Canadian Institute of Mining and Metallurgy, 1988, p. 70.
6. Davies, L.J. and McCollum, J.M., "Refractory Selection for Non-Ferrous Smelting Application," Advances in Refractories for the Metallurgical Industries, Canadian Institute of Mining and Metallurgy, 1988, p. 70.
7. Larson, S.B. and Videtto, R.B., "Improved Brick Performance in Carbon Baking Flues," Proceedings of the International Symposium on Advances in Refractories for the Metallurgical Industries, Canadian Institute of Mining and Metallurgy, 1987, p. 215.

Effect of the piers rigidity on the state of thermomechanical stresses induced in the cathode blocks of alumina electrolytic cells

C. Allaire
Alcan International Limited, Arvida Research and Development Centre
Jonquière, Québec, Canada G7S 4K8

ABSTRACT

In an alumina electrolytic cell, carbon cathode blocks are generally restrained from free expansion along their length, by the action of piers located between their ends and the shell. In such conditions thermal expansion is one of the factors that may contribute to the induced mechanical tensile stresses on top of the blocks, causing their deflection and/or cracking.

A new bench test involving uniaxial heat flow on sample with constrained thermal dilation has been used to verify the effect of the piers rigidity on the state of the induced mechanical stresses in carbon cathode block samples.

The results showed that a pier design involving higher rigidity at the upper part of the blocks may be beneficial since it contributes to induce compressive mechanical stresses on the upper half part of the blocks.

KEYWORDS

Alumina : electrolytic cells : piers design : piers rigidity : cathode stress : thermomechanical stresses : bench test : stress simulation : elasticity equations.

INTRODUCTION

In order to complete the preparation of a mathematical model of the stresses induced in alumina electrolytic cathode cell assemblies, it was decided to undertake physical simulation studies of the mechanical stress component. Generally, workers in that field (1, 2) have no tool to validate their models other than the measurement of the deformations of the shell, which gives informations only about the resultant stresses acting on the shell and not about the locally induced stresses at the level of each element of the cathode assemblies. The set-up used, which will be described in a further paper allows a sample measuring 24 x 24 x 9 inches (60.96 x 60.96 x 22.86 cm) to be subjected to a unidirectional heat flow while preventing its free thermal expansion in one of its orthogonal directions. Measurements of temperatures, deformations and forces permitted the subsequent calculation of the thermomechanical stresses induced in the sample throughout the test. The justification for this study was based on the factors given below, which cover the main advantages of this set-up with respect to the particular problems associated with the mathematical model.

Problems associated with the development of a mathematical model

- Insufficient information concerning the thermomechanical properties of the various components of a cathode assembly (the carbon block, the monolithic mix, the protective and insulating layers etc.).

- Insufficient information concerning the interaction among the various components of the cathode assembly (ex.: interfacial conditions which are difficult to define).

- The interaction of stresses of chemical origin (ex.: corrosion due to cryolite, sodium swelling, etc.) which are difficult to model.

Advantages of the thermomechanical resistance testing set-up

- The determination of the stresses induced in the cathode block requires only the information about its mechanical properties (modulus of elasticity, Poisson ratio, coefficient of thermal expansion and mechanical strength).
- No hypothesis regarding the interaction among the various components of the cathode assembly adjacent to the block is necessary (in particular the materials making up the piers).
- The chemical component of the stresses due to cryolite penetration as well as the effect of the rigidity equivalent to that of the piers might possibly be verified experimentally (with some modifications of the testing set-up in the former case).

It had been agreed that the realization of such simulation studies would depend on the following two principal objectives:

1. The definition of conditions which would permit the attainment of uniformity of the mechanical stresses induced in the cathode block and the reduction of their intensity.
2. The validation of the mathematical model particularly as it pertains to the materials adjacent to the cathode block (ex.: the piers).

However, before being able to describe a research programme based on this subject, it was necessary to first check the real extent of the testing set-up possibilities in relation to these particular objectives. It was therefore decided to carry out preliminary tests utilizing different arrangements of the samples in order to expose either the cooler or the hotter half of the cathode block to the maximum rigidity of the piers. These two extreme cases, for which the results are presented in this paper permitted not only the verification of the suitability of the set-up but also the direct determination of the effect of the piers' rigidity on the stress components induced on the block.

EXPERIMENTAL PROCEDURES

The "sample" cathode assemblies were constructed using material representation of the actual situation in the plant. The cathode blocks measuring 16 x 12 x 5 1/2 inches (40.64 x 30.48 x 13.97 cm) were first of all cut out of a block actually produced in the plant, in such a way that the heat flow during the test was parallel to the direction of pressing (i.e. perpendicular to the direction of alignment of the anthracite grains). The insulating refractories shields beneath the block as well as those covering the two sides, which were free to expand during the test were made up of MOLER SUPRA and SUPEREX 1200 as shown in figure 1. Along the direction in which the block's free thermal expansion was restricted during the test, the attachment of the piers was made up of a layer of CORAL BP (a dense, very rigid alumina brick) covered on the outside by a layer of SUPEREX 1200 (an insulating material possessing very little rigidity) carried out as shown in figure 2.

Temperature measurements at the center of the block were carried out by means of a Chromel-Alumel thermocouple.

Fig. 1
Arrangement of the insulating refractories of the sample cathode assembly along the "non-restrained" direction

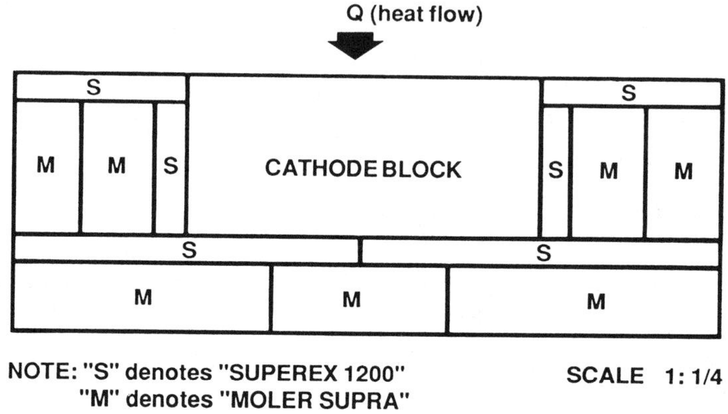

Fig. 2
Arrangement of the insulating refractories of the sample cathode assembly along the "restrained direction"
(in the case shown, the rigidity of the piers affects the colder half of the cathode block. For a rigidity acting on the hotter half of the block the positions of the CORAL BP and SUPEREX 1200 refractories are inverted).

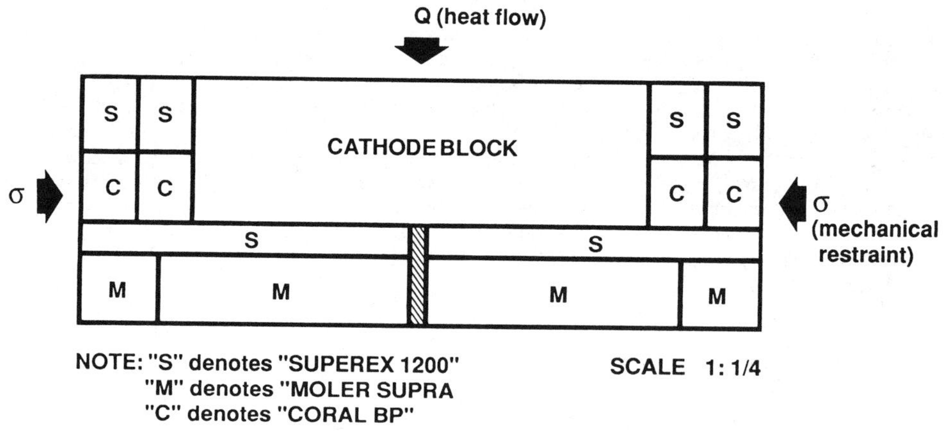

Displacement transducers (LVDT) were placed in direct contact with the block by means of Al_2O_3 rods inserted through holes made in the side shields.

The tests were carried out by exposing the top of the block to a source of heat by means of an electrical heating cover whose temperature increased at a rate of 300°C/hour up to a temperature of 1 000 °C, which was maintained for 20 or 24 hours and was followed by cooling at the rate of 50 °C/hour to ambient temperature. The mechanical component of stress induced in the block was calculated from measurements obtained at 10 minutes intervals throughout the duration of the tests at a distance of 1 inch (2.54 cm) from the hot and cold face in the direction where thermal expansion is restrained.

The calculation of the stresses, based on the generalized elasticity theory applied to two dimensions was carried out according to the following equations (3)

$$\sigma_x = -\frac{E\alpha T}{1-\upsilon} + \frac{E}{1-\upsilon^2}(\varepsilon_x + \upsilon\varepsilon_z) \quad (1)$$

$$\sigma_z = \sigma_{th} = -\frac{E\alpha T}{1-\upsilon} + \frac{E}{1-\upsilon^2}(\varepsilon_z + \upsilon\varepsilon_x) \quad (2)$$

$$\sigma_x - \sigma_z = \sigma_{mech} = \frac{E}{1+\upsilon}(\varepsilon_x - \varepsilon_z) \quad (3)$$

Where σ_x = total stress induced along the axis of the block in which the thermal expansion is restrained.

σ_z = total stress induced along the axis in which the block is free to expand.

σ_{th} = component of the thermal stress induced by the temperature gradient along x and z.

σ_{mech} = component of the mechanical stress induced by the piers along x.

ε_x = deformation of the block along x.

ε_z = deformation of the block along z.

T = temperature of the block.

E = modulus of elasticity of the block.

α = coefficient of thermal expansion of the block.

υ = Poisson's coefficient for the block.

As well, using the TIMOSHENKO equations (3), the induced mechanical stress along x(σ_{mech}) can, in turn, be separated into two new components: an induced compression component (σ_{comp}) and an induced flexion component ($\sigma_{flexion}$). In the case of a total perfect mechanical constraint, the latter two may be written:

$$\sigma_{comp} = -\frac{1}{2C(1-\upsilon)}\int_{-C}^{+C} \alpha TE \, dy \tag{4}$$

$$\sigma_{flexion} = -\frac{3y}{2C^3(1-\upsilon)}\int_{-C}^{+C} \alpha TE \, ydy \tag{5}$$

Where C = the half-thickness of the block in the direction of the thermal flow (the Y-axis).

For the particular case where α and E are independent of the temperature, equations (4) and (5) may be written:

$$\sigma_{comp} = -K_1 \left[\int_{-C}^{+C} T(y)\, dy \right] ; K_1 = \frac{\alpha E}{2C(1-\upsilon)} \tag{6}$$

$$\sigma_{flexion} = -K_2 y \left[\int_{-C}^{+C} T(y)\, ydy \right] ; K_2 = \frac{3\alpha E}{2C^3(1-\upsilon)} \tag{7}$$

In this particular case, the integration products are a function only of the temperature distribution in the material along the direction of the thermal flow. On the other hand, the limits of integration are functions of the region of interaction of the mechanical restraint in relation to the thickness of the material. The Appendix provides, as reference, the products of integration and the expressions for the corresponding induced mechanical stress components associated with two typical temperature distributions: uniform distribution (T (y) = A) and a linear distribution (T(y) = A + By). This has been made by considering three areas of interaction of the mechanical restraint: a restraint acting on the cooler half ($\int_{-C}^{0}$) or the hotter half ($\int_{0}^{+C}$) of the material or on its entire thickness ($\int_{-C}^{+C}$).

During the tests, the component of the flexion induced in the cathode block due to the piers was determined from the difference in the mechanical stresses calculated using equation (3), for the levels situated at 1 inch (2.54 cm) from the hot and cold faces such that:

$$\sigma_{flexion}\,[\text{hot face}] = \frac{\sigma_{mech}\,[\text{hot face}] - \sigma_{mech}\,[\text{cold face}]}{2} \quad (8)$$

$$\sigma_{flexion}\,[\text{cold face}] = -\sigma_{flexion}\,[\text{hot face}] \quad (9)$$

The results obtained were then compared with the theoretical values presented in the appendix. TABLE 1 provides the values for the mechanical properties of the cathode block used for the calculation of the stresses. The modulus of elasticity (E) was considered to be independent of the temperature (in reference to graphite (4)) and equal to the characteristic value of the Arvida cathode blocks for T ambient (parallel to the anthracite grains). For this same category of blocks, the shear modulus (G) at T ambient was used for the calculation of Poisson's coefficient (υ), which was also considered to be independent of the temperature such that: $\upsilon_{//} = \frac{E_{//}}{2G_{//}} - 1$ (the symbol // denotes parallel to the grains of anthracite).

TABLE 1: Mechanical Properties of the Cathode Block Parallel to the Anthracite Grains

PROPERTY	VALUES CONSIDERED FOR 0 < T < 1000 °C
Modulus of elasticity E (GPa)	10.87
Shear modulus G (GPa)	4.28
Poisson's coefficient υ	0.27
Coefficient of thermal expansion 10^{-6} / °C	$0.8102619 + 3.551537 \times 10^{-3}T + 7.553172 \times 10^{-7}T^2 - 1.807961 \times 10^{-9}T^3 + 3.978418 \times 10^{-16}T^4$

The coefficient of thermal expansion of the blocks subjected to the test was determined in the laboratory from a dilatometric test. The results were then subjected to a polynomial regression of the fourth order.

RESULTS AND DISCUSSION

In figure 3, corresponding to the test in which expansion was restricted in the colder half of the cathode block, it is evident from (a) that the hot face, which had initially been subjected to the greater degree of mechanical compression, then became less affected than the cold face (the arrow in the figure indicates the beginning of the cool-down period of the heating cover). As shown in (b), this was caused by an inversion in the flexion component induced by the piers. In other words, to counteract the thermal expansion of the cathode block during the test, the piers had first to transmit a positive moment of flexion (compression on the hot face and tension on the cold face) followed by a negative moment of flexion (tension on the hot face and compression on the cold face) towards the end of the test. The shifting of the rigidity of the piers to the hotter half of the block led to significantly different results as shown in figure 4. In (a), the mechanical compression induced on the hot face remained greater than that acting on the cold face during the entire test. In (b), the absence of inversion signifies that the piers had only to transmit a positive moment of flexion to counteract the thermal expansion of the block. In figure 5, the induced flexion results obtained during the two tests are considered together.

Fig. 3
Evolution of the Mechanical Stress (a) and of the Flexion Component (b) Induced Near the Cold and Hot Faces of the Cathode Block When the Cooler Half Was Restrainded

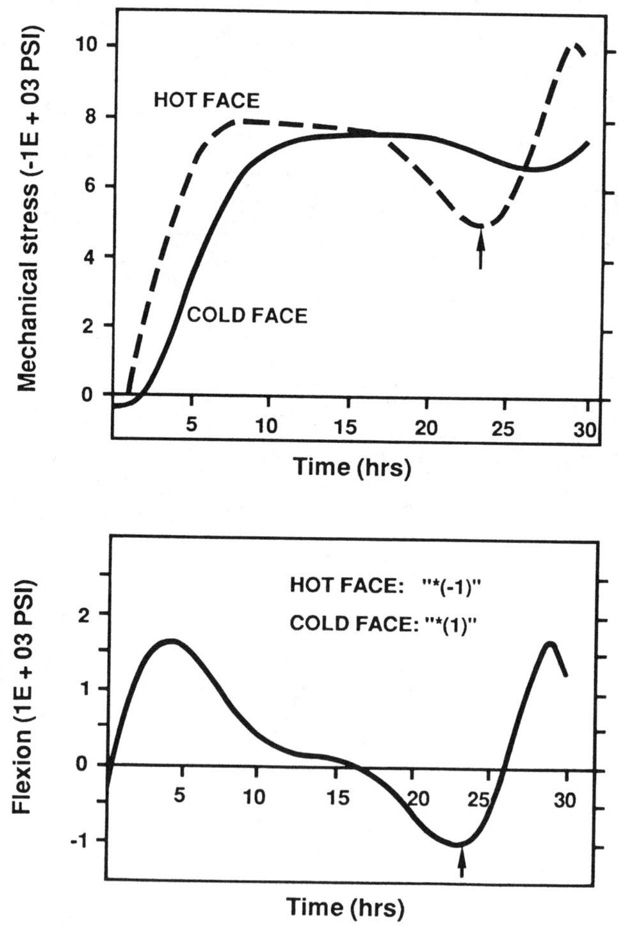

Fig. 4
Evolution of the Mechanical Stress (a) and of the Flexion Component Induced Near the Hot and Cold Faces of the Cathode Block When the Hotter Half Was Restrained

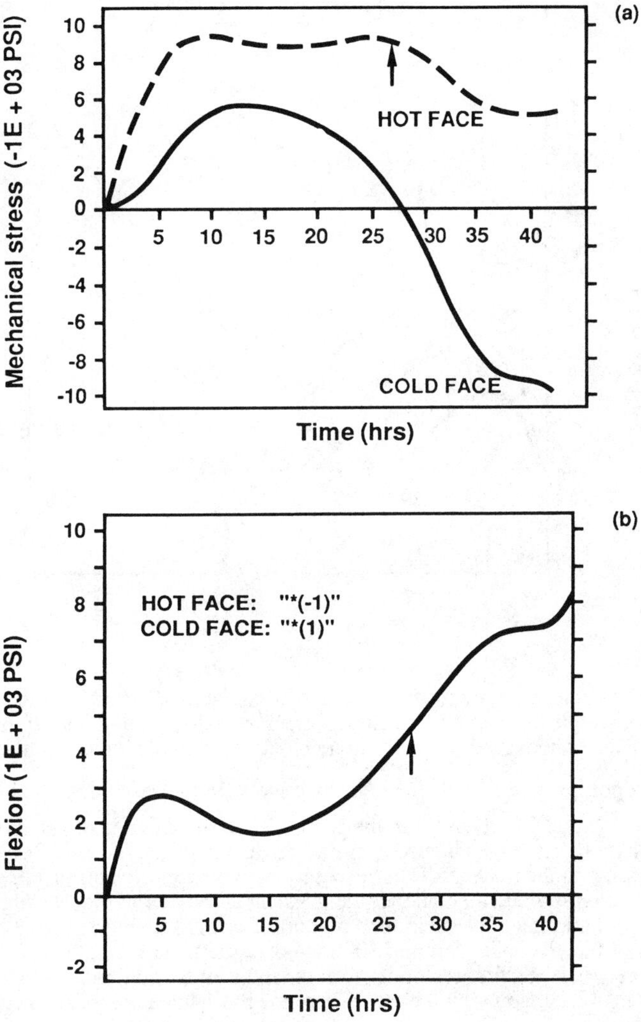

Fig. 5
Comparison of the Degree of Flexion Induced Under the Influence of a Restraint Acting Either on the Colder Half or the Hotter Half of the Cathode Block

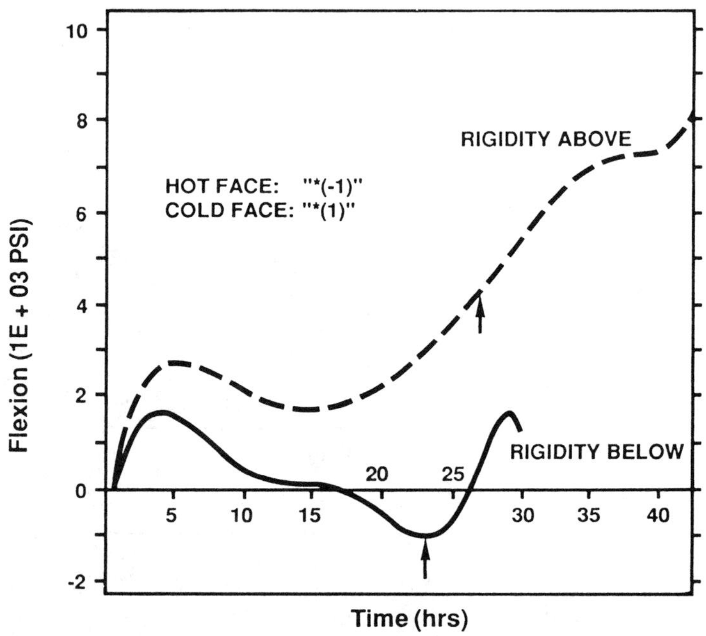

Such differences in behaviour are in accord with theoretical predictions. With reference to the appendix, for the simplified case of a uniform temperature distribution, the equivalent of a rigidity acting on the cooler or hotter half of the cathode block is the induction on its hot face of a tension $(\frac{K_2\,AC^3}{2})$ or a compression $(\frac{-K_2\,AC^3}{2})$, respectively. In the case when the block is under the influence of a temperature gradient (as in the present case) the theoretical treatment based on the linear gradient hypothesis shows in effect, that no inversion of the induced flexion is possible when the rigidity is acting on the hotter half. In the case where the rigidity acts on the colder half, the inversion is possible provided that the thermal gradient at the beginning was high enough (i.e.: B>>A). This is realistic since it took around 5 hours after the soaking period start-up of the tests to bring the center of the block from around 700 to 900°C, the latter being the equilibrium temperature. However, as observed, the stress resulting from the induced flexion under a thermal gradient can only be weaker in comparison with the case where the mechanical constraint acts on the hotter half of the block since:

$$\left| \frac{AC^2}{2} + \frac{BC^3}{3} \right| > \left| \frac{-AC^2}{2} + \frac{BC^3}{3} \right|$$

CONCLUSION

This first series of preliminary studies has permitted to establish that the thermomechanical resistance testing set-up used seems really to be appropriate as a tool for the validation of cathode stress mathematical models, related to alumina electrolytic cells. In effect, by testing two limiting pier configurations, the set-up clearly demonstrated, in conformity with the theory, that piers having greater rigidity in the lower half of the block favour an upward defection of the cathode under the influence of tensile stresses at the hot face. This seems to be similar to the actual design of most of the electrolytic cells in the aluminium industry.

REFERENCES

1. Rolf, R.L. and Peterson, R.W. "Compressible insulation to reduce potlining heaving in Hall-Heroult cells", AIME, pp. 209-213, 1987.

2. Toia, G. et al., "Application of a Mathematical Model to Ensure Optimum Potshell Design", Light Metals,1979, p. 495.

3. Timoshenko, S.P. and Goodier, J.N., "Theory of Elasticity", third edition, McGraw-Hill Book Company, N.Y., 1970.

4. Kirk, R.E. and Othmer, D.F., "Encyclopedia of Chemical Technology", vol. 4, John Wiley & Sons Inc., N.Y., p. 594, 1978.

APPENDIX

Theoretical calculations of the components of induced mechanical stress associated with two temperature distributions and three regions of interaction of the rigidity ot the piers influencing the cathode block.

The calculations are based on the following equations:

$$\sigma_{comp} = -K_1 \left[\int_a^b T(y)\, dy \right] ; K_1 = \frac{\alpha E}{2C(1 - \upsilon)} > 0$$

$$\sigma_{flexion} = -K_2 y \left[\int_a^b T(y) y\, dy \right] ; K_2 = \frac{3\alpha E}{2C^3(1 - \upsilon)} > 0$$

The different cases considered are the following:

1) Temperature distribution:
   i) $T(y) = A$ (uniform)
   ii) $T(y) = A + By$ (linear gradient)

   where "A" and "B" are positive constants

2) Region of interaction of the rigidity of the pier:
   i) $a = -C$ and $b = 0$ (rigidity next to the cooler half of the block)
   ii) $a = 0$ and $b = +C$ (rigidity next to the hotter half of the block)
   iii) $a = -C$ and $b = +C$ (rigidity along the overall thickness of the block)

   where "C" represents half the thickness of the block.

Component of mechanical stress	Region of interaction of the rigidity of the piers	Temperature distribution	
		$T(y) = A$	$T(y) = A + by$
Compression *	$-K_1 \left[\int_{-C}^{+C} T(y)\, dy\right]$	$-K_1 [2AC]$	$-K_1 [2AC]$
	$-K_1 \left[\int_{-C}^{0} T(y)\, dy\right]$	$-K_1 [AC]$	$-K_1 \left[AC - \frac{BC^2}{2}\right]$
	$-K_1 \left[\int_{0}^{+C} T(y)\, dy\right]$	$-K_1 [AC]$	$-K_1 \left[AC + \frac{BC^2}{2}\right]$
Flexion*	$-K_2 y \left[\int_{-C}^{+C} T(y) y\, dy\right]$	0	$-K_2 y \left[\frac{2BC^3}{3}\right]$
hot face: $y = +C$	$-K_2 y \left[\int_{-C}^{0} T(y) y\, dy\right]$	$-K_2 y \left[\frac{-AC^2}{2}\right]$	$-K_2 y \left[\frac{-AC^2}{2} + \frac{BC^3}{3}\right]$
cold face: $y = -C$	$-K_2 y \left[\int_{0}^{+C} T(y) y\, dy\right]$	$-K_2 y \left[\frac{AC^2}{2}\right]$	$-K_2 y \left[\frac{AC^2}{2} + \frac{BC^3}{3}\right]$

* Note: - a negative component of stress indicates a compression
- a positive component of stress indicates a tension

Design, operation and electrochemical aspects of Hall-Héroult cells containing solid wetted cathodes: a 30-year chronicle

R.C. Dorward and J.R. Payne
Center for Technology, Kaiser Aluminum, Pleasanton, California 94566, U.S.A.

Abstract

The investigation of solid cathodes for aluminum reduction cells dates back to the late 1950's. These early tests showed that substituting a wetted stable cathode for the turbulent metal cathode allows the anode-cathode distance (ACD) to be decreased with an attendant voltage reduction of about 1V, depending on current density. Since high efficiency is retained, specific energy consumption is decreased by about 20%. These savings are not as great as anticipated, however, because the effective electrical resistivity of the electrolyte increases as the ACD is lowered. This effect is related to the void fraction of anode gas in the interpolar space. Although TiB_2 materials are available that withstand the harsh chemical environment, cell working operations result in an inordinate amount of part breakage. From an operational viewpoint the ideal cathode material would be a durable TiB_2 composite coating that could be troweled onto conventional cathode (carbon) blocks.

Introduction

The Hall-Heroult smelting process, in which alumina is dissolved in molten NaF-AlF_3 salt at 940-980C and electrolytically decomposed with direct current, is universally used to make aluminum metal. A typical aluminum smelter has an energy efficiency of about 40%. A major portion of the energy consumed in the process can be attributed to the voltage drop between the anode and cathode. The spacing between the anode and cathode (the anode-cathode distance, or ACD) must be maintained at 4-5 cm to prevent electrical "shorting" between the carbon anode and the molten aluminum pool, which serves as the cell cathode. Therefore, if it were possible to replace the turbulent metal cathode with a dimensionally stable cathode, the ACD could be reduced significantly with a concomitant saving in energy.

The potential benefits of using electrically conductive titanium diboride (TiB_2) for this application have been recognized for over 30 years (1,2). During the 1956-64 time period for example, Kaiser Aluminum and the British Aluminium Company actively pursued the activities of producing TiB_2 powder, forming cathode shapes, and evaluating them in pilot and industrial cells. The pilot cells, of nominal 10 KA capacity, were of the "Ransley-well" design, in which the metal deposits on a slightly sloped cathode surface and drains into a center collection trench. The pilot cell tests demonstrated that the ACD could be reduced to provide an energy savings of about 20% (3). However, the development program had to be terminated when it became increasingly apparent that the cathode materials then available (carbothermic TiB_2, TiC and TiB_2/TiC-carbon mixtures) were prone to intergranular disintegration by molten aluminum, which led to premature failure.

In the 1970's, the advent of the energy crisis and improvements in materials technology led to a reassessment of the practical merits of the solid wettable cathode concept. In the first of these campaigns, a 15 kA cell of the Ransley-well design, in which four varieties of TiB_2 were bonded to a graphite substrate, was operated for 66 days. One of the materials was clearly identified as satisfactory (chemically stable) for its intended purpose. Although the cell had to be shut down before attaining the goal of 180 days of operation, it confirmed the promise of TiB_2 cathodes as a means of lowering the anode-cathode distance (4). The premature shutdown of the cell was due to the detachment of the TiB_2 tiles that were used to pave the cell bottom. The cell autopsy revealed that all of the TiB_2 tiles under the anode shadow were displaced toward the metal well and the cell periphery; most were found broken.

The majority of the breakage was attributed to mechanical forces arising from the daily cell servicing operations. These findings lead to a reassessment of the design philosophy for a wettable cathode cell as described below.

Horizontal Electrode Concepts

During the 30 odd years that TiB_2 has been considered for use as a reduction cell cathode, many different concepts have been proposed (5). Designs have ranged in complexity from "candle leads" to 3-dimensional cathode shapes. The cathode compositions have varied from pure, high-density TiB_2 to composites containing titanium carbide, carbon, graphite, silicon carbide, aluminum nitride and boron nitride. Reasons for the multi-component systems have included cost reduction, improved structural integrity, and designed coefficients of thermal expansion. Patents have also been granted for designs and processes that deposit TiB_2 coatings on high alloy steels or carbon substrates. All of these considerations must be viewed in the context of retrofitting the cathodes into existing cells.

Design Considerations

Two general cell design strategies can be envisioned: (a) a large mass inventory of simple TiB_2 shapes that last a long time and could be recovered at overhaul, (b) a minimal inventory of more complex TiB_2 shapes that are discarded at overhaul time. From basic material considerations, the last alternative requires the utmost quality of TiB_2 shapes. The attributes and disadvantages of several candidate cathode concepts are discussed below.

Packed Bed Cathode. This design is the most elementary and requires the simplest TiB_2 shapes. Nominal 1-3 cm diameter balls or oblate speroids could be cold pressed in a pill press and sintered. Shrinkage and its effect upon dimensional control is not important. The balls would be randomly poured into the cell cavity and leveled out to a bed depth of 8 to 15 cm. The random packing fraction for spheres is 0.64, but other shapes could reduce the inventory considerably. The spaces between the spheres would fill with metal, creating a continuous aluminum film at the bed surface. At least one retaining wall would be required to separate the bed from a metal accumulation well. One such design is based on a forehearth created by a transverse TiB_2 wall, which also serves as a weir to maintain a constant metal level in the electrolytic region. Other attributes of this design are

- tolerance to physical abuse,
- the TiB_2 inventory could be reclaimed and recycled, and
- an expensive variety of pure TiB_2 may not be required,

while the disadvantages include:

- a large inventory of TiB_2
- a susceptibility to cathod contamination by undissolved alumina ("mucking"), and
- questionable integrity of the weir or retaining wall.

Piecewise Cathode. This design involves embedding individual components such as rods or cylinders into the cell bottom. In the case of cylinders, the part fills with metal so that the cathode surface is primarily molten aluminum (6). Although a cursory examination suggests that such a design would minimize TiB_2 inventory per unit anode area, this is not necessarily the case. It is also important that the vertical cylinders be crack-free to prevent the internal aluminum volume from seeking the level of the metal pool. As for the packed bed concept, the design is also susceptible to mucking.

Planar Cathode. A slightly sloped planar system of flat tiles fixed to a graphite underlayment requires the least amount of TiB_2 per unit anode area, although large tiles may be expensive due to the stringent requirements of soundness. The long-term survival of large tiles has not been established, so it may be necessary to utilize smaller parts to increase survival probability.
It is also significant that reliable fastening techniques would have to be developed for an extremely hostile environment (7).

Another planar cathode system is based on a TiB_2-carbon composite mix that can be troweled onto a conventional cathode block (8). As with the original TiB_2- TiC-carbon cathodes (9), the long-term survivability of these materials has yet to be demonstrated (10). A similar composite material of TiB_2-graphite (11) has also been the subject of intensive investigation recently (12,13).

The Replaceable Cathode Module. A number of the shortcomings of the previously discussed cathode designs can be alleviated by the use of a replaceable cathode module (RCM) concept (14). The basic philosophy behind this approach is the recognition and acceptance that TiB_2 parts will fail in service. A shortcoming of earlier approaches is that local failures of a few TiB_2 parts could disable a cell prematurely, requiring that it be taken out of service for repair.

The RCM design is achieved by making cathode assemblies correspond on a one- to-one basis with the anodes. Ideally, individual cathodes should be physically separate and also separable from the cell lining. The following attributes are important features of this concept:

- The cathode can be assembled under controlled conditions outside of the cell to accommodate the special requirements of bonding, if necessary.
- The cell can be started up without the cathodes in place.
- The cathode assembly can be slowly heated in a controlled atmosphere prior to installation in cell.
- The cell lining life and cathode life are decoupled, i.e., cathodes can be installed at any time in a functioning cell lining, and functioning cathodes can be removed from a failed lining for reuse in another cell.

Pilot Cell Operations

A modified 15 kA cell was operated for ten months in 1979 to evaluate the replaceable cathode module concept (15). In the course of testing 20 cathode modules of six generic designs (flat plates, open-ended cylinders and cones, closed-end cylinders, packed bed, vertical and horizontal rods), a simple procedure was perfected for installing replaceable cathode modules. This method effectively eliminated thermal shock and start-up concerns with TiB_2 cathodes. However, these activities also identified two major problems that had to be resolved before proceeding to a commercial-size test cell. First, cell working operations resulted in an inordinate amount of TiB_2 part damage. Second, only 60% of the expected energy reduction was achieved upon decreasing the ACD from 4.5 to 1.6 cm, i.e., there was not a one-to-one correspondence between the interpolar bath resistance and ACD (3,15).

The interpolar voltage was measured between a probe in the anode about 2 cm from the sole and another probe in the metal pool below the cathode. Voltage- current (V-I) data were generated by reducing the electrolysis current incrementally from about 3000 A per anode (~1.6 A cm^{-2}). Ignoring the small voltage drop between the probe and the bottom of the anode, the bath resistance was then determined from the slope of the V-I plot, i.e.,

$$V = V_d + IR_b \qquad (1)$$

The plots were linear above about 1000 A (~0.5 A cm^{-2}), allowing good comparative estimates of interpolar resistance to be made. Figure 1 gives the results of three separate tests, two on a close-packed bar assembly, and the third on a flat plate assembly. The resistance values obviously do not extrapolate to zero at O cm ACD. Furthermore, the resistances are considerably greater than those depicted by the dashed line, which were calculated from the specific bath resistivity and estimated interpolar cross-sectional area (taken to be the average of the anode and cathode areas, with no fringing effects). The discrepancy between the estimated (calculated) and the apparent (measured) values increased as the ACD decreased. Figure 2 shows how $\bar{\rho}/\rho_5$ (ratio of the effective bath resistivity at reduced ACD to the apparent resistivity at 5 cm ACD) for these data changed with ACD.

One possible explanation for the increase in effective bath resistivity at reduced ACD is related to anode gas evolution. Considering the large rate of gas generation in a reduction cell (~0.25 cm^3 $A^{-1}s^{-1}$), the total gas volume fraction must rise as the ACD is decreased, unless the transport velocity increases correspondingly. Soviet investigators have reported specific gas volumes (ratio of gas volume in the interpolar space to the area of the anode bottom) of 0.3-0.6 cm^3cm^{-2} under pilot cell anodes, depending on current density and depth of anode immersion (16). The effective resistivity of the interpolar gap is obviously affected by the amount of gas in this region. The simple case of a uniformly distributed gas phase can be described by a number of relationships, including the Bruggemann equation (17)

$$\bar{\rho} = \rho(1 - \epsilon)^{-3/2} \qquad (2)$$

where $\bar{\rho}$ and ρ are the resistivities of the gas-containing and gas-free electrolyte, respectively, and ϵ is the volume fraction of gas in the electrolyte. Therefore, for a constant volume of gas under an anode, the effective resistivity ($\bar{\rho}$) must increase as the ACD is decreased. A uniform gas dispersion in the interpolar space as the above model assumes is unlikely; a more probable situation is a layer of gas-rich electrolyte near the anode surface. This latter case has been mathematically represented by a number of colloca- tions (3,18-20), which qualitatively predict the observed effects, given the number of assumptions that have to be made. These implications will be discussed in greater detail later.

In an attempt to negate the effect of the adverse "polarization" effect, a pilot cell containing a near vertical electrode system was designed, built and operated. The following section describes this cell and its electrochemical performance.

Near Vertical Electrode Concept

The concept of steeply sloped electrodes for an aluminum reduction cell is not new; several European companies investigated the notion between 1930 and 1950, but were defeated by the low current efficiency that occurs when carbon is used as a cathode surface. In 1957 Ransley proposed a steeply sloped TiB_2 cathode design (1), which was advocated for several reasons, including more electrode area per unit floor area and improved cathode life. However, the concept as envisioned by Ransley had several drawbacks: poor electrolyte circulation path, high anode wastage (large butts), and difficult ACD control. An alternative highly sloped design that minimizes these drawbacks, while being compatible with the RCM concept, is described below.

The basic feature of the near-vertical electrode system is the use of cathode surfaces arranged in back-to-back pairs as depicted in Figure 3. The associated anodes for paired cathode surfaces are suspended from an overhead structure located in the same vertical plane as the cathode structure. As the anodes are consumed in the electrolysis process, they are rotated toward the fixed cathode faces. Predicted advantages of the design included:

- Installation of a cathode structure or module while the cell is in operation.
- Rapid upward motion of the two-phase mixture of bath and anode gases.
- A fairly uniform concentration of alumina throughout the interpolar space, thereby minimizing local anode effects.
- Continuous or semi-continuous alumina feeding at the top of each cathode structure where an agitated bath surface of reduced alumina concentration should minimize mucking.
- No requirement for cell-bottom support or backup material since the TiB_2 parts are free bodies.
- Much lower cathode voltage drop than a conventional pot lining/collector bar system.

Although the sloped, top-entering cathode concept offers a number of technical and operating advantages, they are not obtained without some major impact on cell construction. The superstructure allowed for insulated covers, which were required to prevent cathode damage from frozen crust trapped between an anode and a module. The use of TiB_2 top-entering cathode collector bars also required that the cell temperature be lowered to about 875C. This was accomplished with a 1.35 ratio bath containing 3% CaF_2, 5% Al_2O_3, 8% MgF_2 and 8% LiF, which has a primary freezing point of about 855C.

Electrochemical Performance

Voltage-current (V-I) data were obtained on two anode-cathode pairs; the active surfaces of both anodes were found to be smooth and "spikeless" upon removal, indicating good cathode integrity. The voltage was measured between the anode rod and the cathode hanger. First, data were obtained at two ACD's that differed by 2.5 cm at the center of the anode's active face (~5 cm and 2.5 cm, respectively). Although the ACD's were estimated no better than $\pm$0.6 cm in this instance, it is instructive to compare the reduction in voltage with the change in ACD, i.e.,

$$\Delta V = I\rho\Delta(ACD)/A. \qquad (3)$$

Ignoring current fringing effects, the discrepancy between the measured voltage reduction at 4000 A (1.35 A/cm^2 anode current density) was about 0.7 V, i.e., only 55% of the "expected" voltage reduction was achieved.

In a similar test, data were obtained as an anode was lowered in five measurable increments down to what appeared to be about a 2 cm ACD as estimated at the top. The V-I data are given in Figure 4 and the slopes of the plots (ohmic resistance of bath + anode + cathode circuits) are shown in Figure 5 as a function of ACD. The reduction in interpolar resistance with change in ACD ($dR/d(ACD) = \rho/A$) was calculated as 150 $\mu\Omega$/cm. This compares with a measured value of 92.5 $\mu\Omega$/cm, i.e., the reduction in resistance with decreasing ACD was only 60% of the theoretical value. Assuming the ACD estimates are correct, this results in an extra 0.9 V at a 1.6 cm ACD for a current of 3840 A (1.35 A/cm^2) -- not significantly different from the horizontal electrode situation.

We note that the theoretical dR/d(ACD) value of 150 $\mu\Omega$/cm is in reasonable agreement with that determined from the slope of the line joining the zero ACD resistance (R_{anode} + R_{cath})* and the high ACD resistance (140 $\mu\Omega$/cm). It is also noteworthy that the bath resistivity back calculated from the measured resistance at the large ACD is close to the assumed value (0.44 vs. 0.48Ω - cm). Another interesting feature of the results is the upward curvature in the V-I plots at the lower ACD's. The ohmic resistance, therefore, appears to be dependent on current density in the near vertical situation.

Current efficiency for a 7-day operating period was 86% based on the amount of metal tapped plus that removed from the cell after shutdown. Although this is a reasonably high value for a new cell, we note that it was not obtained under low ACD conditions. Assuming the present data to be representative of a production cell, the specific energy requirement is 13.6 Kwh/kg at the retrofit anode current density of 1.35 A/cm^2.

Discussion

The basic premise of the wettable cathode concept was that the resistive interpolar voltage would decrease on a one-to-one basis with reductions in the interpolar gap. The test data from four pilot cell campaigns demonstrated that the "expected" decrease in interpolar voltage is not achieved. One reason for this observation in cells with large planar cathodes is related to current fringing effects, i.e., the effective cross sectional area of the interpolar space is greater than the projected area of the anode sole. The larger the ACD, the greater the effective cross-sectional area, and the smaller the anode the larger the effect. From reference electrode measurements, Haupin estimated a "fanning" factor (21), which is defined as an extension in each direction to be added to the dimensions of the anodes. Even after allowing for current fringing, however, a "polarization" effect still exists (3,15).

The results can be reduced to two observations: (a) the dependence of ohmic resistance (and cell voltage) on ACD is too small, and (b) the ohmic resistance at reduced ACD is too high. Table I lists a number of conceivable reasons for these observations, and whether they are consistent with observations (a) and (b) above. Two candidates seem most plausible, both involving gas fraction effects:

- increasing gas fraction as the ACD is reduced
- constant gas fraction and lower bath resistivity than assumed.

TABLE I. Possible Reasons for Observed Electrochemical Results

	Consistent with	
	High V, R	Low dR/d(ACD)
1. High external ohmic resistance	Yes	No
2. Bath resistivity higher than assumed	Yes	No
3. Bath resistivity lower than assumed	No	Yes
4. ACD higher than estimated	Yes	No
5. High cathodic/anodic polarization	Yes	Maybe
6. Constant (ACD independent) gas fraction effect	Yes	No
7. ACD dependent gas fraction effect	Yes	Yes
8. Lower ρ + constant gas fraction effect	Yes	Yes

If the gas bubbles are confined to a layer near the anode, and the bubble dynamics are not affected

* R_{anode} (~65 $\mu\Omega$) was measured; R_{cath} is estimated (~35 $\mu\Omega$).

by the ACD, there should be a constant (ACD independent) bubble resistance. Referring to Figure 6, the bath resistance is then comprised of two components in series:

$$R_b = R_\delta + R(ACD\text{-}\delta) \qquad (4)$$

and from Eqn (2)

$$R_b = \frac{\rho(1-\epsilon_\delta)^{3/2}\,\delta}{A} + \frac{\rho(ACD\text{-}\delta)}{A} \qquad (5)$$

Since δ and ϵ are assumed independent of ACD, then

$$dR/d(ACD) = \rho/A \qquad (6)$$

and $$\Delta R = \frac{\delta\rho}{A}\,[(1-\epsilon_\delta)^{-3/2} - 1] \qquad (7)$$

In other words, the only way to account for the observed behavior is a considerably lower gas-free bath resistivity than that estimated from the usual compositional relationships (22), i.e., case 8 in Table I. However, if the bubble dynamics are ACD dependent, e.g., $\epsilon_\delta = f(1/ACD)$, then one would not conclude that ρ is lower than assumed.

Although the void fraction theory qualitatively describes the type of behavior observed, the linearity of the typical V-I plot does not fit the model. Since the rate of gas evolution is proportional to current, the resistance of the interpolar space (slope of the V-I plot) should be dependent on current. It is possible that the expected upward curvature in the plots could be minimized by a current (or gas volume) dependent transport velocity, i.e., the higher the current, the greater the rate of bubble removal from the interpolar space. Although such behavior has been observed in commercial smelters (23) and laboratory $PbCl_2$-KCl cells (24), we cannot rule out other current inhibiting factors such as concentration polarization, or reaction and charge transfer over- voltage, which in turn could be dependent on bubble effects (25,26).

The fact that gas bubbles can have a significant effect on the electrical resistivity of other electrolyte systems is well recognized (27). However, the issue of anode gas void fraction has not until recently attracted serious attention in the American-European aluminum industry, probably because of the large interpolar gap in commercial Hall-Heroult cells,. Only a few research investigations, reflecting different concerns, have been reported. The first, by Haupin, involved *in situ* measurements of the interpolar voltage distribution (21). Reference electrode scans of a 7.6 mm interpolar gap in a pilot cell revealed linear voltage relationships except within 0.25 cm of the anode face. A second feature of the scans was an erratic or noisy character of the voltage vs. distance trace within 2.2 cm of the anode face, which was attributed to a "bubble layer thickness" effect. Voltage drops due to the gas layer were 0.09-0.34 v, depending on bath composition and current density.

Dernedde and Cambridge physically modeled the hydrodynamic behavior of the fluids in a reduction cell under the driving force of the anode gas (28). Air was passed through a porous plate to simulate the anode process; water and an organic fluid simulated the electrolyte and the aluminum pool, respectively. The purpose of the work was to clarify certain design and operational features of Soderberg-type cells, but provided limited direction to the problem of gas effects on interpolar resistance. This work was extended by Fortin et al (29), who modeled the effects of current density, electrode angle, electrolyte velocity and ACD on gas film and bubble behavior. ACD was found to have little effect on the observed response. A similar study by Solheim and Thonstad (20) showed that the specific volume of gas increased about 10% upon decreasing the ACD from 5 to 2 cm. In addition, the average resistivity of the two-phase layer appeared to increase with increasing gas volume.

There is also a considerable collection of papers by Soviet authors dealing with the interpolar gas content of electrolytic cells for magnesium and aluminum production (16,23,24,30,31). A number of these authors have also addressed the related issue of cell hydrodynamics.

Lastly, there has been an extensive body of work done in aqueous systems to investigate the resistance of bubble layers on electrodes and the increase of electrolyte resistivity due to entrained bubbles. The situation with either horizontal or near-vertical electrodes generally consists of two or possibly three regions of different bubble regimes:

- Small (0.5 mm) nascent bubbles adhering to the electrode surface.
- A two-phase layer of bubbles that have detached from the electrode.
- A variable bubble dispersion in which the gas content diminishes to zero near the cathode.

The principal difference between a horizontal downfacing anode and a near- vertical anode is the nature of the detached-layer flow. In the horizontal case an individual bubble starts in motion toward the electrode boundary and by a "snow plow" effect grows into a large flat bubble or film. This results in the stochastic expulsion of large gas films rather than bubbles. In the near-vertical case the bubbles rise in a swarm. When a bubble detaches from the anode, it attains a single bubble terminal velocity in a distance of a few bubble diameters. As a given bubble rises, it encounters other bubbles that are accelerating. From some of these collisions, mergers occur and the bubble diameter increases which in turn, causes the rise velocity to increase. Certainly for electrode heights less than 50 cm no large gas films develop. The void fraction at a given point is an integration of local gas generation from the bottom to that point. The local current density is therefore dependent on the electrode height (consumable anodes "burn" non-uniformly to counter this effect).

As the bubbles rise, they are compressed toward the anode surface due to the buoyancy component normal to the surface. This situation enhances the probability of collisions between free-rising bubbles and those attached to the surface. Thus, bubble growth occurs and the rise velocity increases. A qualitative estimate of the bubble size disengaging at the bath surface of the 40 KA cell was 3-5 mm. If the flow is one-dimensional from bottom-to-top, and there is not re-entrant flow at the free surface, then a net throughput of bath must occur. Ideally, it should be equal to the volumetric flow of anode gas. However, slip or entrainment may occur and cause the liquid flow rate to be greater or less than the gas flow rate.

The liquid entering the bottom of the electrode gap will exhibit a velocity distribution across the gap, which changes with height because liquid flow is induced by gas bubble flow in the layer adjacent to the anode. This situation can be termed as "2-region, 2-phase flow". An energy balance of the interpolar gap includes an entrance loss, a friction loss, and an exit loss balanced against the differential fluid head. A quantitative predictive model has not been derived to include these factors. An overall calculation based upon a void fraction necessary to satisfy the V-I relationship resulted in a differential pressure of 1600-2000 Pa, while the pressure drop estimate was less than 200 Pa. A discrepancy this large cannot be attributed to errors or uncertainties in fluid properties. The reason is more likely related to an incomplete understanding of the bubble size, spatial distribution, and/or the velocity profile.

A computer code was written to calculate local current density based upon the simple consideration that the gas evolved at a given unit area moves upward to the next unit area. In doing so the bubbles attain the single-bubble rise velocity. A void fraction estimate can be made by utilizing a residence time computed from the rise velocity plus liquid velocity. The liquid velocity was taken as a ramp function in the liquid zone and was assumed constant across the bubble layer. Furthermore, the slope of the ramp was increased as the analysis progressed up the interpolar space.

This model automatically creates a stagnant liquid layer adjacent to the cathode. Also, the thickness of this layer increases with electrode height. The liquid flow, induced by the gas flow, has the maximum velocity in the 2- phase anode layer. A continuously varying bubble size distribution was replaced by a single bubble size uniformly incremented as the calculation moves up the electrode height. A bubble layer thickness was estimated by close packing the bubbles in a square array and in multiple layers. The procedure resulted in a close approximation to the experimental results at interpolar gaps less than 2.0 cm, but undercalculated the current at large gaps.

In summary, no engineering design method exists for predicting the terminal voltage of a given cell as a function of its physical dimensions, current density, and interpolar gap. A comprehensive mathematical model needs to be constructed and verified with carefully designed experiments to provide an engineering design procedure for electrolytic cells and their optimization. Recent physical and mathematical models, which recognize the importance of gas driven flow in Hall-Heroult cells (32-35), are important first steps in such an endeavor.

Summary and Conclusions

The operation of a number of pilot Hall-Heroult cells over a period of 25 years has demonstrated that

1. The use of dimensionally stable TiB_2 cathodes allows the interpolar spacing to be reduced substantially while still retaining high current efficiency.

2. Productivity increases of 20-30% can be achieved at somewhat lower specific energy consumption than a conventional pilot cell with a liquid metal cathode.

3. The interpolar resistance can be reduced about 40% by decreasing the ACD from 4.5 cm to 1.6 cm. For a typical electrolyte composition this corresponds to a voltage reduction of about 1 V at a current density of 1.35 A/cm^2, or a specific energy savings of ~3.3 kWh/kg.

4. The reduction in cell resistance is not as great as anticipated because the apparent electrical resistivity of the interpolar space increases as the ACD is lowered.

5. The adverse resistivity effect is consistent with an increased void fraction of anode gas in the interpolar space, although the ACD dependency of this effect remains unclear.

6. A highly sloped electrode system does not afford significant relief from the void fraction problem.

7. A highly sloped electrode cell with top-entering cathode leads is capable of operating at a specific energy rating of 14 kWh/kg (at 1.35 A/cm^2).

8. Titanium diboride cathode materials are available that withstand the chemical environment of the reduction cell.

9. An inordinate amount of TiB_2 part breakage occurs as a result of normal cell working operations in a conventional down-fed anode system.

10. The replaceable cathode module concept effectively eliminates thermal shock and cell start-up concerns with TiB_2 cathodes, and failed cathodes can be replaced in an operating cell without shutting it down.

11. From an operational viewpoint, the most desirable cathode would be a relatively inexpensive, but chemically and physically durable, trowel-on TiB_2 composite material.

Acknowledgements

The efforts of D. W. Dow, W. H. Goodnow, R. A. Lewis and R. D. Hildebrandt are gratefully acknowledged for their extensive conceptual and/or experimental contributions to this study. The US Department of Energy (Contract DE-AC03- 76CS40215) partially funded a portion of the program (M. McNeil and W. Thielbahr technical contracting officers). Pure TiB_2 and TiB_2 composite materials were supplied by PPG Industries, Kawecki Berylco, Carborundum, Union Carbide and Elektroschmelzwerk.

References

1. C. E. Ransley, U.S. Patent 3,028,324 (1962).
2. C. E. Ransley, Extractive Metallurgy of Aluminum (ed. G. Gerard), Vol. 2, p. 487, AIME, New York (1963).
3. R. C. Dorward, J. Appl. Electrochem, 13 569 (1983).
4. W. H. Goodnow, Interim Report, ERDA Contract E (04-3)-1257 (1977). 5. K. Billehaug and H. Oye, Aluminium 56, 642 and 713 (1980).
6. P. A. Foster and S. C. Jacobs, U.S. Patent No. 4,071,420 (1978).
7. C. H. Schilling, D. I. Hagen and P. E. Hart, Report PNL-6144, DOE Contract DE-AC06-76RLO 1830 (1987).
8. L. G. Boxall, W. M. Buchta, A. V. Cooke, D. C. Nagle and D. W. Townsend, U.S. Patent 4,466,996 (1984).
9. R. A. Lewis and R. D. Hildebrandt, U.S. Patent 3,400,061 (1968).
10. R. C. Dorward and J. R. Payne, Final Report, DOE Contract DE-AC07-76CS40215-2 (1985).
11. L. A. Joo, K. W. Tucker and F. E. McCown, U.S. Patent 4,376,029 (1983).
12. C. H. Schilling and G. L. Graff, Reports PNL-6593 and PNL-6594, DOE Contract DE-AC06-76RLO 1830 (1988).
13. L. Joo, K. Tucker, J. Gee and J. Shaner, CMP Report 86-10, Center for Metals Production (1986); also Light Metals, p. 345, AIME (1987); p. 413 (1990).
14. W. H. Goodnow and J. R. Payne, U.S. Patent 4,349,427 (1982).
15. R. C. Dorward and J. R. Payne, Topical Report, DOE Contract DE-AC07-76CS40215-1 (1983).
16. W. S. Siraev, D. V. Forsblom and D. Ya. Khalpakchi, Tsvet. Met. 49, 33 (1976).
17. R. E. DeLaRue and C. W. Tobias, J. Electrochem. Soc. 106, 827 (1959).
18. H. Vogt, J. Appl Electrochem 13, 87 (1983).
19. W. Haupin, The 3rd Int. Course on the Process Metallurgy of Aluminium (1984), Chapter 7.
20. A. Solheim and J. Thonstad, Light Metals, p. 397, AIME (1986).
21. W. E. Haupin, J. Metals 23, No. 10, 46 (1971).
22. G. J. Houston, M. P. Taylor, D. J. Williams and K. Grjotheim, Light Metals, p. 641, AIME (1988).
23. A. I. Begunov, Izv Vus Tsvet. Met. 19, 29 (1976).
24. V. A. Kryukovskii, P. V. Polyakov, G. V. Forsblom, A. M. Tsyplakov and V. V. Burnakin, Tsvet. Met. 45, 62 (1972).
25. E. W. Dewing and E. Th. van der Kouwe, J. Electrochem. Soc. 122, 358 (1975).
26. P. J. Sides and C. W. Tobias, ibid. 127, 288 (1980).
27. R. B. MacMullin, Chlorine, Its Manufacture, Properties and Uses (ed. J. S. Sconce), p. 127, Reinhold Publishing Corp. New York (1962).
28. E. Dernedde and E. L. Cambridge, Light Metals, p. 111, AIME (1975).
29. S. Fortin, M. Gerhardt and A. J. Gesing, Light Metals, p.721, AIME (1984).
30. M. K. Kulesh, A. A. Dmitriev and V. O. Volodchenko, Tsvet. Met. 43, 23 (1970).
31. V. V. Nerubashchenko, L. N. Antipin and A. V. Kuleshova, ibid. 40, 53 (1967).
32. A. Solheim, S. T. Johansen, S. Rolseth and J. Thonstad, Light Metals, p. 245, AIME (1989).
33. D. C. Chesonis and A. F. LaCamera, ibid., p. 211 (1990).

34. K. J. Fraser, M. P. Taylor and A. M. Jenkin, ibid., p. 221 (1990).
35. R. Shekar and J. W. Evans, ibid., p. 243 (1990).

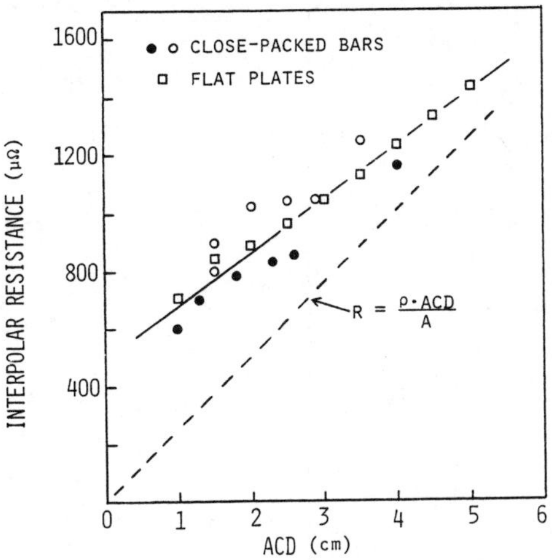

FIG. 1. Effect of ACD on interpolar resistance of 10 kA pilot cell with 5° sloped electrode system.

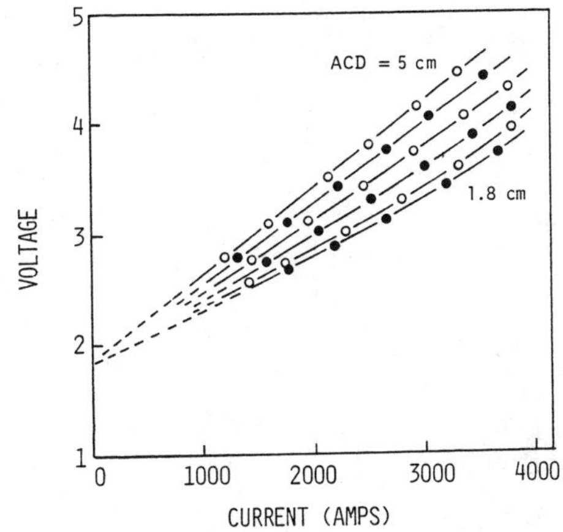

FIG. 2. Effect of ACD reduction on effective bath resistivity of 10 kA pilot cell with 5° sloped electrodes. Symbols same as Fig. 1.

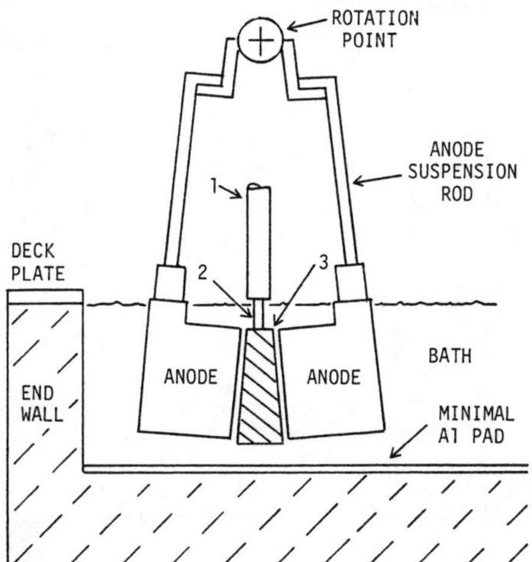

FIG. 3. Schematic representation of near-vertical electrode system with top-entering cathode lead.
1. Cathode lead
2. Cathode collector bar
3. Near-vertical cathode

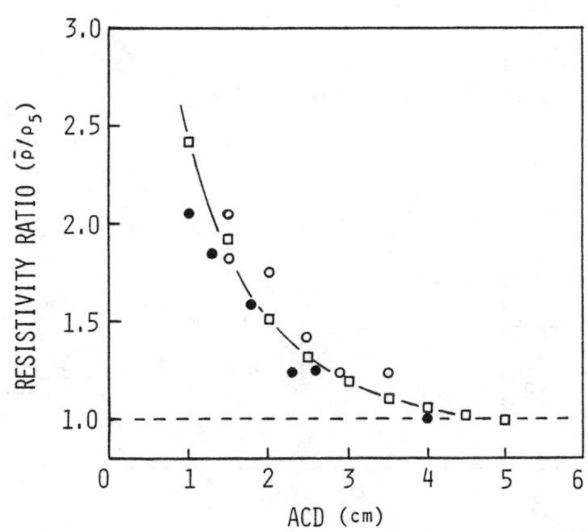

FIG. 4. Voltage-current plots for near-vertical anode that was lowered incrementally by controlled adjustments from about 5 cm to 1.8 cm.

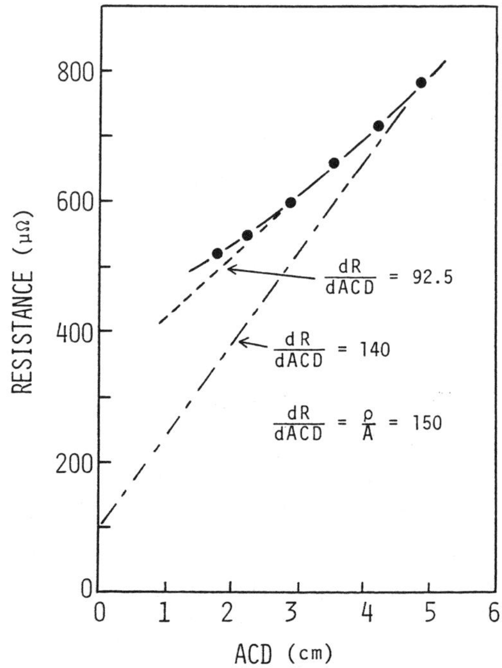

FIG. 5. Measured cell resistance vs. ACD for near-vertical anode/cathode pilot cell. Resistance gradients are μΩ/cm, ρ = .48 Ω-cm, A = 3160 cm , R(cathode + anode) = 100μΩ.

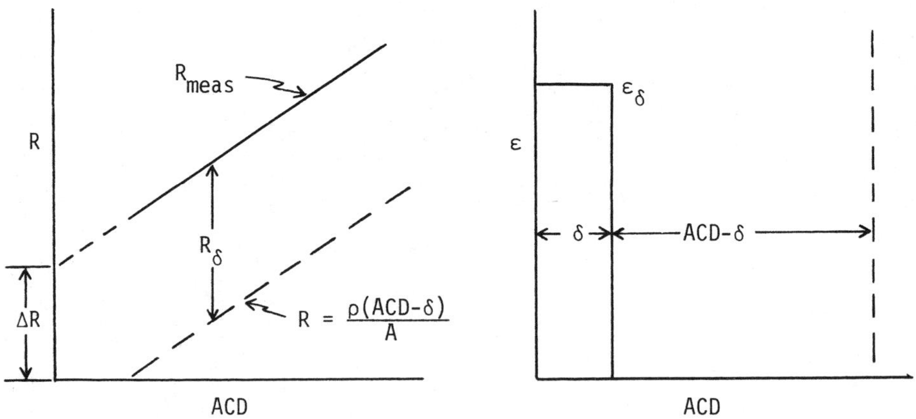

FIG. 6. Schematic representation of interpolar resistance, which is comprised of two components: R_δ due to a gas-containing layer, and $R_{(ACD-\delta)}$, the resistance of the gas-free region (see Equations 4 and 5)

Intalco potlines' operation with lithium-modified bath chemistry

R.A. Hawkins
Intalco Aluminum Corporation, P.O. Box 937, 4050 Mt. View Road, Ferndale, Washington 98248, U.S.A.

Abstract

The first large-scale operation on lithium-modified bath at Intalco began in 1978. The conversion of the first potline was the subject of a paper presented at the 1983 annual meeting of TMS/AIME. The present paper deals with plant operation and technical results for the years 1985 through 1987 when the remaining two potlines were converted and the lithium fluoride level was increased to 2% (+8% free aluminum fluoride). Bath temperatures averaged less than 947°C, and plant current efficiency averaged 92.4% over the three years.

Keywords

Aluminum fluoride; bath chemistry; Intalco; lithium carbonate; lithium-modified bath; and TAC.

Introduction

Intalco is located on Puget Sound 100 miles north of Seattle and 50 miles south of Vancouver, B.C. Figure 1 shows the plant's location outside Ferndale, Washington. The plant is situated on the coast with a deep water port allowing direct alumina delivery by ocean-going vessels. Figure 2 points out the alumina unloading facility with ore being transferred from the pier silos to the dry scrubbing system. Power is supplied by the Bonneville Power Administration which operates a series of hydroelectric dams on the Columbia River and other area rivers.

The Intalco facility consists of three potlines with 240 pots per line. Originally built in the late 1960's, the plant utilizes Pechiney technology with sideworked pots and prebaked anodes.

Each potline is comprised of four rooms of sixty pots each as seen in Figure 3.

FIGURE 1

The reduction cells at Intalco were originally designed to run at 115 ka. Improvements in emission collection, pot design, and operating procedures allowed increases in amperage to 137 ka by 1987. One of these operating changes was the conversion to lithium-modified bath.

FIGURE 2

History

Intalco's experience with lithium started in the summer of 1974. Two pots in Potline B were operated with a 1% lithium fluoride (LiF) bath. In November of 1974, Potline A joined the test by adding an additional two pots.

A third phase of the testing began in August, 1976, with the addition of ten test pots in Potline C. Based on encouraging results, the remainder of the 240 pots in Potline C began normal operation with a 1.5% LiF bath in April, 1978. The procedures used for the conversion and success of the Potline C operation were reported by Cheney(1).

FIGURE 3

Potline C was operated with 1.5% LiF bath for three and one-half years. The operating characteristics and technical results were monitored closely during this period. It was apparent by this time that the productivity of the Intalco potlines could be improved by including lithium in the bath. Potline A was converted to lithium in June, 1982.

During the initial period of operation with two of three potlines at 1.5% LiF, it was learned that lithium in the aluminum could create problems in the cast house and for certain customers. Conversion of Potline B was delayed so its uncontaminated metal could be used for existing orders with critical lithium specifications.

Lithium in the Metal

With two of the three potlines using a 1.5% LiF bath, the metal leaving the lines contained 12 to 14 ppm lithium but arrived in the cast house with approximately 8 to 10 ppm lithium. The reason for the decrease, even without treatment, was due to normal lithium oxidation at ladle temperature. If the transport ladle could be held for 14 to 16 hours, the problem would correct itself.

Lithium in the metal created problems in several areas. When the lithium content in molten aluminum was 5 ppm or greater, a tendency to oxidize at the float/spout interface was encountered. This oxidation resulted in metal flow control problems. Inconsistent flow coupled with a tendency to form a crust in the ingot head, particularly with magnesium-containing alloys, led to surface quality problems and scrap. It has been widely reported for a number of years that lithium at the 3 ppm level was responsible for a blue haze corrosion in foil products. More recently, lithium at 1 to 2 ppm has been shown to create problems with lack of adhesion in foil/paper or foil/plastic laminates which are used in the packaging industry.

FIGURE 4

With those concerns, cast house personnel undertook a program to develop practices to control lithium content in ingot products. The result of this program led to the decision to purchase the rights to use a treatment system developed by Alcan. The TAC system, "Treatment of Aluminum in Crucibles" reported by Dube and Newberry(2), was installed in June of 1984. With the TAC installation, Potline B was placed on lithium in June of 1984. Figure 4 is a picture of the Intalco TAC facility. The TAC operating results were discussed in a 1986 AIME paper by Freudenberger(3).

Operating Results

The effect the conversion to lithium had on the operating parameters for each potline was in most cases similar, the differences being due primarily to the operating conditions surrounding the time the conversions were made.

Potline C reached its LiF target in one month. Several alumina changes and adjustments to the lithium carbonate (Li_2CO_3) addition table had it operating at a 2.0% LiF level. Potline A moved to its target more slowly, taking 24 weeks. Potline B showed the effect of the tramp lithium in the system from the recycling of anode crust and ladle cleanings before the line started regular lithium carbonate additions. See Figure 5.

Bath temperatures in all three potlines dropped an expected 15°C as the lithium carbonate additions were made. Potline B's aluminum fluoride level remained higher due to the different make-up table in effect at the time. This difference resulted in the additional 4° to 6°C temperature decrease indicated. See Figure 6.

The decrease in the excess aluminum fluoride due to the chemical combination of Li_2CO_3 and AlF_3 was evident in Potline A and Potline C. Changes in the AlF_3 addition table stabilized both lines at around 7.8%. Potline B was using a more aggressive addition table going into the conversion and did not experience a sharp drop in the free aluminum fluoride. In 1986, that addition table was changed, and the line moved toward an 8.0% target. See Figure 7.

Potline C was increasing the metal height as the line came off a power curtailment. A cooling trend was countered with a tapping table change in 1979. This same method was used to control Potline A. The tapping table for Potline B was not changed and the metal pad remained constant. See Figure 8.

The steady decrease in the bath height of all three potlines was the result of a change in operating philosophy. The pots performed better with bath levels below six inches. In 1986, the plant began using 100% Kwinana alumina. The higher sodium content of this ore required regular bath tapping and storage of excess rod shop crust to maintain these targets. See Figure 9.

A decrease in the number of anode effects in both Potline B and Potline C corresponded to their lithium conversion. The Potline A conversion preceded a slight increase in their anode effect frequency. Several projects designed to reduce anode effects, including the installation of a computer anode effect prediction system, mask the actual effect the lithium conversion had on each potline. The results of the anode effect prediction system were reported by Meyer and Earley in 1986(4). See Figure 10.

The correlation between pot voltage and lithium content was also hard to pinpoint as computer parameters and work rules were being changed during the conversion of both Potline A and Potline B. Potline C results were masked by the return from a power curtailment in 1977. See Figure 11.

Potline C returned from the power curtailment 2100 amps above their previous operating level. The amount this increase can be credited to lithium was placed in question by the Potline A operating level prior to its lithium conversion. The slight drop in Potline B amperage was in controlled response to poor market conditions at that time. See Figure 12. A steady improvement in the specific power consumption following the lithium conversion was evident in all three potlines. See Figure 13.

Intalco potlife took several excursions in part due to power variations and operating level changes. Potlife in Potline C and Potline B improved after their conversion to lithium. The drop in potlife for Potline A rebounded and continued to improve. See Figure 14. The improvement in current efficiency that took place in all three potlines during the lithium conversion continued to be evident. See Figure 15.

Intalco has operated twelve years with one or more potlines on lithium-modified bath. Table I lists the plant annual technical results for 1985 through 1987. The reference to "free" AlF_3 accounts for the combination of aluminum fluoride and lithium fluoride to form lithium cryolite. Intalco uses the convention: (% free AlF_3) = (% excess AlF_3) - 1.077 x (% LiF)

TABLE I. Combined Technical Results

	1985	1986	1987
Current efficiency (%)	92.49	92.46	92.35
Amperage (ka)	135.1	135.9	136.9
Power consumption (dckwh/lb)	6.134	6.143	6.138
AlF_3 consumption (lbs/1000 lbs)	19.71	21.62	22.09
Li_2CO_3 consumption (lbs/1000 lbs)	1.53	1.47	1.46
Net carbon (lbs/1000 lbs)	404.3	404.3	404.7
AlF_3 (free %)	8.25	7.92	8.09
CaF_2 (%)	4.71	4.81	4.87
LiF (%)	1.85	1.94	1.95
Temperature (°C)	942.7	945.7	946.8
Metal Height (inches)	12.18	12.22	12.23
Bath Height (inches)	5.68	5.75	5.74

Conclusion

Intalco potlines continue to operate very well with a LiF bath. A sustainable improvement in current efficiency and specific power consumption continues as amperage increases have pushed plant output to above 300,000 tons per year.

References

1. R. Cheney, Light Metals, p. 519 (1983).
2. G. Dube and V. J. Newberry, Light Metals, p. 991 (1983).
3. W. Freudenberger, Light Metals, p. 913 (1986).
4. H. J. Meyer and D. G. Earley, Light Metals, p. 365 (1986).

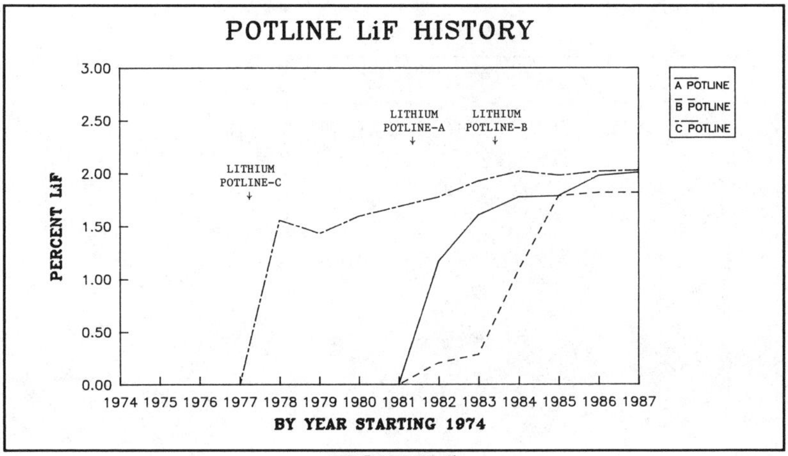

FIGURE 5

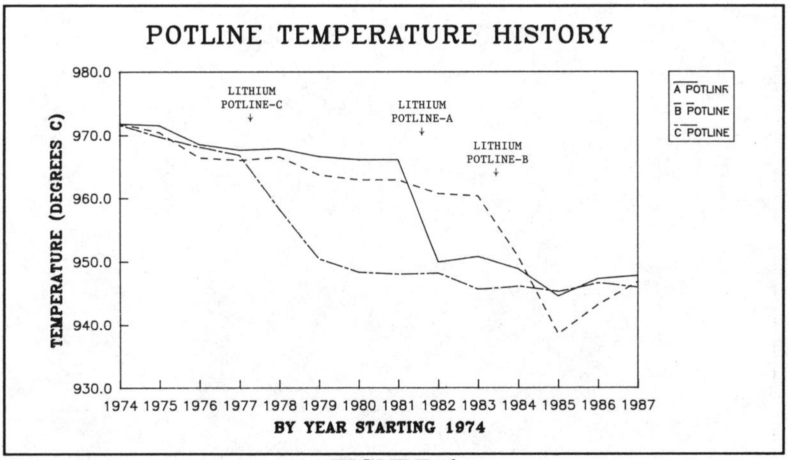

FIGURE 6

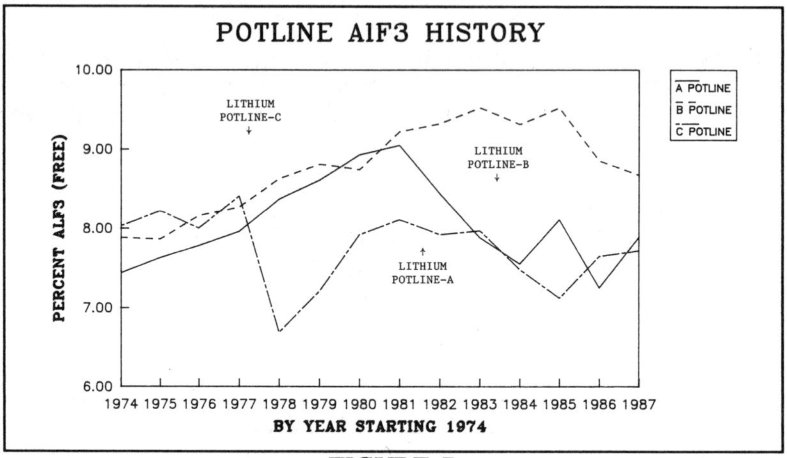

FIGURE 7

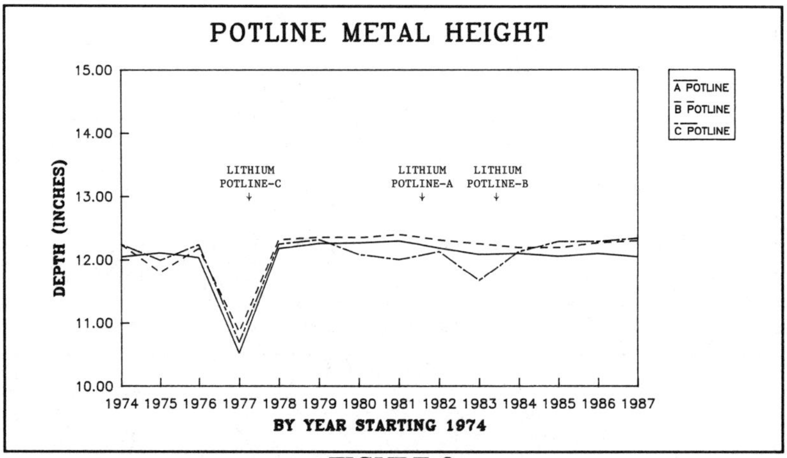

FIGURE 8

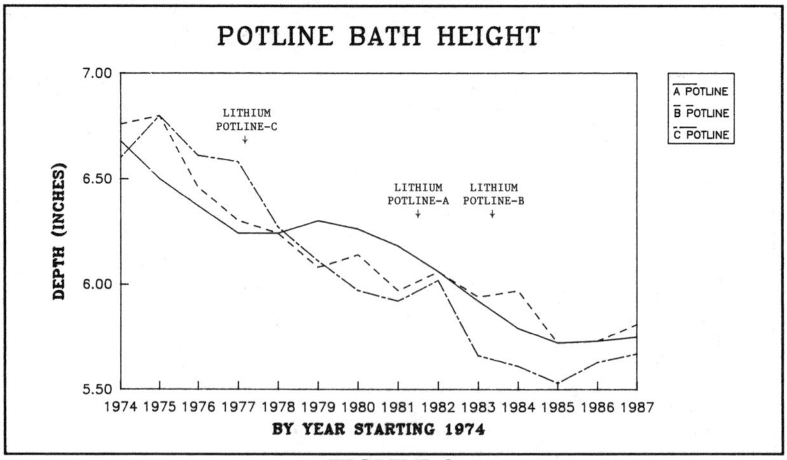

FIGURE 9

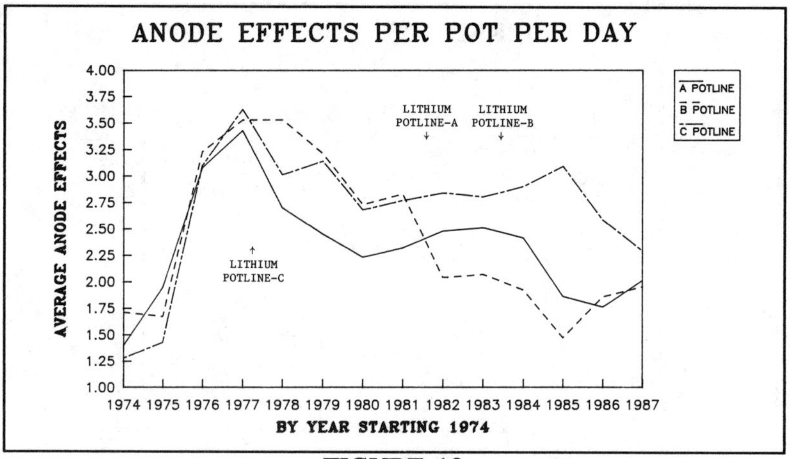

FIGURE 10

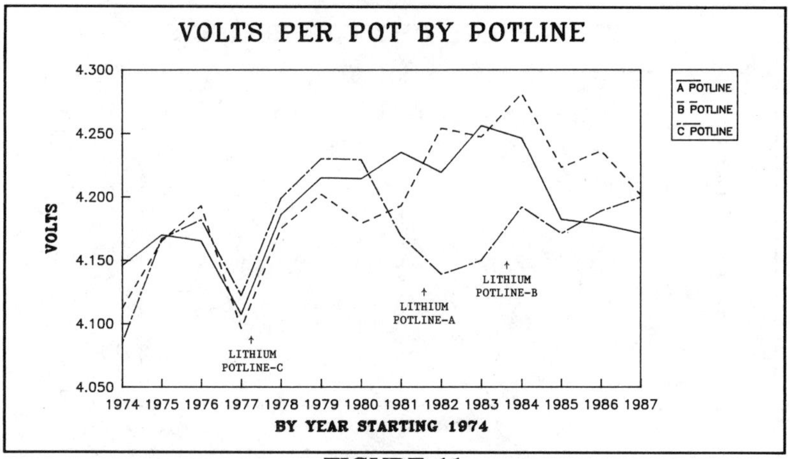

FIGURE 11

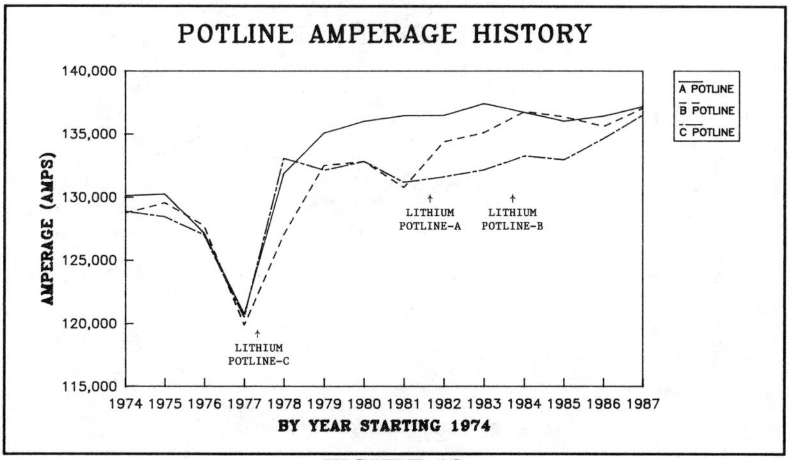

FIGURE 12

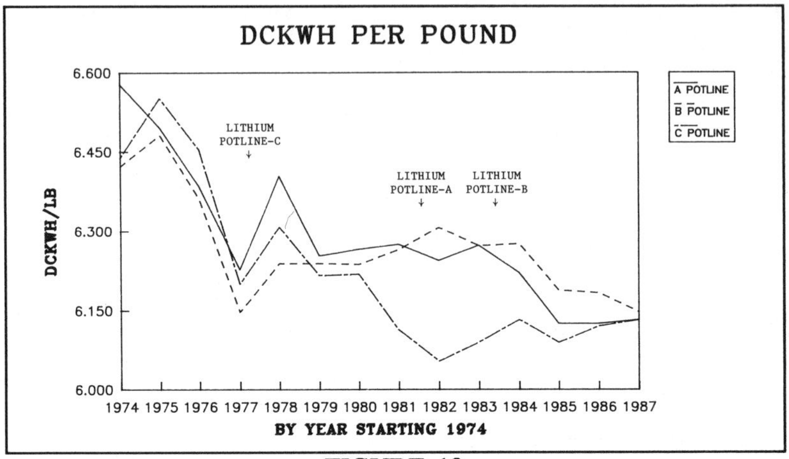

FIGURE 13

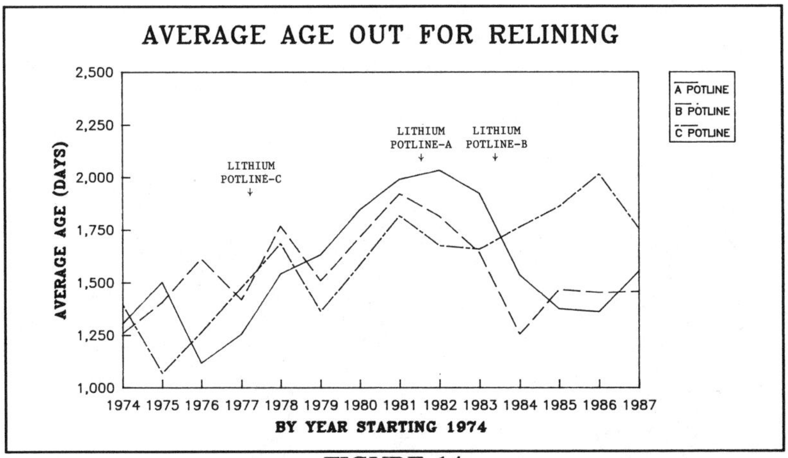

FIGURE 14

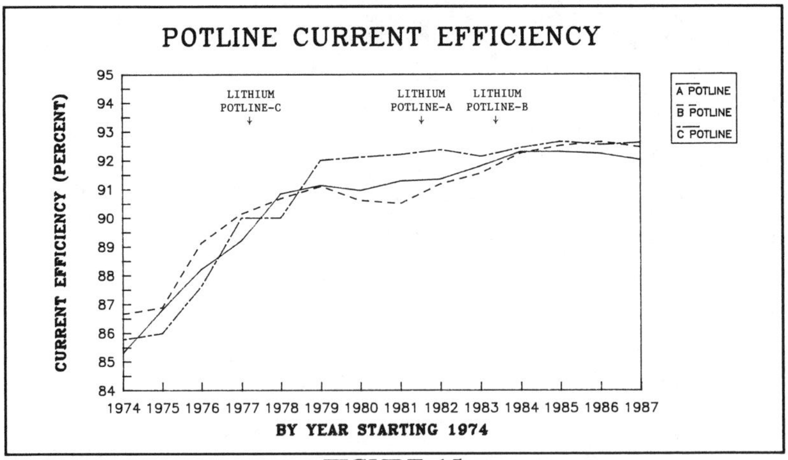

FIGURE 15

The production of Al-B, Al-Ti, and Al-Ti-B alloys by electrolysis

Wei Qing Bin, T.A. Utigard and J.M. Toguri
Department of Metallurgy and Materials Science, University of Toronto, 184 College Street, Toronto, Ontario, Canada, M5S 1A4

ABSTRACT

The production of Al-B, Al-Ti and Al-Ti-B alloys by fused salt electrolysis were investigated in a laboratory scale Hall-Heroult cell at 970°C. The maximum content of boron in the Al-B alloy was about 0.1 mass % and that of titanium in the Al-Ti alloy was about 1.5 mass %. However, when B_2O_3 and TiO_2 were added to the bath simultaneously, the maximum contents of boron and titanium in the Al-B-Ti alloy were 0.8 mass % and 4.5 mass % respectively.

The surface phenomena between the Al alloy and electrolyte as well as the Al alloy and the graphite crucible were observed by x-ray radiography. When a conventional electrolyte is used, the aluminium does not wet the graphite and forms a contact angle of approximately 140-160°. On the addition of TiO_2 and B_2O_3, the wetting behaviour changed noticeably, leading to an improved wetting of the cathode by the aluminium drop.

The microstructures of the three alloys were examined by scanning electron microscopy.

INTRODUCTION

Aluminium-titanium-boron alloys have been found to be excellent grain refining agents (1). The addition of only one mass percent of titanium to pure aluminium refines the grain size by a factor of 10^{-1} for all cooling rates (2). Further, the addition of titanium and boron to aluminium has a significant effect on its mechanical properties. The tensile strength increases, and both the percent elongation and the hardness decrease with increasing contents of Ti and B (3,4).

There have been a number of studies on effective methods of adding both titanium and boron to aluminium. Presently, Al-Ti-B alloys are produced either by thermal reduction of KBF_4 and $TiBF_7$ using aluminium or by heating a mixture of pure Al, Ti and B to temperatures above 1200°C.

In the present investigation, the possibility of directly producing Al-Ti-B alloys by the electrolysis of Al_2O_3, TiO_2 and B_2O_3 added to a Hall-Heroult cell was studied. This procedure could have both a cost and quality advantage as compared to the traditional ways of making these alloys.

In the Hall-Heroult process, liquid aluminium is formed and collected at the cathode. The cryolite based electrolyte contains normally 2 to 5% aluminium oxide. In this investigation, TiO_2 and B_2O_3 were added to the electrolyte in addition to Al_2O_3. The following chemical reactions with the corresponding standard Gibbs energy changes and equilibrium constants at 970°C indicate that both TiO_2 and B_2O_3 are thermodynamically less stable than Al_2O_3.

	ΔG°(kJ/mol)	Equilibrium Constant	
$4Al + 3TiO_2 \rightarrow 2Al_2O_3 + 3Ti$	-365.7	2.3×10^{15}	(1)
$2Al + B_2O_3 \rightarrow Al_2O_3 + 2B$	-336.7	1.4×10^{14}	(2)

Ti and B may also form by electrolysis according to the following reactions:

$$Ti^{4+} + 4e^- = Ti \text{ (cathode)} \qquad (3)$$
$$B^{3+} + 3e^- = B \text{ (cathode)} \qquad (4)$$

At 970°C, the decomposition voltage of Al_2O_3, TiO_2 and B_2O_3 are 2.21 V, 1.87 V and 1.65 V, respectively. Since the decomposition voltages of TiO_2 and B_2O_3 are lower than that of Al_2O_3, these oxides will decompose during the electrolysis of Al_2O_3 leading to the formation of Al-Ti-B alloys.

Based on the above thermodynamic analysis, a direct production of Al-Ti, Al-B and Al-Ti-B alloys by fused salt electrolysis in a cryolite solvent is a promising alternative.

EXPERIMENTAL

A. Alloy Electrolysis

For these experiments approximately 60 g of electrolyte containing 88 mass % Na_3AlF_6, 5 mass % Al_2O_3, 5 mass % MgF_2 with additions of 1 mass % TiO_2 and/or 1 mass % B_2O_3 was used. The cryolite ratio was 2.7 and approximately 2 g of aluminum was added.

The electrolysis was performed at 970°C at a cell voltage of 4.0 V and cell current of 2.0 A. The cathode-anode distance was maintained at 2 cm.

B. Radiographic Studies

The different phenomena occurring in the cell were observed using an x-ray radiographic system. This system is shown in Fig. 1 and is identical to that used in previous investigations (5-7).

The experiments were carried out in graphite crucibles (ID = 23 mm), both with and without a BN lining. The graphite crucible worked as the cathode and a graphite rod (OD = 12 mm) immersed in the melt from above worked as the anode.

The standard electrolyte used contained 88 mass % Na_3AlF_6, 3.0 mass % Al_2O_3, 5.0 mass % AlF_3, and 4.0 mass % CaF_2. Different amounts of TiO_2 and B_2O_3 were added to this mixture. The mass of electrolyte used for each experiment was 15 g. The mass of aluminum used varied between 2-5 g.

RESULTS AND DISCUSSION

Series A: Alloy Electrolysis.

The results of the electrolysis of the standard bath containing 1 mass % TiO_2 is shown in Fig. 2 where the Ti content in the metal phase is plotted as a function of the duration of electrolysis at 970°C. The Ti content increases almost linearly with time up to a maximum of 1.5 mass % at which point it becomes constant. Under the experimental conditions, the time to reach saturation is 120 min. The rate of Ti transfer is 1.49×10^{-2} mass %/min. The value of 1.5 mass % Ti corresponds to the solubility limit of Ti in Al at 970°C(8).

The addition of 1 mass % B_2O_3 to the electrolyte results in an electrolytic transfer of approximately 0.1 mass % to the aluminum phase. This corresponds to the saturation value at 970°C as reported by Hansen(9). The time to reach this value is about 210 min as shown in Fig 3. The rate to transfer is 6×10^{-4} mass %/min which is much slower than that of Ti.

When both 1 mass % TiO_2 and 1 mass % B_2O_3 are added to the electrolyte, there is an enhanced rate of electrolytic transfer of both Ti and B to form an Al-Ti-B alloy as shown in Fig. 4. The rate of transfer of Ti and B are 3.4×10^{-2} and 6.5×10^{-3} mass %/min respectively. The maximum solubility of Ti and B is 4.5 and 0.8 mass % respectively.

It is interesting to compare the hydrogen content of the Al-Ti alloy produced by electrolysis and by the standard method of either the thermal reduction of KBF_4 and $TiBF_7$ or the direct melting of pure Al with Ti and B. As seen in Table 1, the H_2 content of the final Al-Ti alloy is significantly lower with the present method as compared to the standard methods. This may be partly due to the low temperature (970°C) and the presence of the electrolyte which protects the alloy from the atmosphere.

Series B: Cathode Phenomena.

As shown by the x-ray image in Fig. 5a, pure aluminum (99.7% Al) did not wet the graphite cathode at all. However, when TiO_2 and B_2O_3 were added to the electrolyte, the wetting of the cathode by the liquid aluminium improved slowly as electrolysis was carried out, as shown in Table 5 and Figs. 5b and 5c.

The contact angle before electrolysis was 156° (Fig. 5a), while after 30 min it was 144°. After 60 min of electrolysis (Fig. 5c), the aluminum had completely wetted the graphite crucible. The contact length between the aluminum drop and the graphite was initially 5.8 mm and after 60 min it was 14 mm.. These observations can be accounted for by either a decrease in the surface tension of the aluminum as the content of Ti and B increases as electrolysis proceeds, or by chemical interactions with the graphite substrate.

It was observed that as the TiO_2 content in the electrolyte was increased to 3 or more mass %, the aluminum drop tended to float up along the side wall as shown in Fig. 6. This phenomenon was also observed when 2 mass % TiO_2 and 2 mass % B_2O_3 were both added to the electrolyte. It is difficult to explain this behaviour based on density changes alone. The x-ray images show a foamy layer of electrolyte and gas surrounding the aluminum drop. One possible explanation is that the electrolyte is supersaturated in TiO_2. This would lead to the formation of a solid suspension which could stabilize the foam. This interesting observation is presently being studied further.

Radiographic observations of the alloys in the presence of the electrolyte, but not under conditions of electrolysis, indicated that the type of substrate significantly influenced the contact angle of the alloy. The results are summarized in Fig. 7. In all cases, when the substrate was alumina, the Al-Ti, Al-B and Al-Ti-B alloys did not wet the alumina substrate even after contact for 1 hr. The contact angle varied between 140-150°, while that of pure aluminum was 150°. In the case of the graphite substrate, the Al-Ti and Al-Ti-B alloys were found to wet the substrate. Within 40 min of contact time, the wetting angle approached zero. However, aluminum and the Al-B alloy did not wet the graphite substrate.

Microscopic observation of the solidified aluminum phase in graphite crucibles are also shown in Fig. 7. Pure aluminum shows large crystals while the addition of ~ 0.2 % B decreases the grain size by almost a factor of 10. Addition of ~ 5.0 % Ti leads to a fine grained structure and shows the presence of two layers. These layers were also observed radiographically at 970°C. In the case of the Al-Ti-B alloy, three layers are observed. The structure is fine grained.

When the alloys are formed in alumina crucibles, only one layer is observed microscopically. The Al-Ti and Al-Ti-B samples show large needlelike crystals distributed in a fine grained aluminum matrix while the Al-B samples do not contain such structures. These needlelike structures are probably the equilibrium Al_3Ti phase in the Al-Ti and Al-Ti-B alloys. This type of structure has been observed previously in the Al-Ti system (10).

The presence of carbon prevents the formation of the needlelike Al_3Ti phase. As shown in Fig. 7, the Al_3Ti phase appears in the scanning electron micrograph as white granules. The exact composition of these white particles has not been defined, however, it appears that they are more dense than the aluminum phase which sits on top of it. A typical SEM micrograph of the Al-B alloy is shown in Fig. 8.

In the case of the Al-Ti-B alloy in contact with a graphite crucible, SEM micrographs of each of the three layers are shown in Fig.8 . As in the Al-Ti alloy, a white granular phase exists towards the bottom of the alloy sample. Between the Al-rich Ti-B phase and the white granular particles, there is a region which displays large petal-like and "Wang"-like structures. This latter term refers to the shape of the chinese character "Wang",王. The effect of carbon is demonstrated by the small white particles, while the presence of B and carbon leads to the "Wang" structure. Further studies on the chemical composition of these various structural units are being investigated.

CONCLUSIONS:

1. TiO_2 and B_2O_3 were successfully reduced to Ti and B by electrolysis in a laboratory Hall-Heroult cell using cryolite as the electrolyte. The maximum content of boron in the Al-B alloy was about 0.1 mass %, and that of titanium in the Al-Ti alloy about 1.5 mass %. The maximum contents of boron and titanium in Al-Ti-B alloy were 0.8 mass % and 4.5 mass % respectively.

2. Titanium containing alloys formed in graphite crucibles showed a complete wetting behaviour. On the other hand, Al-B alloys did not wet graphite. Al-Ti, Al-B and Al-Ti-B alloys do not wet alumina.

3. The microstructures of the alloys depend on both the alloy composition and the type of substrate.

4. The addition of Ti and/or B to aluminum results in grain refinement.

ACKNOWLEDGEMENT

The work was performed during one of the authors' (Wei Qing Bin) stay at the University of Toronto. He gratefully acknowledges the help that he received from Dr. Sai Wai Ip.

REFERENCES

1. M. Vader, J. Noordegraaf and E. Klein Nagelvoort, Proceedings 8th International Light Metals Congress, Leoben-Vienna, 1987, edited by F. Jeglitsch and R. Ratzi, published by Aluminium Verlag, Dusseldorf, Germany, 1988, pp. 464-467.

2. S. Hori and S. Saji, Report of Research Group, The Light Metal Educational Foundation, 1983, pp. 23-24.

3. Wei Qing Bin, unpublished results, 1989.

4. D.R. Sigler, C.L. Gibbons II, and W.L. Haworth, Met. Trans. A, 12A, pp. 905-907 (1981).

5. T. Utigard and J.M. Toguri, Met. Trans. B, 16B, pp. 333-338 (1985).

6. T. Utigard and J.M. Toguri, Light Metals, 1986, edited by R.E. Miller, published by Metallurgical Soc. of AIME, Warrendale, PA, 1986, pp. 405-413.

7. T. Utigard, J.M. Toguri and T. Nakamura, Met. Trans. B, 17B, pp. 339-346 (1986).

8. W.G. Moffatt, Binary Phase Diagrams Handbook, Genium Publishing Corp, Schenectody, NY, 1985.

9. M. Hansen, constitution of Binary Alloys, McGraw-Hill, 1958, p. 71.

10. S. Hori, H. Tai and Y. Narita, J. Japan Inst. Light Metals, 32, pp. 596-603 (1982).

Table 1: H_2 Content in Al-Ti Alloy

	Ti mass %	H_2 $cm^3/100g$
Alloys formed by Electrolysis	0.54	0.41
	0.80	0.59
	1.13	0.37
	1.58	0.44
Other Methods	0.6 - 1.5	0.7 - 1.0

Table 2: Change in contact angle between aluminum drop and graphite substrate during electrolysis.

Time (min)	0	15	30	60
Wetting angle (θ°)	156	149	144	0
Contact length between aluminium drop and graphite substrate (mm)	5.8	7.1	9.6	14

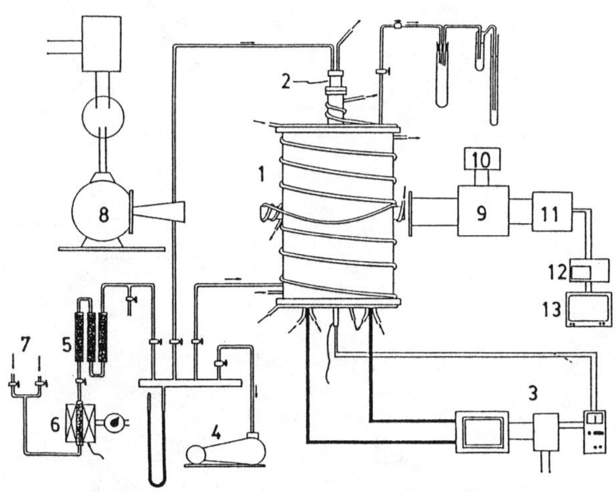

Fig. 1: Experimental apparatus. 1: Water-cooled resistor furnace, 2: Reaction tube, 3: Power and temperature controller, 4: Vacuum pump, 5: Drierite, 6: Copper getter furnace, 7: Inlet gases, 8: X-ray source, 9: Image intensifier, 10: Film camera, 11: T.V. camera, 12: Video recorder 13: TV monitor.

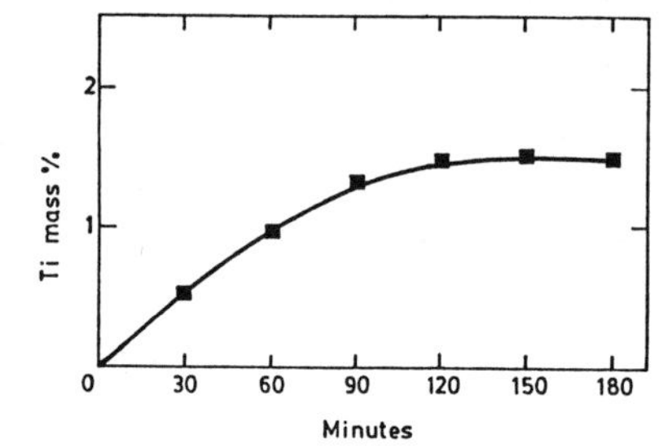

Fig. 2: Rate of Ti transfer to Al during electrolysis of TiO_2.

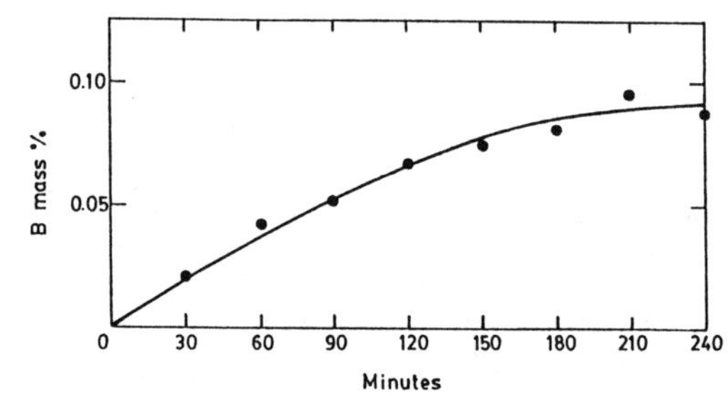

Fig. 3: Rate of B transfer to Al during electrolysis of B_2O_3.

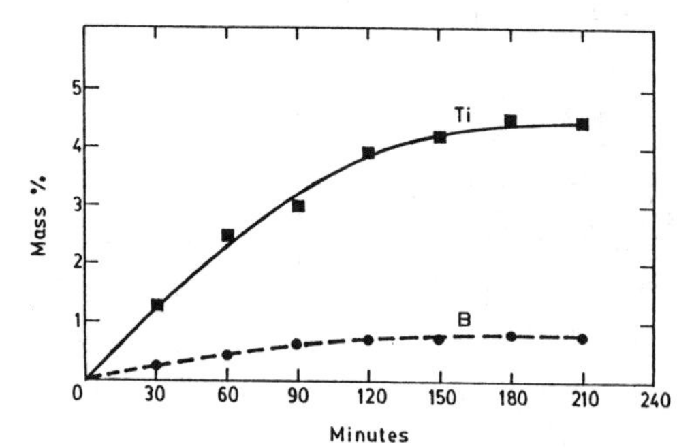

Fig. 4: Rate of Ti and B transfer to Al during electrolysis of TiO_2 and B_2O_3.

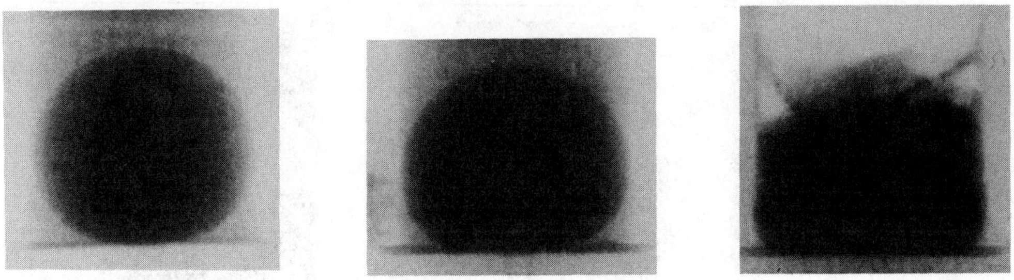

Fig. 5: X-ray radiographs taken at 970°C during electrolysis in a graphite cell.

a) Al drop at the begining of the experiment, contact angle 156°, contact length 5.8 mm.

b) After 30 min, contact angle 144°, contact length 9.6 mm.

c) After 60 min, contact angle 0°, contact length 14 mm.

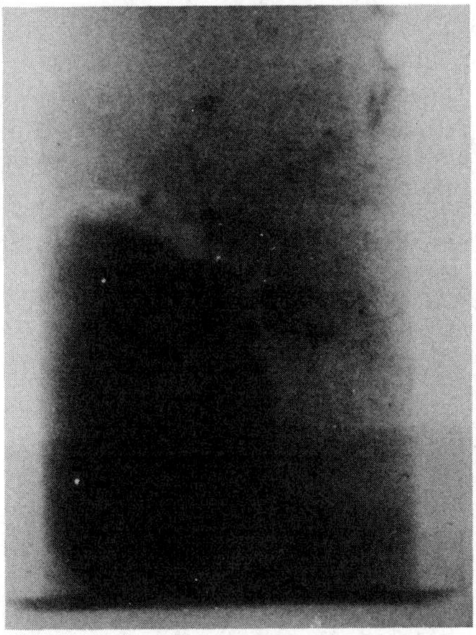

Fig. 6: X-ray radiograph taken after 20 min of electrolysis. Note the presence of a foam on the top surface.

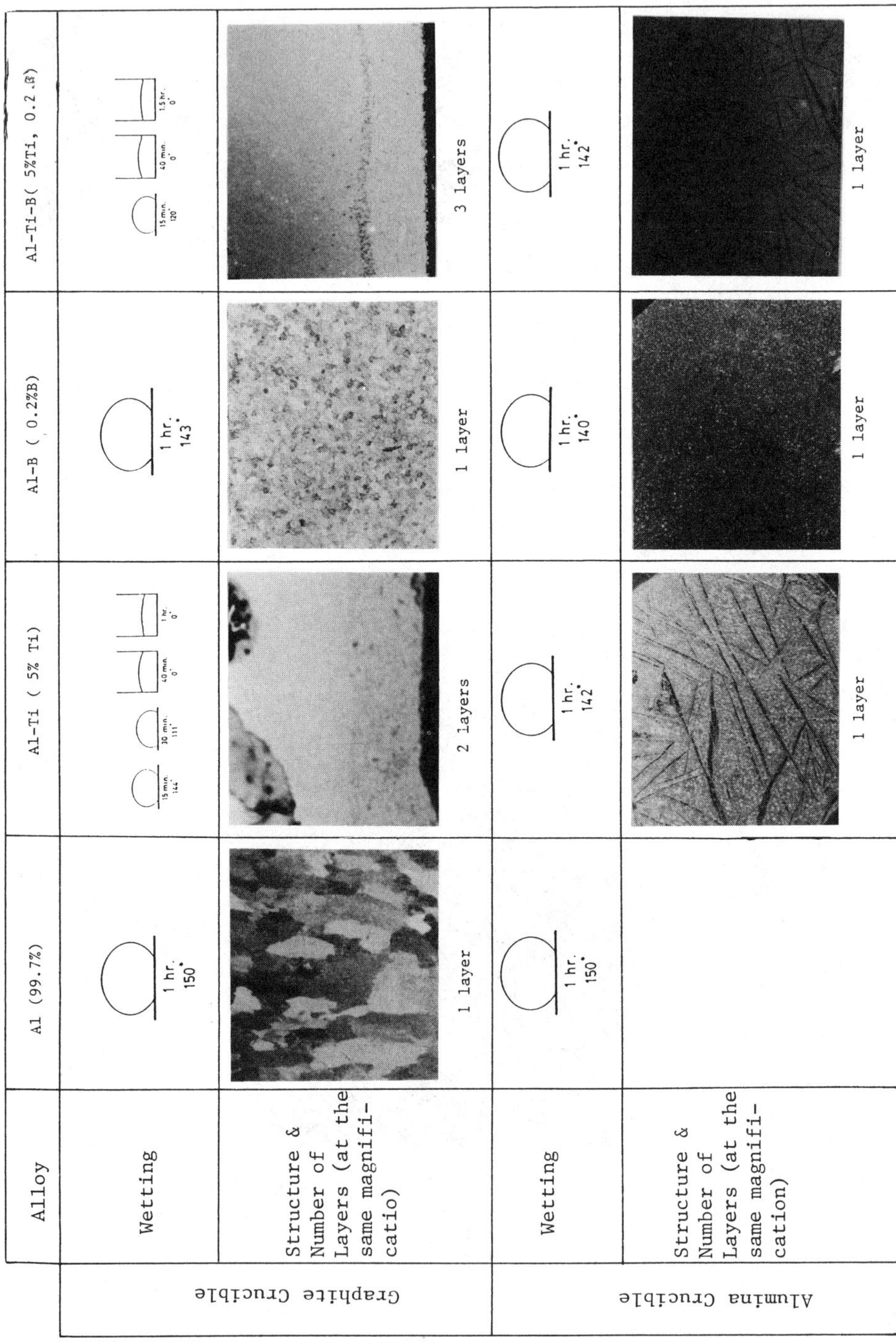

Fig. 7: Summary of the physical form of Al-Ti, Al-B and Al-Ti-B alloys in contact with electrolyte at 970°C.

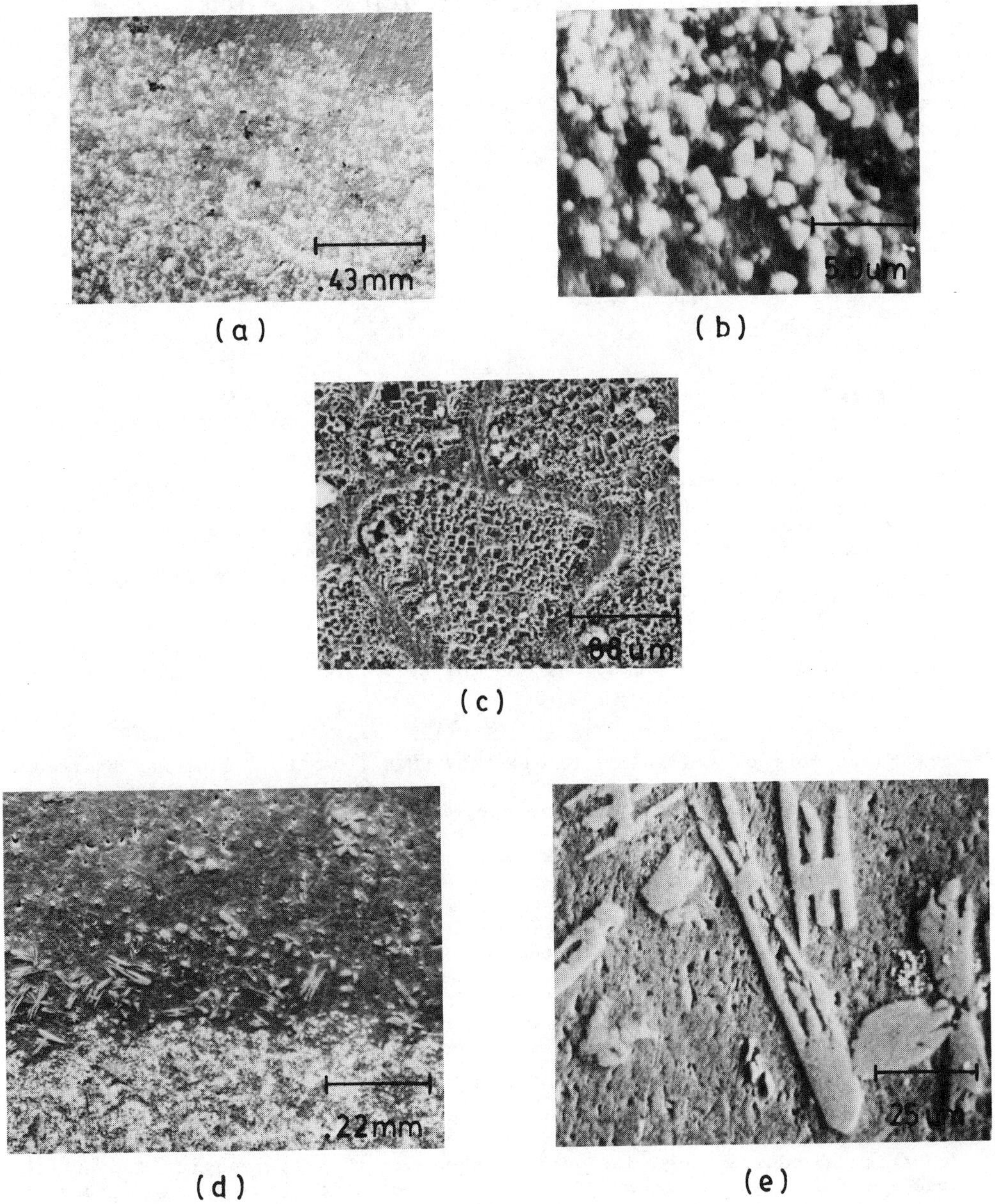

Fig. 8: SEM photo micrographs (a) Al-Ti alloy in graphite crucible, (b) white granular particles, (c) Al-B alloy in graphite crucible, (d) Al-Ti-B alloy in graphite crucible, (e) "Wang" shaped particles.

F-line in Saint-Jean-de-Maurienne, ten years after

A. Crapart and J.M. Peyneau
Aluminum Pechiney, Plant and LRF, B.P. 114, St-Jean-De-Maurienne, Cedex, 73303, France

Abstract

The first ALUMINIUM PECHINEY potline of AP18 electrolytic cells was started-up 10 years ago in the alpine site of SAINT-JEAN-DE-MAURIENNE (FRANCE).

It can be said that the general design and operating principles of this type of high amperage point-fed electrolytic cells were definitely marked by a considerable improvement of magneto-hydrodynamical and thermo-electrical models, the minimizing of cathode expansion, the bath alumina content control and the controlling of the operation parameters in an optimal restricted range.

The technical results of the F-line (especially the current efficiency and specific D.C. power consumption) and the cell-life of the first generation are outlined in this paper.

Introduction

The first cell of the SAINT-JEAN-DE-MAURIENNE (FRANCE) F-line was started-up on October 11th 1979. More than ten years later, a review of the technical results of this potline comprising 60 ALUMINIUM PECHINEY 180 kA cells is given in this paper. These 60 cells are of the AP18 generation (1) developed at the ALUMINIUM PECHINEY research centre in SAINT-JEAN-DE-MAURIENNE (FRANCE). A great deal of effort was invested in magnetic, electrical and technical design calculations, advantage was taken of the latest technological advances in computerised control to achieve a cell designed for stable and optimized operation. Today, more than 2,000 cells of this generation are in operation throughout the world.
After start-up, the F-line benefitted from the latest developments made by ALUMINIUM PECHINEY's researchers on the new AP28 generation (2) in terms of current efficiency (3 and 4), process control (5, 6 and 7) and lining (8).

Current Efficiency

ALUMINIUM PECHINEY's research engineers have studied the main phenomena governing the current efficiency (4). A number of fundamental parameters capable of considerably influencing the current efficiency of industrial cells have been isolated. So, it is possible to guarantee current efficiency results of above 94 % on industrial cells.

One important objective is to obtain a low aluminium dissolution speed by operating cells in a stable way at a reduced differential velocity between the bath and the metal, with weak horizontal currents and a high excess AlF_3. In addition, to keep the reoxydation of aluminium at a minimum, the cells must operate with a very low alumina content. Cell operation procedures should lead to these favorable working conditions by promoting a balanced anode current distribution, correct ridge positions, a good thermal balance of the cell and a suitable chemical composition.

Figure 1 gives the yearly current efficiency of the F-line. The low value of 1980 is due to the start-up period, some teething problems with the alumina feed devices and a bath aluminium fluoride excess still too low. After 1981, cell operation improved and the bath composition was changed (see figure 2). As a consequence the current efficiency increased and stabilized around 94 %, which represents the average for the last nine years.

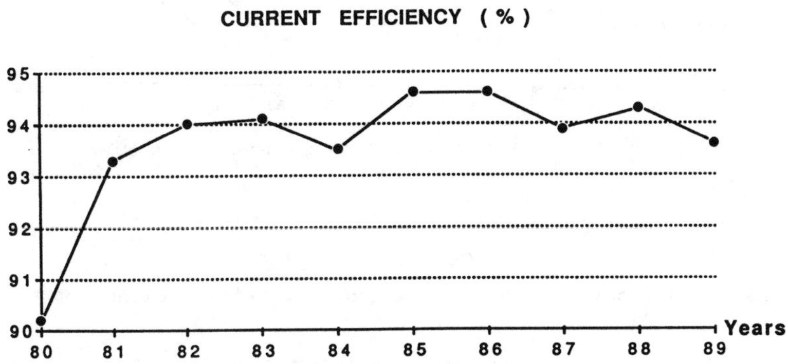

FIG. 1. Current Efficiency Versus Time.

It is to be noted that in 1989, the current efficiency was seriously affected by strikes in the plant and the start-up of 21 new cells.

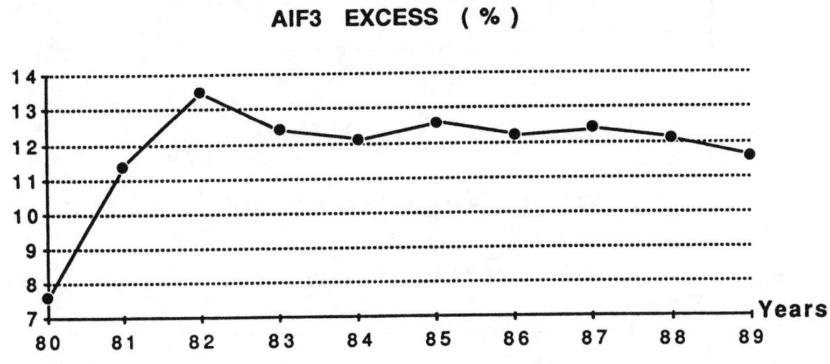

FIG. 2. AlF_3 Excess Versus Time.

D.C. Power Consumption

Figure 3 gives the yearly D.C. power consumption of the F-line. Apart from the first year, D.C. power consumption has been varying between 13.17 and 13.44 kWh/t Al. The average value for the last nine years is less than 13.30 kWh/t Al.

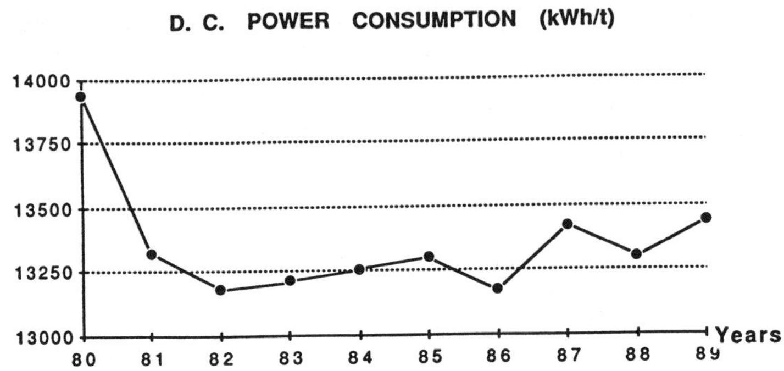

FIG. 3. D.C. Power Consumption Versus Time.

Same remark as for current efficiency results in 1989.

Current And Voltage

Figures 4 and 5 give the yearly current and yearly voltage of the F-line.

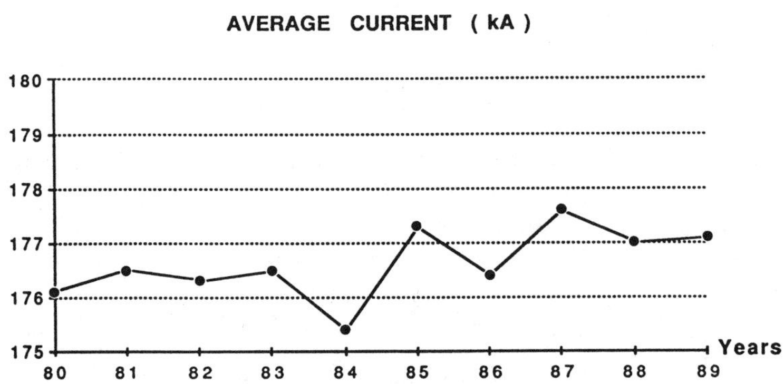

FIG. 4. Average Current Versus Time.

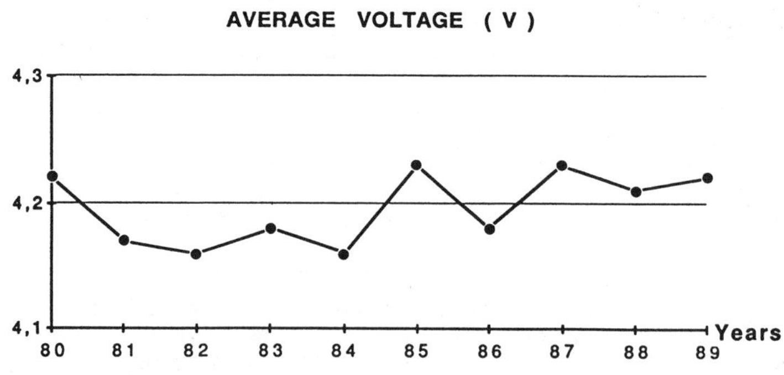

FIG. 5. Average Voltage Versus Time.

Other Technical Results

Figure 6 gives the yearly anode effect rate per cell and per day. Once the alumina feed device problems of the first year had been overcome, an average anode effect rate of 0.15 anode effects per cell and per day was obtained for the last eight years and 0.11 for the last four years.

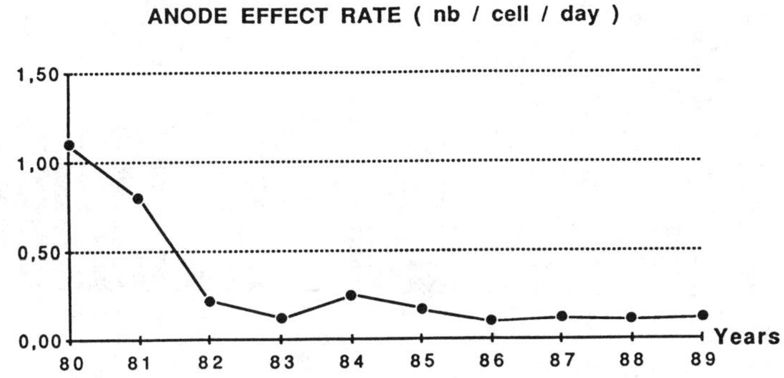

FIG. 6. Anode Effect Rate Versus Time.

Figure 7 gives the yearly iron and silicon content of the metal produced.

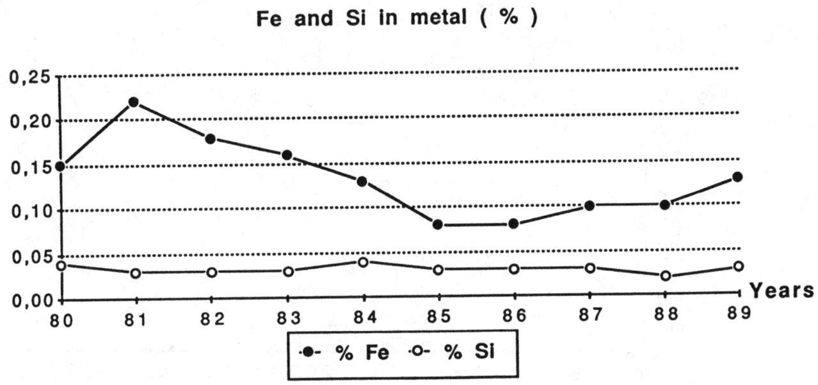

FIG. 7. Iron And Silicon In Metal Versus Time.

The iron content increase in the last three years is due to the stepping-up of the anode change cycle in 1987.

Cell Life And Cathode Drop

To have a good cathode life, the lining must be resistant to infiltrations, resistant to erosion by liquid products and resistant to impregnation by fluorinated and sodium products. These properties are obtained by a suitable cell lining design and more generally by optimizing cell design together with stable cell operations.

The AP18 cell design follows these principles and the cell life results of the first generation of the F-line confirms this.

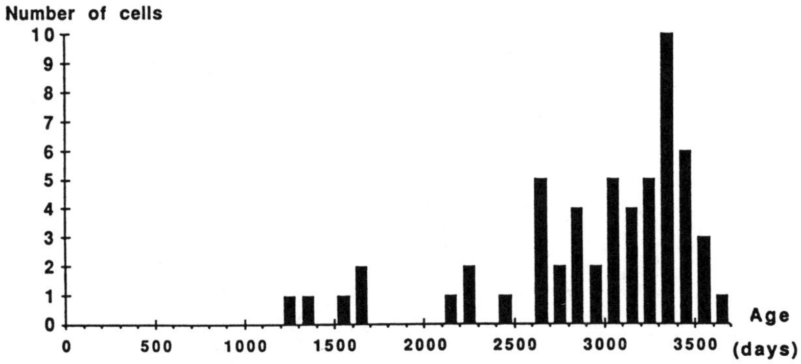

FIG. 8. Histogram of Age of Dead Cells.

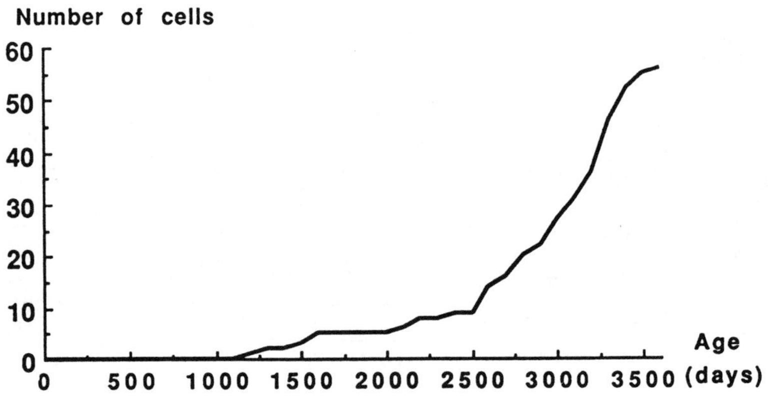

FIG. 9. Number of Dead Cells Versus Age.

As of January 1st 1990, 56 of the 60 first generation cells have died. The last four cells have an average life of 3,687 days i.e. more than 10 years. The average life expectancy is close to 3,100 days.

It is to be noted that the five first cells were stopped preventatively to carry out trials on lining and would normally have had a longer life.

Figure 10 gives the yearly cathode drop of the whole line. The decrease observed after 1986 is due to the new cells coming into operation.

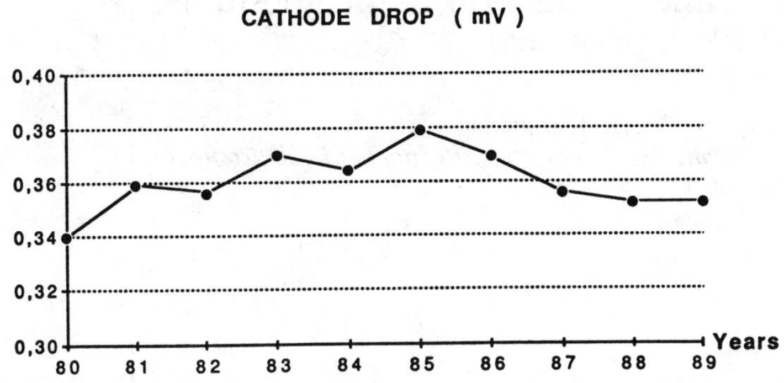

FIG. 10. Cathode Drop Versus Time.

Conclusion

As of January 1st 1990 the F-line results are very good, and even after 10 years of operation the technical results continue in the same way. The current efficiency (average of 94 %), D.C. power consumption (below 13.3 kWh/t Al) and the cell life of the first generation (more than 3,100 days average life expectancy) are to be specially noted. All the F-line results are confirmed by the results of the 7 other lines of the AP18 generation (4 and 8) in operation.

References

1. M. Keinborg and J.P. Cuny, ALUMINIUM PECHINEY 180 kA Prebake Pot, from Prototype to Potline, p. 449-460, Light Metals (1982).
2. B. Langon and P. Varin, ALUMINIUM PECHINEY 280 kA Pots, p. 343-347, Light Metals (1986).
3. M. Leroy, T. Pelekis and J.M. Jolas, Continuous Measurement of Current Efficiency, by Mass Spectrometry, on a 280 kA Prototype Cell, p. 291-294, Light Metals (1987).
4. B. Langon and J.M. Peyneau, Current Efficiency in Modern Point Feeding Industrial Potlines, p. 267-274, Light Metals (1990).
5. Y. Macaudière, Evaluation of Process Control of the Potline, CIM (1987).
6. M. Reverdy, Technological Advances in High Amperage Aluminium Smelting, 8th International Light Metals Congress, Leoben, Vienna (1987).
7. J.M. Peyneau, The Automated Control of Bath Composition of High Amperage Cells, CIM (1988).
8. J.M. Peyneau, Design of Highly Reliable Pot Linings, p. 175-181, Light Metals (1989).

Aluminothermic production of magnesium at the 100 kVA pilot scale

A.F. Saavedra and N.E. Richards
Manufacturing Technology Laboratory, Reynolds Metals Company, Sheffield, Alabama, U.S.A.

ABSTRACT

Through a campaign of tests using a 100 kVa submerged arc furnace, key parameters were determined for producing magnesium at atmospheric pressure from dolomite and magnesite using selected sources of scrap aluminum as reductant. Conversion efficiencies ranged from 37 to 90%. Starting from a premelted slag, Al_2O_3/CaO, molar ratio 1:2, we discovered the practical relationships dependent upon preconditioning the reaction zone chamber to, and management of, the condenser/receiver. Yields and operation improved with increasing electrode separation when the voltage was about 25 V and current 4000 amps, while feed rate and form of the reductant strongly affected the process. Although not designed for thermal efficiency, magnesium was produced at about 22 kwh/kg and, while coalesced metal was recovered, the main problem unresolved was the sustained coalescence of the condensed magnesium.

KEYWORDS

Magnesium production; metallothermic reduction; aluminothermic reduction; electric furnace.

INTRODUCTION

A few years ago, when the combination of availability, sourcing, and pricing of magnesium, essential for the production of aluminum can, body, and end stock, had the potential for negatively impacting our future strategies, we initiated several studies of magnesium production. From the technical aspect, those improvements of the "Magnetherm Process," published by Avery (1) were attractive because, for an aluminum company, that metal was the preferred reductant and the process could operate at atmospheric pressure.

An ensuing internal economic analysis for a plant with 5500 tonnes Mg/yr capacity, for an aluminothermic plant, showed that it would cost about $7.2 million dollars (1980) and yield a pretax return of 50%.

Consequently, we conducted two series of evaluations to provide a technical base against which business decisions could be made: Firstly, laboratory tests at the 20 kw scale to check the chemical and kinetic feasibility of the favorable thermodynamics. We reported on these in last year's CIMM meeting (2). Secondly, field tests at the 100 kVa scale to establish operability, methodologies, and how well we could manage the enhancements discovered from the laboratory studies in one order of magnitude scale up. This report covers the equipment, exploration of its operation, manifestation of the chemistry, material balances, and our efforts, at that time, for recovering coalesced magnesium.

EQUIPMENT

The experiments were conducted in a semi pilot scale apparatus which consisted of three main parts: a carbon-lined furnace vessel, a refractory-lined top, and a steel condenser. This apparatus is shown in Figure 1. The furnace was single-phase, top-entry electrode, bottom-current return, and rated at 100 kVa. The refractory-lined top accommodated the orifice and water-cooled seal for a 20-cm electrode, the vapor outlet duct, and feed port.

Fig. 1. View of 100 kw Furnace and Condenser for Aluminothermic Magnesium Production

The condenser was fabricated in two sections from heavy wall steel tubing. The upper section was connected to the lined top outlet duct with a flanged joint. This portion of the condenser was equipped with heating elements to preheat it to a predetermined temperature. The lower, or bottom, section of the condenser was detachable to allow cleanup of the condenser after tests were terminated. In addition, the condenser had an internal finger to increase its condensing surface; this finger had heating capability to help preheat the condenser to operating temperatures.

All the furnace components, reactor, lid, and condenser, were sealed together using high temperature gaskets ensuring that there was complete electrical isolation between components to prevent current flow through the outer steel shell.

The apparatus was set on a steel platform equipped with a tilting mechanism to tap slag from the reactor vessel.

Power to the furnace was provided by single-phase, water-cooled, transformers, which had the capability for different voltage outputs ranging from 24 volts to 96 volts. The top entry electrode was supported by a water-cooled copper clamp attached to an extending mast that could be adjusted vertically to establish the working distance between electrode tip and furnace bottom. The current was drawn from the furnace bottom by four graphite electrodes embedded in the rammed carbon furnace bottom.

Feed materials were delivered to the furnace by a 10-cm diameter screw conveyor, driven by a timer-controlled air motor. The premixed feed was placed in a cone shaped hopper located at one end of the screw conveyor; the other end was located directly above the feed delivery port.

A control panel housed instruments such as air meters, volt meters, power meter; also, switches for furnace tilting, electrode positioning, and feed control were located there.

Slag temperature was monitored through an observation port on the furnace top and measured by an optical pyrometer. A two-color pyrometer was used in some experiments, but its reliability was questionable. The condenser temperature was monitored by thermocouples located inside this compartment.

EXPERIMENTAL

Feed Preparation: Material used as components of the feed were calcined dolomite, calcined magnesite, and finely divided aluminum metal. In one experiment, aluminum dross concentrate was used instead of pure aluminum.

The components of the feed were mixed thoroughly in predetermined amounts. The amount of magnesite was determined to keep a dolomite/magnesite ratio which varied from 2.5 to 3.0 and to maintain an Al_2O_3/CaO ratio close to 1 in the resulting slag. The aluminum needs were determined by the MgO in the mixture of dolomite and magnesite and it was near the stoichiometric requirement for the reaction --

$$3MgO + 2Al \rightarrow 3Mg + Al_2O_3$$

or, more specifically,

$$3MgO + 2Al + 1.82CaO \rightarrow 3Mg + Al_2O_3 \cdot 1.82CaO$$

Table I summarizes the charge composition for the experiments conducted. The mixed feed was kept heated in a drying oven to avoid moisture pickup, and it was removed from the heated container only when it was required by the feeder demand.

Table I. Reduction Charge Summary

Test No.	Oxides kg	Dol/Mag Ratio	Reductant kg	MgO/Al Ratio	Reductant Type
1	165	2.86	42.5	2.19	Aluminum Scrap*
2	210	2.85	53.9	2.20	"
3	100	2.87	25.9	2.18	"
4	88	2.85	22.7	2.19	"
5	92	2.66	26.6	1.96	"
6	86	2.80	29.1	1.66	"
7	113	3.00	25.0	2.53	"
8	152	2.98	37.7	2.35	"
9	53	3.03	19.5	2.44	Dross Conc.**

* Scrap Al Content 97%; ** Dross Al Content 60%

Furnace Operation: Initial Stage -- Prior to the start of a test, prefused calcium aluminate slag was placed around the electrode which was shorted to the furnace bottom. The nominal composition of the starting slag was 50 w/o Al_2O_3 and 50 w/o CaO, which corresponds to the composition $7Al_2O_3 \cdot 12CaO$, and it falls near the eutectic point of the CaO-Al_2O_3 system. The calcium aluminate slag was prepared before the run, and in some instances the slag from a previous experiment was used. The initial power input was low, 20-25 kw, to avoid burning the electrode and furnace bottom by excessive open arc; on the other hand, too low a power would cause the electrode to freeze at the bottom. The power was gradually increased as the temperature of the initial charge increased. When the size of the pool was large enough to keep the electrode tip submerged, the power was increased to 50 kw and eventually to 70 kw. The power increments were accomplished by increasing the current input as well as by repositioning the electrode. As the slag became more fluid, the voltage dropped gradually, thus requiring the electrode to be repositioned. The slag was superheated to reach 1650°C, or over, as this was considered the working temperature for the process.

In some runs, the initial stage was extended for some time in order to get the condenser preheated to the required temperature. Other experiments had the condenser detached from the furnace for the slag melting phase to minimize the buildup of dust from slag melting operation in the condenser. The dust was a product of carbothermic reduction of magnesium that back reacted with CO and the product was a fine mixture of MgO and carbon.

In most tests, there was frozen slag left in the bottom of the furnace from the previous test, and additional crushed slag was added to build a molten pool 13 to 20 cm deep; this was considered necessary to achieve the resistance required for stable operation. Additional lime was used in some experiments to bring the Al_2O_3/CaO ratio close to 1.

Reduction Stage: When the starting slag was molten and superheated to over 1600°C, a rod sample of the slag was taken through the inspection port to determine the pool composition and depth. Argon was introduced at a rate of 24 l/min to purge the remaining air and to check for leaks in the reactor; then the argon flow was reduced to a rate of 1.5 to 2.5 l/min to ensure positive pressure inside the vessel.

The feed mixture, described previously, was introduced in the furnace via the screw type feeder at a pre-established rate which was maintained constant through the reduction stage. In some experiments feed rate was varied, but in most tests it was kept at 18 kg/hr.

The voltage and current were adjusted as the feed started to reach the furnace until a constant power input was established. To maintain a constant voltage, the electrode was repositioned until the desired voltage was achieved. Current input was maintained nearly constant.

The condenser was preheated during the initial stage and by the time the reduction period started, it had normally reached temperatures above 500°C. In the initial test, the condenser temperature only reached about 300°C; this situation was corrected by properly insulating the condenser and providing the necessary auxiliary heat to preheat the condenser to 700°C before the reduction stage began. When the partial pressure of magnesium had increased sufficiently, there was a rapid increase of temperature in the condenser. At this point, the auxiliary heat for the condenser was removed.

Feeding of the reduction mixture was continuous until it was exhausted or the pool had reached a level at which it was necessary to tap the furnace. Afterwards, operation of the furnace resumed as before.

Before tapping the slag, usually there was a holding period at low power to reduce the magnesium vapor pressure. If the tap were an intermediate one, only a portion of the slag was extracted and reduction operations resumed as soon as conditions inside the furnace were back to those before the tap.

During the final tap, as much slag as possible was poured out of the furnace, then the furnace was sealed again and left to cool with argon gas flowing to protect the magnesium produced. Attempts to tap molten magnesium from the condenser were unsuccessful and in all tests the magnesium produced was removed only after the apparatus had cooled sufficiently.

In order to produce an accurate mass balance after each experiment, the apparatus was taken apart and all products left inside after the final tap were removed and carefully weighed and sampled.

The coalesced magnesium was remelted in a gas fired furnace. For this purpose, fluxes normally used in magnesium foundries were used and the magnesium metal was cast into ingots.

RESULTS

Slag Pool Building: An average of 13 hours was required to melt the prefused slag and frozen skull. Figure 2 shows the specific power consumed to melt the slag as a function of the Al_2O_3-CaO ratio. The data indicated that slags with an Al_2O_3-CaO ratio close to 1 required less power to melt regardless of the MgO content in the slag. Adding lime to the slag left in the furnace brought the ratio close to 1 which otherwise could have been over 1.1 (test 8). Slags with excess CaO also required more power to melt, as seen in the same figure.

Feed System Operation: The initial test showed that magnesium metal vapors could be contained inside the apparatus without excessive leakage. In this test, the feed was placed in a hopper directly above the charging port, but the unexpectedly high angle of repose of the charge left the feed port open allowing air to enter the furnace.

A new feed system was designed using a screw conveyor to deliver the feed to the furnace. This allowed for a more positive sealing of the furnace as well as an accurate way to control the feed rate. A timer-actuated solenoid valve controlled the air motor driving the screw conveyor.

Preheating the feed to approximately 200°C before charging to the furnace reduced its moisture content and the amount of air introduced in the reactor. Preheating the charge also improved its flow characteristics.

Reduction Stage: Table II summarizes the results of the reduction stage. Feed rates ranged from 12.7 to 26.1 kg/hr of preheated charge. The average power consumption for this stage ranged from 75.3 kw to 96 kw and the corresponding specific power varied from 3.66 kwh/lb to 6.92 kwh/lb.

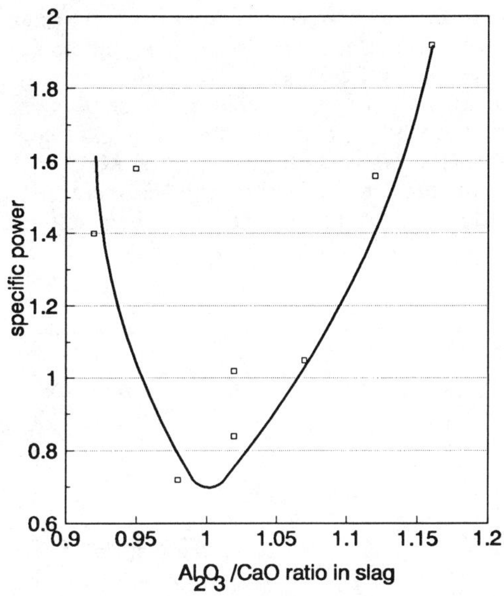

Fig. 2. Effect of Al_2O_3/CaO Ratio on Melting Calcium Aluminate Slags

Table II. Reduction Stage Summary

Test	Feed Rate kg/hr	Feed Ratio MgO/Al	Power Input Average kw	Percent Conversion Efficiency Mg	Percent Conversion Efficiency Al	Slag Ratio Al_2O_3/CaO
1	14.3	2.20	75.3	53.5	52.4	1.05
2	17.0	2.21	89.6	65.1	64.1	1.10
3	14.1	2.20	80.4	37.4	37.3	1.18
4	12.7	2.23	86.8	41.2	42.1	1.09
5	19.7	1.96	89.1	74.9	65.6	1.07
6	18.6	1.67	96.7	90.4	76.2	1.02
7	26.1	2.50	95.3	62.1	70.1	0.96
8	17.3	2.35	88.5	64.7	67.8	1.00
9	18.1	2.42	81.7	55.6	61.4	0.96

Mass balance calculations were performed for each test that helped understand the effects of the operating variables on the results. The magnesium conversion efficiency varied from about 37% to slightly over 90%. In most cases, the corresponding aluminum conversion efficiency matched closely that for magnesium, but best magnesium conversion efficiency was achieved with charges that had excess aluminum and high power input. The worst recoveries resulted in a test in which the Al_2O_3-CaO ratio of the slag was considerably over the preferred ratio of 1.

Some of the tests explored the effect of electrode position on the efficiency of the process. Operating the furnace with the electrode tip deep in the molten slag resulted in slow reactions and poor conversion efficiencies. According to mass balance results, better conversion efficiency of MgO to Mg was achieved when the reduction operations were conducted with the electrode tip high in the molten slag. Figure 3 shows the effect of voltage across the slag on the conversion efficiency of MgO to Mg. The tradeoff from operating the electrode tip high in the molten pool is increased electrode consumption. Maintaining the electrode tip near the furnace bottom resulted in the lowest electrode consumption. This is presented in Table III.

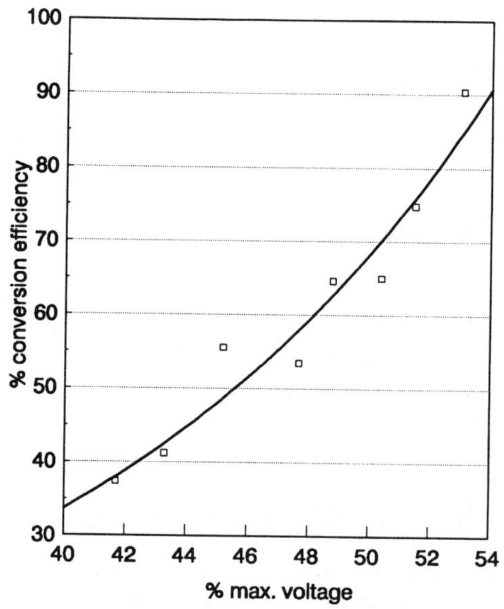

Fig. 3. Effect of Operating Voltage on Magnesium Conversion Efficiency

Table III. Conversion Efficiency and Electrode Consumption

Test	kw	% Maximum Volt.	MgO Conversion Efficiency	Elect. Consumption kg/hr
1	75.3	47.7	53.5	0.71
2	89.6	50.4	65.1	0.49
3	80.4	41.7	37.4	0.04
4	86.8	43.3	41.2	0.10
5	89.1	51.5	74.9	0.20
6	96.7	53.1	90.4	0.29
7	95.3	54.2	62.1	0.22
8	88.5	48.8	64.7	0.24
9	81.7	45.2	55.6	0.30

In some tests, slag was tapped from the reactor vessel and operations were resumed after the slag level was within the desired range.

Condenser Operation: The condenser temperature increased gradually during the pool building stage, but it did not exceed 300°C without additional heat provided by external heaters and proper thermal insulation. Without these provisions, the product accumulated was mostly finely divided powder. It was observed that tests in which the condenser temperature did not exceed 800°C, the condenser product was mostly powder, as shown in Table IV. The condenser temperature increased noticeably when the reduction stage started and there were temperature surges every time feed was introduced in the reactor. The rate of temperature increase of the condenser was affected by the feed rate. This is shown in Figure 4. Additionally, it was critical to maintain the apparatus properly sealed to prevent air leaks that resulted in production of MgO from powdered Mg.

Table IV. Condenser Operation and Results

Test	Condenser Temp. °C Start	Condenser Temp. °C End	Product % Coalsc.	Product % Powder	kwh/kg Mg Produced
1	342	835	62.5	37.5	46.1
2	450	860	82.7	17.3	34.2
3	350	580	0	100.0	57.5
4	540	730	0	100.0	62.4
5	520	900	70.0	30.0	26.2
*6	700	917	81.0	19.0	29.3
7	645	942	70.5	29.5	22.0
8	572	879	33.2	66.8	33.7
9	605	775	0	100.0	23.1

*Placed flux in condenser before test.

Typically, the material deposited in the condenser consisted of an initial layer of powdered magnesium covering the walls and bottom of the lower section. The next layer was mostly Mg droplets, some of which were partially fused together and some powder. The inside layer consisted of mostly large metallic buttons and only small amounts of powder.

Efforts to minimize the production of fine powder were unsuccessful as were efforts to tap the molten magnesium metal directly from the condenser. For this reason, the apparatus was left to cool with an argon cover before the condenser was removed and cleaned.

Use of dross concentrate as reductant resulted in a product that was contaminated by sodium and magnesium nitride. All product was finely divided.

DISCUSSION OF RESULTS

Based on findings in the bench scale experimentation (2), the experiments at the 100 kw scale were conducted following the general conditions observed earlier. Among these were temperature of the molten slag, preheating the condenser, slag composition as a result of reduction charge.

The magnesium conversion efficiency is affected by feed composition, feed rate, power input, and electrode tip position during the reduction stage.

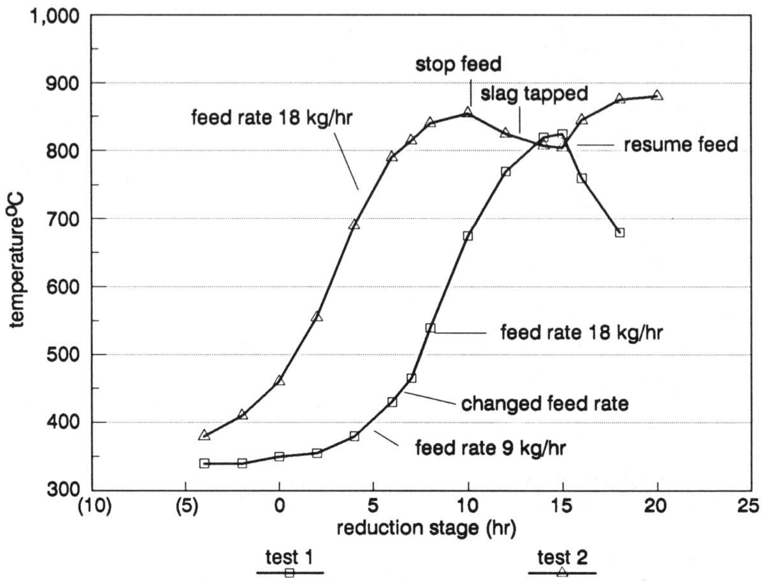

Fig. 4. Effect of Feed Rate on Condenser Temperature

Experiments where the MgO-Al ratio of the reduction charge was below 2 had the best magnesium conversion efficiency. Experiment 6 had a reduction charge MgO-Al ratio of 1.66 with a magnesium conversion efficiency of 90%, although the aluminum conversion efficiency only reached 76%.

Experiments with reduction charge composition near the stoichiometric ratio of MgO-Al of 2.24 had conversion efficiencies ranging from 37% to 65% for both magnesium and aluminum. Average power input varied from 75 kw to 90 kw and, in general, efficiencies were higher for tests conducted with high power input. However, the efficiency of the process is affected more by the electrode tip position than by the power input. It has been stated (3) that reaction in the Magnetherm Process takes place at or near the surface of the molten slag.

It was established that when the electrode tip was near the furnace conductive bottom, conversion efficiencies were low. In tests conducted with the electrode tip high in the molten slag pool, the conversion efficiencies were higher. As in the Magnetherm Process, in the aluminothermic process, the reactions seem to take place near the surface of the slag. This was also evidenced by the pressure increase in the vessel immediately after each charge reached the molten pool and by the temperature fluctuation in the condenser a few instants later.

The electrode tip position can be related to the voltage across the molten slag since it increases with the distance between the electrode tip and the conductive bottom.

Another important factor considered in the operation was the electrode consumption. Results indicated the consumption of electrode is affected by the voltage across the molten slag which is established by the electrode tip position. In some instances, excessive electrode consumption during the reduction stage led to contamination of the magnesium produced.

While preparing the reduction charges, it was necessary to consider the final slag composition to maintain the Al_2O_3-CaO ratio close to 1. This delivered a slag that melted at lower temperature resulting in energy savings. By allowing the slag to become richer in either Al_2O_3 or CaO, the process became less efficient and the slag required higher temperatures to remain fluid. A risk of operating at higher temperatures was that of reaching conditions where some carbothermic reduction took place resulting in contamination of the product due to back reaction between magnesium and carbon monoxide.

It was possible to remove part of the slag during a run with minimum losses if the tap time and amount of air introduced to the system were minimized.

The most important problem remaining is that of removing molten magnesium metal during furnace operation. Attempts were made in several tests to remove magnesium metal from the condenser while it was molten. However, because of the reactive nature of magnesium, these attempts failed.

The use of dross concentrate as reductant in this process has the drawback of excessive contaminants being present in the reducing agent. Results were affected by the lower-than-expected aluminum content, presence of sodium salts, iron, and aluminum nitride.

Thermal balances indicated that the system operated with a thermal efficiency of only about 25%. A large amount of energy was lost to the cooling water used to keep the joints of the furnace components at temperatures that prevented the seals from failing. Nevertheless, the energy demand per kilogram of magnesium produced reached 22 kwh and it is certain this figure can be reduced further. The theoretical energy required was calculated to be 4.5 kwh per kilogram of magnesium. It is believed that in an industrial scale operation the energy demand will be near 11 kwh per kilogram of magnesium.

CONCLUSIONS

1. It is possible to produce pure magnesium metal by aluminothermic reduction with MgO conversion efficiency near 90%. Careful control of furnace atmosphere is needed to avoid contamination.

2. The process favors electrode tip operation high in the pool. Electrode tip position near the bottom resulted in low conversion efficiency. Reactions take place more efficiently at or near the surface of the molten pool. Maintaining the electrode tip at or too close to the pool surface can result in excessive electrode consumption.

3. Charge preparation should take into account the MgO/Al ratio and the resulting slag, Al_2O_3/CaO, ratio. The former should be below the stoichiometric value of 2.24 and preferably around 2. The latter ratio should be maintained near 1.

4. The condenser should be preheated to temperatures over 750°C to promote the coalescence of magnesium metal. Excessive air leaks or contaminants from side reactions yield impure product, mostly in powder form. Means to enhance the collection of magnesium vapors need to be further investigated.

5. The use of aluminum dross as reductant resulted in a product contaminated with sodium and magnesium nitride.

6. Magnesium metal was produced with an energy consumption of 22 kwh per kilogram.

REFERENCES

1. J. M. Avery, U.S. Patents 3,761,247; 3,782,922; 3,994,717.

2. N. E. Richards, A. F. Saavedra, R. L. Porter, and L. M. Ruch, "An Evaluation of the Aluminothermic Production of Magnesium," The Metallurgical Society of CIM, 1989.

3. R. A. Christini, "Equilibria Among Metal, Slag and Gas Phases in the Magnetherm Process," Light Metals, TMS-AIME Proceedings, Las Vegas, 1980.

Mathematical modelling and computer simulation of aluminum fabrication processes

The paramount importance of thermal properties and coefficients in thermal process modelling

L.I. Kiss, A. Charette and R.T. Bui
Université du Quebec à Chicoutimi, Chicoutimi, Québec, Canada, G7H 2B1

Abstract

For a long time the progress made in the quantitative description of complex industrial processes had been slow due to the lack of means to solve the appropriate mathematical models. Although the difficulties associated with the handling of mathematical models for arbitrary 3D geometries, complicated boundary conditions, nonlinearities etc. have not yet been eliminated, it becomes more and more evident, that significant research efforts have to be made to strengthen physical modelling, for providing reliable input data base.

While - using advanced computer technology and sophisticated numerical methods - we can solve our basic set of equations with an accuracy of say 0.1 percent or less, the input data like thermal conductivities, specific heats, emissivities, surface heat transfer coefficients, contact resistances etc. are known only with large uncertainties like ±20% - if available at all.

A review is given to demonstrate the efforts to reach a well-balanced situation between the accuracy of numerical solving methods and that of the input data which describe the physical behaviour of substances in the system, and the interactions between the system and its environment.

Aspects of various problems that will be touched upon in this article include: critical evaluation of literature data, discovering the sources of scatter in results due to deviations in model concepts; the use of sensitivity analysis to clarify the effect of various input data on the overall performance of the simulation; the proper selection of set of properties to characterize the behaviour of matter under given load patterns; and finally, the efforts to develop new measurement methods and devices using microcomputer control and evaluation, making possible the study of thermal properties even in factory environments.

Keywords

Simulation; thermal processes; modelling; material properties; heat conduction; heat transfer coefficient; thermal measurements.

Introduction

Although simulation methods have shown significant, impressive development in the past decades, we do not possess procedures for treating industrial problems which require merely fundamental physical constants as input data. Building up solutions from very basic parameters like the mass and charge of an electron, the velocity of light, Planck's and Boltzmann's constants etc. is not only irrational in terms of the economy of labour and time,

---

* Technical University of Budapest, presently Invited Professor at the University of Québec at Chicoutimi

but also impossible due to the incompleteness of our knowledge.
Either to save work and time, or to fill the gaps in our ability to describe complex industrial processes in their totality, we have to generally use a large number of input data like coefficients, material properties, source terms, coupling or interchange factors etc. during modelling.
These quantities come from several branches of the science, reflecting the results of various physical models and hypotheses, or they can be of an empirical nature, in cases where theory has been unable to provide satisfactory quantitative description of a certain phenomenon.
As advanced mathematical models of industrial technologies are highly complex, with several dozens or hundreds of input data like material properties and other parameters in the differential equations and boundary conditions, the effect of a single parameter is often obscured. In other words, the sensitivity of the results of a model to changes in certain input data is low in a given situation. This fact is an advantage on the one hand, as errors in the estimation of not well-known data do not significantly affect the whole simulation (this makes sensitivity analysis a necessity), but, on the other hand, the efforts are utopian to determine those properties by parameter indentification using the given mathematical model. (In other words, we can not use that model to find a stable solution of the inverse problem.)
If a complex mathematical model is applied to a set of input data with significant - often unknown - uncertainties, it creates a situation when even a false model can be "verified" with a "proper" selection of parameters, when, for example the mathematical model is "tuned" to approximate a few experimental results by assigning a thermal conductivity value to a refractory brick which is characteristic for aluminium alloys.
The development of science has always produced new and attractive theories, offering more complex models to substitute the older, simpler ones. The appeal to apply them is strong, but it is very important to maintain harmony among the different aspects of a model, to ensure a balanced situation in accuracy, labour and time requirements. The application of a sophisticated theory with several parameters instead of a single empirical constant will not increase the accuracy of the whole simulation if adequate reliability of the new parameters cannot be ensured.
The introduction of new technologies and materials is a challenge for thermal process modelling. The application of plasma and laser beams for cutting and shaping, high-velocity extrusion, ultra-fast cooling and heating processes to produce new quality materials (glass-metals, microcrystalline structures) - all result in thermal loads and heating rates unusually high compared to those obtained in traditional applications. This makes it necessary to check the validity not only of the corresponding equilibrium or transport properties, but even that of the physical models.
The introduction of new materials (ceramics, composites etc) occurs generally faster than the ability to supply all the necessary data to describe their behaviour under various load conditions. These new materials also require changes in design and modelling philosophy having qualitatively new material laws and application criteria.
This paper does not attempt to give a systematic treatment of all the problems associated with the selection, availability and measurement of material properties and coefficients in thermal process modelling. It can merely call the reader's attention to this field, using examples to illustrate the difficulties and ways to overcome them.

Modelling examples

The amount of difficulty associated with supplying thermal data for simulation is basically determined at the very start, by the selection of the model itself, its complexity and adequacy. This is true even for a relatively simple and much studied process such as heat conduction.

Heat conduction

The simplest linear transient conduction model

$$\frac{\partial T}{\partial t} = a \nabla^2 T \qquad (1)$$

requires the knowledge of one single parameter (namely, the diffusivity of heat - $a = k/\rho c$, where k is the thermal conductivity, W/ms, ρ is the density in kg/m^3 and c is the specific heat, J/kgK) in case of boundary conditions of the first kind.
But in the case of second or third kind boundary conditions, the number of necessary independent properties increases to two, as the conductivity plays an explicit role in the corresponding equations. Also, in case of heat conduction with internal sources, a knowledge of the ρc volumetric heat capacity is required.
Handbook data show significant scattering especially in case of conductivities and diffusivities. The reasons are multiple. Material properties are functions of composition and temperature. Sensitivity to compositional changes is especially large in the high purity domain, which results in greater scattering of measured data for "pure" metals than for well defined alloys.
As heat capacity is an equilibrium property, it can be determined without measurable spatial temperature difference inside the sample, while for measuring thermal conductivity the creation of spatial temperature differences is an absolute necessity. This can result in systematic deviations among various measurement methods, when measurement results are assigned to temperatures. The situation is further complicated by the fact that certain measurement methods are primarily sensitive to conductivity, while others are sensitive to diffusivity. The systematic errors are principally different for the two cases. Conductivity measurements always require the direct or indirect (e.g. by using etalons) determination of heat fluxes, which relies on the perfectness of passive or active thermal insulation methods. The determination of diffusivity can be based purely on length and time measurements (as it is reflected by its dimension L^2T^{-1}), which have accuracies superior to those of temperature or calorimetric measurements.
Transient methods for the determination of thermophysical properties have another advantage, namely, the possibility of determining more than one parameter from a single experiment. Obviously, this practice leads to a more coherent data base, than using, for example, specific heat data from classical calorimetry, conductivity from a steady-state apparatus and finally, density from mass and volume measurements.
On the other side, in case of strong temperature dependence, the concept of diffusivity loses its original attractiveness. It is evident, from the quasi-linear heat conduction equation

$$\rho c \frac{\partial T}{\partial t} = \nabla [\, k(T)\ \nabla T \,] \quad , \tag{2}$$

that the simple substitution of a with a(T) in Eq.1 can lead to an error depending on dk/dT and on the square of the temperature gradient:

$$\rho c \frac{\partial T}{\partial t} = k\, \nabla^2 T + \frac{dk}{dT} (\nabla T)^2 \tag{3}$$

From a theoretical point of view, the $q=-k\nabla T$ Fourier law is only a first-order, linear approximation of the more general $q=f(\nabla T)$ function. At thermal equilibrium ($\nabla T=0$), the heat flux is zero, and from basic symmetry assumptions it is also evident that the McLaurin expansion of the above function cannot contain scalar terms. Such a higher order approximation cannot contain terms of even order, as follows:

$$q = -k_1 \nabla T - k_3 (\nabla T)^3 - \ldots \tag{4}$$

According to this theory, in case of high heat fluxes (gradients) we can expect the onset of a new, "second" conductivity parameter. Although there are no convincing experimental facts regarding the magnitude of k_3, it seems reasonable to suppose that it plays a role only in

the early time, near-surface, extra high-intensity heat conduction processes.
In the past few decades more attention has been paid to the modified, hyperbolic type heat conduction equation where also a new physical property, the relaxation time t_r (or in an equivalent formulation, the so-called velocity of propagation) has been introduced:

$$t_r \frac{\partial^2 T}{\partial T^2} + \frac{\partial T}{\partial t} = a \nabla^2 T \qquad (5)$$

Similar to the previous case, significant effect of the relaxation term can be expected only at extremely fast temperature changes and correspondingly, in small spatial domains.
For modelling traditional thermal technologies, any pair of properties from conductivity k, volumetric heat capacity ρc, and diffusivity of heat a, can be selected as independent parameters. But even for this simple model the availability and reliability of data are restricted, especially if one desires to take into account their temperature dependence (as a minimal requirement in most of the cases) to improve the accuracy of modelling.

Convective heat transfer

In modelling industrial processes, the description of heat transfer via solid-fluid or liquid-gas interfaces is often required. It is traditionally done by using the concept of surface heat transfer coefficients. This concept has proven its usefulness in various fields of thermal engineering applications, but can cause serious difficulties when applied to non-simple geometries, mixed-type flows or elevated temperatures.
As the surface heat transfer coefficient reflects in itself not only the material properties, but also the structure of the whole flow and temperature fields in the fluid, extra care should be taken when using literature data. Handbook sources are frequently incomplete in the description of the validity ranges, or the experimental conditions. For example, dimensionless relations giving the heat transfer coefficients for the very basic case of inside flow in a duct, are sometimes not accompanied by the information about the means of heating during the experiments, the length-to-diameter ratio or the entering conditions. Electrical resistance coil or condensing steam heating produces two significantly different boundary conditions; fully developed flow or a sharp cornered entrance to the heated part affects the distribution of local heat transfer coefficients etc.
Without this supplementary information, the comparison of various literature data does not have much sense.
In the recent stages of the development of modelling thermal processes, the uncertainties due to the use of heat transfer coefficients can be eliminated in many cases by treating the convective transfer numerically in its totality. Undoubtfully this is the most promising approach, but the economy of simulation work often requires that we treat convection separately, generating heat transfer coefficients as input data for the main model. In such an approach, the errors due to the inadequacy of the flow conditions during measurement to meet those in the simulation are eliminated, since the numerical method for the determination of heat transfer coefficients uses geometrical data and material properties as its own input.

Sources and availability of input data

With the improvements in modelling techniques, the demand for constitutive laws and properties is increasing. Usually handbooks, standards and product information leaflets serve as sources for these data. As we tried to illustrate before, the critical evaluation, the "expert-screening" of input data is a very important and organic part of modelling, but cannot fill the gaps where measurement data are not available at all.
The "measurement culture" in thermal engineering is well behind that in the electrical, optical or mechanical fields. This is true for the accuracy as well as for the number of parameters which are measured, and for the offer and availability of thermal property measurements.

Practically, the only specific thermal parameter which is measured widespread in industry is temperature. Electrical or mechanical properties (density, resistance, hardness, tensile strength etc.) are measurable even on the factory floor, while thermal conductivity, specific heat or diffusivity measurements belong mostly to the activites of research laboratories and academic institutions.
Standardization also reflects the above mentioned situation. While it is "evident" that density, elasticity modulus, strength, specific electrical resistance etc. are specified for structural materials, the only thermal parameter specified is probably the coefficient of thermal expansion. Inside the materials which are applied in heat exchangers, moulding-forms, furnace-walls important heat transfer processes take place, and their thermal performance can be described by conductivity, diffusivity, emissivity etc.
These tendencies have led to the situation, that even though we have well-founded theoretical proposals to improve the description of heat transfer processes (an example is the suggestion of adding a relaxation term to the conduction equation), there still remain several problems : besides the unanswered question about the magnitude of relaxation times between 10^{-6} and 10^{-13} seconds, we do not know even the room temperature conductivities for many materials.
The authors are convinced that not only is it necessary, but of late, also possible to make significant developments - perhaps even a breakthrough - in thermal measurements, applying the achievements of computer technology extensively, and combining them with new measurement principles.
Traditionally, measurement methods for thermal properties are highly labour and time consuming and expensive procedures. Conventional techniques are based on the few famous cases which could be solved by the tools of classical mathematical analysis. Temperature dependence of the thermophysical properties is usually determined by tiresome, lengthy, serial, point-by-point measurements.
To overcome the difficulties, new measurement concepts coupled with automatic, intelligent measurement systems are necessary. Traditionally the measuring instrument was fabricated to meet the restrictions of the simple, analytic mathematical models used for evaluation. Applying microcomputer control and numerical evaluation, the philosophy in the planning of experiments can be changed: one can tailor the evaluation software to the requirements of practical use, and that of the physical model.
These developments, at the same time, can lead to the construction of cheaper, faster, easy-to-operate measurement devices, which do not require high-level expertise and can be operated even at factory level to supply thermal properties for new products or for computer control of thermal technologies. Such research and development activities have been initiated at TUB and UQAC in the field of ceramics, metals and high temperature refractories.

Conclusions

1. Research efforts are necessary to strengthen physical modelling with respect to the demands raised by the introduction of new materials (engineering ceramics, composites, glass-metals etc) and by new technologies (extra-high thermal loads, ultra-fast temporal changes) and to check the validity ranges of new theoretical models.

2. In thermal process modelling special attention should be paid to critical evaluation, expert screening of input data and to the application of sensitivity and error analysis when data are available only with large uncertainties.

3. New measurement technologies should be developed which are capable of determining more than one properties under the same conditions, and can be applied in a wide temperature range including elevated temperatures (ceramics, refractories).

Hydrodynamics, heat and particle mass transfer phenomena in reverberatory furnaces—mathematical modelling and experimentation

R.I.L. Guthrie, D. Frayce
McGill Metals Processing Centre, McGill University, Montreal, Quebec, Canada
J.-P. Martin
Alcan Research Laboratories, Jonquière, Quebec, Canada

Abstract: Following melt alloying and chlorine fluxing of liquid aluminium in reverberatory-like aluminum holding furnaces, it is common practice to allow the melt to settle for about an hour prior to casting operations. The purpose of this procedure is to allow inclusions to settle to the furnace bottom or to float out to an overlying layer of dross. Mathematical modelling of this process reveals that significant natural convection currents are generated, particularly during the early part of this holding period, and that as these currents gradually diminish, so does the rate of precipitation of inclusions. A 6.25-ton pilot scale holding furnace was used to test the model's predictions of changes in metal quality with holding time. Satisfactory agreement was observed between predicted and "*LiMCA*"* measured changes in inclusion density levels of $(Ti\text{-}V)B_2$. Both approaches revealed exponential-like decays in inclusion density levels, with time constants in the order of 10 to 60 minutes.

Keywords: Aluminum holding furnaces, mathematical/physical modelling, inclusions, *LiMCA*/Podfa measurements, natural convection, liquid metal quality.

INTRODUCTION

Until recently, it was commonly supposed that molten aluminum holding furnaces of the tilting variety were intrinsically superior to stationary, bottom tapped, furnaces in that they were able, on average, to supply higher quality metal. This argument rests on the presumption that strong vertical temperature gradients within a melt, induced by burners above the metal, ensure stagnant conditions within the metal bath. For the majority of inclusions which are heavier than the liquid aluminum they

---

* *LiMCA* -- ***Li****quid* ***M****etal* ***C****leanliness* ***A****nalyzer*

displace, analyses show that a stagnant melt will become progressively cleaner from the top down. Notwithstanding this long held belief in the superiority of tilting furnaces over their bottom tapped, stationary counterparts, on-line metal quality measurements made possible by the *LiMCA* technique (1) have now revealed that provided appropriate measures are taken to control the level of melt flowing through the trough, there are no clear differences in metal quality between such furnace types. This is shown in Figures 1 and 2. Metal quality, in this sense, is defined as the number density level of inclusions greater than 20 μm effective diameter per 1 kg of melt. Thus, it was shown (2) that the concentration levels of inclusions within a melt decayed in an exponential like fashion, but that longer and longer time constants were needed to fit the observed values of decay in inclusion level densities.

More recently, a mathematical modelling paper by C. Sztur et al. (3) of Pechiney, France, on this specific problem has confirmed the importance of natural convection phenomena in holding furnaces in affecting the settling trajectories of denser inclusions. This latter approach, while revealing and complementary to our previous work (2), uses a Lagrangian frame of reference, thereby predicting the motions of individual inclusions and the role of convection currents in modifying the trajectories of vertically settling particles. The purpose of our present work is to extend the empirical plant metal quality data presented in reference 2, by modelling induced natural convection currents in a 6.25-ton pilot scale vessel from first principles, and then going on to predict their effect on changes in the bulk density levels of inclusions within a melt. For these studies, a Eulerean frame of reference was required, in order to be able to compare metal quality measurements obtained in the actual furnace, with those predicted.

EXPERIMENTAL PROCEDURES

Figure 3A provides a cross-sectional view of the experimental 6.25-ton vessel at the Alcan Research and Development Center, at Jonquière, Québec, while Figure 3B provides a view of the actual furnace facility. The furnace was first charged with commercially pure liquid aluminium (99.7% Al), at 750°C, using adjacent pot-line facilities for metal supply. The bath was then stirred with a preheated impeller, fixed to a rotor shaft inserted through one of the longitudinal side-walls of the furnace. Next, an argon-chlorine (10% Cl_2) gas mixture was bubbled through the base of the submerged impeller. The purpose of this procedure was to create a fine dispersion of reacting bubbles and thereby remove impurities, such as oxide films and any dissolved alkalis, as salts (NaCl, $CaCl_2$ . . .) from the melt, into an upper layer of dross. This dross layer was then skimmed off as far as possible, so as to create a clean bath surface. A *LiMCA* melt quality analyzer system (comprising a submerged borosilicate glass tube fitted with a small hole (electric sensing zone) through which metal is drawn and analyzed for size and frequency of inclusions) was then set up at a given depth within the furnace (15 to 75 cm) so that metal quality could be monitored every minute using 5-ml samples of liquid metal. Once the background level of inclusions had diminished and stabilized to around 5,000 per kg of aluminum, or less, known particles (TiB_2) were generated within the melt by making suitable additions of a rod-shaped grain refiner alloy comprising Al-5% Ti-1% B. Once this grain refiner melts and becomes mixed in with the bath of aluminum, titanium diboride particles are precipitated in the 2-3 μm diameter. However, agglomerates of such particles will report in the range 10 μm diameter to as much as 100 μm in diameter.

An alternate procedure for generating inclusions of known chemistry was to add a "master alloy" containing 4% Boron. This boron, once dissolved in aluminium, acts as a getter for residual titanium within the melt to form (Ti-V) B_2 inclusions. Again, once such precipitates are formed, they will settle to the furnace floor during the "furnace settling period".

Figures 4A, 4B and 4C show the results of these experiments, wherein the *LiMCA* measurements, normalized to unity, of such particles are plotted versus settling time. During the addition/generation of these TiB_2 or (Ti-V) B_2 particles, the melt was first stirred with an impeller so as to achieve good mixing in of the TiB_2 or (Ti-V) B_2 .inclusions The melt was then allowed to rest. Reference to Figure 4 shows that the N_{20} *LiMCA* readings (depicted as solid squares) appeared to drop off in an exponential manner with time. The best fit exponential curves are shown as thin continuous lines, whereas the theoretically predicted curves, to be explained in the next section, are shown as double-thickness curves. One should note the close similarity between predicted and fitted curves in 4A and 4AB, while the discrepancies in 4B will be explained in the discussion section.

THEORY

The governing differential equations describing fluid motion within the bath are the Navier Stokes and Continuity Equations. Turbulence was simulated using a two-equation differential model for the kinetic energy of turbulence (k) and its rate of dissipation ($\dot{\varepsilon}$). These have already been presented for this particular system in a previous work by the authors (4) for two-dimensional natural convective flows within a container of high aspect ratio (i.e., a long furnace [~ 4 m], of finite width [~ 1 m] and finite depth of metal [~ 0.9 m]). Thus, heat losses from the refractory side-walls and refractory floors of these trapezoidal furnaces extract heat from adjacent metal. The resulting increases in liquid density arising from cooler metal around the interior surface walls of the furnace will generate natural convection currents. The authors have already proposed (2) that such currents could be of such a magnitude as to affect the motions of entrained inclusions.

BOUNDARY AND INITIAL CONDITIONS

On the basis of previous embedded thermocouple data, steady state side-wall and floor heat losses were known for this experimental furnace, *a priori*, and could therefore be inserted as boundary conditions into the mathematical model (h ~ $2 w\, m^{-2} K^{-1}$, [T metal - T ambient] = ΔT). For the top surface of the liquid metal, domain, conditions were set so as to maintain the top surface of the metal bath at a constant (and higher) temperature than the underlying, cooling, metal. Initial conditions for the melt were a uniformly high temperature and zero bulk velocities (i.e., residual stirring effects from the impeller were not accounted for). A cruciform array of thermocouples were put in place (see Figure 3A), and power to the top heater was continuously adjusted so as to maintain a constant melt surface temperature as far as practicable. In treating particle/side-wall interactions, these were regarded as sinks to particles advecting to these surfaces during a given time step. Further, it was assumed that their Stokes settling velocities in a stagnant medium applied within the concentration boundary layer.

For the turbulence model boundary conditions, a Neumann boundary condition was set up for the kinetic energy of turbulence at the walls ($\partial k/\partial n = 0$) while a Dirichlet boundary condition was applied to $\dot{\varepsilon}$ at the first node, P, of the mesh close to the wall. There, convective terms of turbulence energy are negligible. Given that the production of k is balanced by its rate of dissipation, the appropriate boundary condition, $\dot{\varepsilon}_P = U^3/yK$, where K is the Von Karman constant.

NUMERICAL PROCEDURES

The major part of the simulation work described in this paper was carried out using the Phoenics Version 1.4 software package of CHAM. The hardware consisted of a Tektronix workstation linked to a VAX 8600 machine via a Datapac communication system. A second alternative for protracted runs was a CRAY XMP machine at the Cray Research Center, Minneapolis.

A non-uniform grid with a fine mesh adjacent to the side-walls was required in order to be able to resolve the momentum and thermal boundary layers. It is worth noting that the latter were computed to be significantly thicker (~ 10 x) than their corresponding momentum boundary layers (which were in the order of 1 mm), as would be expected for such a low Prandtl number liquid (~0.01). Space and time grid independence studies were carried out, as were runs for turbulent thermal natural convection in enclosed rectangular cavities to check the correctness of the Phoenics' application to this problem. Special subroutines had to be written for the software package in order to take into account the role of natural convection on flow, thermal and particle concentration fields.

RESULTS

Figures 5A and 5B show predicted velocity vector flow fields in the 6.25-ton experimental furnace after 310 and 910 seconds of settling. It is immediately apparent that a strong initial recirculatory flow, predicted to be set up near the melt surface at the start of the settling period, damps down quite significantly with time. Simultaneously, thermal stratification of the melt begins, with steep temperature gradients close to the top surface expanding downwards with time as heat is lost from the bulk of the melt to the base and side-walls.

In the absence of any natural convection currents, the time required for purely conductive losses from the melt to result in a fully thermally stratified bath is readily calculated, and can be shown to be in the order of one to two hours for a metre-deep bath of aluminum.

Referring to the enthalpy (temperature) profiles (Figures 6A and 6B) across the trapezoidal cross-section of the experimental furnace, one sees that iso-enthalpy lines are largely horizontal across the width of the furnace apart from distortions adjacent to the side-walls and within the recirculating vortex on the northwest side of the metal bath. The explanation for this relative insensitivity of enthalpy profiles to these natural convective flows (whose maximum velocities are predicted to be ~ 1-5 cm/s adjacent to the side-walls at the start of the settling period) is, of course, molten aluminium's high thermal conductivity.

This insensitivity contrasts with the complex iso-concentration contour plots for 25 micron inclusions shown in Figures 7A and 7B, again after 310 s and 910 s of settling. One will note that the transient concentrations of inclusions are both time and position sensitive, with the concentration levels deep within the bath, registering much lower values (~ 0.1) (C normalized to unity at $t = 0$) compared to the higher levels (~ 0.9) in the top section of the furnace.

It is interesting to note that this result is opposite to that which would be expected for a stagnant melt, for which the cleanest melt during transient settling phenomena would be near the top of the melt. The explanation is the relatively strong mixing in the top half of the furnace maintaining uniformly higher levels versus lower concentration levels adjacent to the side-wall concentration boundary layer "sinks" for these inclusions. As noted, the flux of inclusions, $\dot{n}''$, to the furnace side-walls and floor were modelled according to an equation of the form

$$\dot{n}'' = C\, U_S \operatorname{Sin} \theta \tag{1}$$

where C is the number density of inclusions, and θ is the angle subtended between the vertical and a furnace side-wall. For the horizontal furnace floor, $\theta = 90°$. U_S, the Stokes settling velocity of the particles, was calculated according to

$$U_S = \frac{\Delta\rho g d^2}{18\mu} \tag{2}$$

where Δρ is density difference, d particle diameter and μ melt viscosity. Referring once more to the predicted and observed local changes in inclusion density levels of (Ti-V) B_2 particles at three different grid points within the melt (Figures 4A, 4B and 4C), one sees that two of the three predictions compare very favorably with observed experimental data. The discrepancy between the third set [Location B in Fig.3A]), is thought to have arisen over difficulties with maintaining the top surface of the melt at 760°C during the run, owing to problems with the temperature controller. As seen from Table 1, the maximum temperature difference recorded between thermocouples in the submerged grid was 30°C (the surface and furnace floor thermocouples in Figure 3A) for the first two runs, but only 17°C for the third test 45cm below the melt line..

Unfortunately, only one *LiMCA* system was available, so these runs had to be reproduced sequentially for the three locations A, B and AB1 shown in Figure 3. In each case, the metal bath was probed with long sensing tubes, so as to allow transient changes in metal quality to be measured.

DISCUSSION

The work that most clearly bears on the results now reported, is that of Sztur et al.[3] As noted, those authors used a Lagrangian approach to predict the motions of variously sized inclusions settling in a convecting flow field. What is not clear from their study is their appreciation that the flows within their furnace would have been transient in character, and that as stratification of the melt proceeds, the induced natural convection currents within their melt would have diminished.

It is interesting to note from the present work, that as a result of the particular geometry of the furnace, naturally convecting flows down the two sidewalls of the furnace were quite asymmetrical. Thus, the downflow along the short vertical sidewall on the east side of the furnace is rather weak as compared to flows down the longer west side wall. This leads to the latter's dominant effect on flow recirculation patterns within the bulk. However, as just noted, after 910 seconds of settling, it is seen from Figure 5B that the currents are strongly suppressed by a thermally stratified field (Figure 6B), and that only a small amount of convection of any significance then persists near the top left side surface of the melt. Under these conditions, maximum temperature differences of 30°C between melt surface and furnace floor were recorded (Table 1). As such differences are not uncommon with one metre deep baths of liquid metal, the pilot scale facility proved to be invaluable for allowing realistic simulations of typical industrial operating furnaces.

In order to take into account the role of turbulence on particle/ inclusion trajectories, Sztur et al. imposed stochastic velocity fluctuations on their time averaged flow fields. The result is a suite of possible trajectories for a particle located at a specific point within the melt at time zero of the computational scheme. Unfortunately, neither the magnitude of these fluctuations, nor a rationale for their choice, is provided in their text. For the present computations, the well known $k - \varepsilon$ model provided the necessary "blurring", or rather turbulent diffusion, of particle concentration fields for our Eulerian framework.

The present transient concentration fields given in Figure 7, refer to a mono-sized population of 25 micron diameter inclusions. *LiMCA* data revealed that the relative fraction of inclusions in the 30 to 40 micron size range, did diminish versus the 20 to 30 micron size range during the settling period, as would be expected. However, being in a much smaller population group, their nett effect on the raw *LiMCA* data, would have been marginal to the computations presented in Figures 4 and 7.

Given the trends and good overall agreement between observations and predictions, it is considered that the results of the mathematical model developed to describe the complex transient flows, temperatures and inclusion concentrations developed within such holding furnaces are consistent. It is felt that the model can now be usefully extended to predict the performance of existing industrial furnaces, and to explore the merits of other possible designs or operating procedures that would enhance melt quality during the settling period. It is appropriate to note that previous work (2) has shown that this step in the operation needs to be optimized for cost-effective operations.

CONCLUSIONS

It has been shown that a mathematical model of transient natural convective flows within a bath of molten aluminum is capable of predicting the exponential decreases in inclusion density levels during the furnace holding period, and that improvements to melt quality can be satisfactorily accounted for. The work shows that neither the assumption of a stagnant melt, nor the opposite condition of a well stirred melt are legitimate approximations in treating the phenomena involved.

ACKNOWLEDGEMENT

One of the authors, D. Frayce, would like to express his appreciation to Alcan International and the Ministère de l'Enseignement supérieur et de la Science for financial support during the course of these studies. The authors acknowledge the help of Mr G. Dubé, of Alcan International, in providing technical support and guidance to this project.

REFERENCES

1. D. Doutre, R.I.L. Guthrie, "On-line measurements of inclusions in liquid metals", *Proceedings of an International Seminar on Refining and Alloying of Liquid Aluminum and Ferroalloys*, Trondheim, Norway, 1985, pp. 145-163.

2. J.-P. Martin, G. Dubé, D. Frayce, R.I.L. Guthrie, "Settling Phenomena in Casting Furnaces: A Fundamental and Experimental Investigation, *Light Metals Proceedings 1988*, TMS of AIME, pp. 445-455.

3. C. Sztur, F. Balestreri, J.L. Meyer, B. Hannart, "Settling of Inclusions in Holding Furnaces: Modelling and Experimental Results", *Light Metals 1990*, Ed. Christian M. Bickert, the Minerals, Metals and Materials Society, pp. 709-716.

4. D. Frayce, R. Guthrie, J.-P. Martin, "Mathematical Modelling of Flows in Holding Furnaces Containing Molten Aluminum, *Proceedings of the International Symposium on Reduction and Casting of Aluminum*, Ed. Christian Bickert, Vol. 8, 27th Annual Conference of Metallurgists, 1988, Pergamon Press, pp. 103-114.

Table 1 SUMMARY OF EXPERIMENTS

EXP. NUMBER	FURNACE TYPE	DEPTH LIMCA PROBE	PARTICLE TYPE	τ_{mn} [20; 300]	σ_{mn} [20; 300]	N°_{20} (Kpart/Kg)	σ (N_{20})	$\overline{N_{20}}$ (Kpart/Kg)	TEMP. MEASURED ?	TMAX PLANE (1) °C	TMAX PLANE (2) °C	ΔTMAX PLANE (3) °C
G2006	CRUCIBLE 140 Kg	40 cm	TiB2 (rod)	11.4	0.87	39.8	2.4	6.75 (8.3)	N	740		
G2106	CRUCIBLE 140 Kg	40 cm	Tib2 (rod)	15.4	1.44	9.20	0.64	2.31 (2.3)	N	740		
G2306	CRUCIBLE 140 Kg	40 cm	Ti-V-B (master)	68.0	22.5	33.8	2.04	27.3 (4.7)	N	740		
G2906	BICKELEY 750 Kg	50 cm	Ti-V-B	20.3	4.75	94.2	12.2	19.0	N	740		
G1208	6.25 TON	75 cm	Ti-V-B	52.7	3.81	43.8	1.72	16.5	Y	760	740	30
G1508	6.25 TON	15 cm	Ti-V-B	50.0	1.56	47.6	0.79	19.6	Y	760	740	30
G1608	6.25 TON	45 cm	Ti-V-B	52.6	6.46	61.2	3.76	26.7	Y	744	729	17

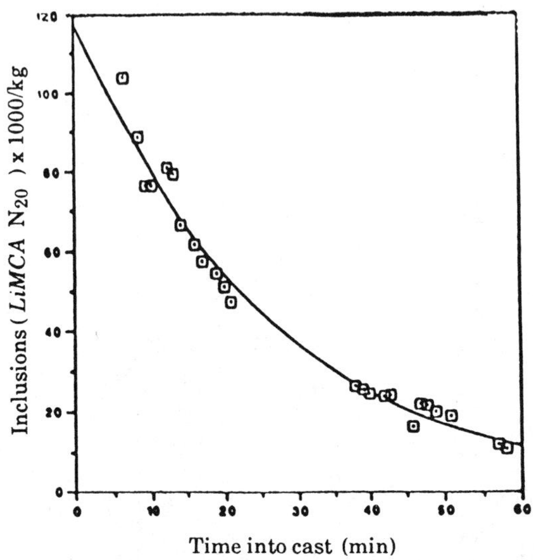

Figure 1: Change in launder metal quality (inclusion counts x1000/kg) versus time into cast during tapping operations from a tilting furnace (lifetime $= 1/k =$ 26 min)

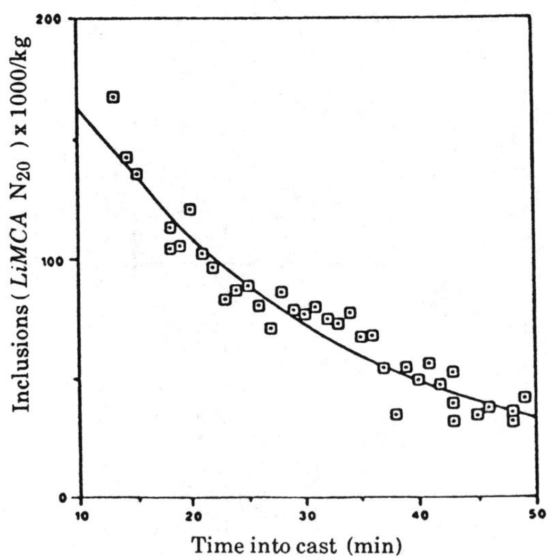

Figure 2: Change in launder metal quality (inclusion counts x1000/kg) versus time into cast during tapping operations from a stationary furnace (lifetime $=$ 25 min)

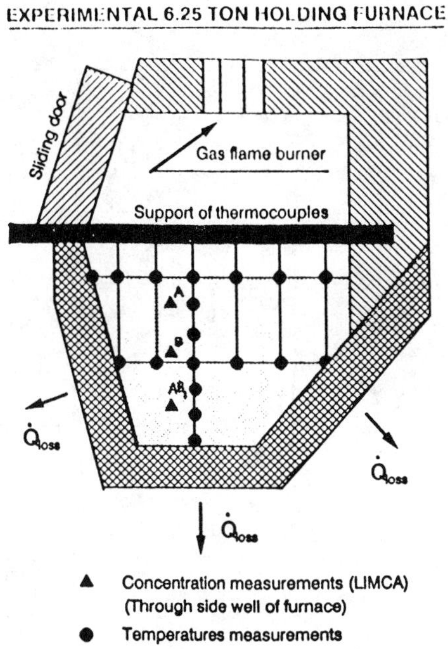

Figure 3A Experimental set-up

Figure 3B View of furnace facility

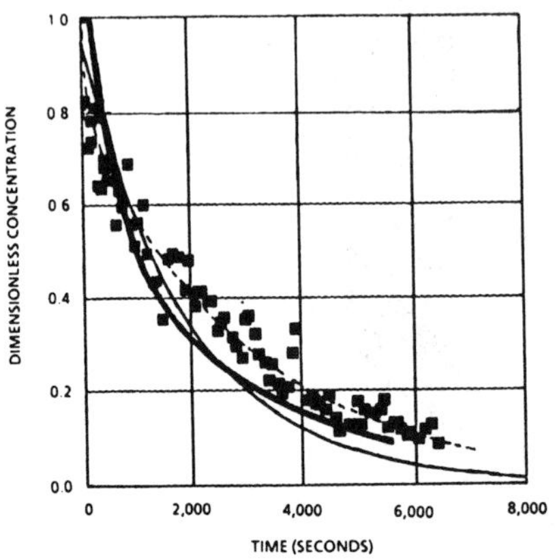

Figure 4A:

Relative change in the level of inclusions at position A1 versus settling time (secs), (— — — exptl fit, ▬▬▬ predicted, ——— "theory," expon. best fit)

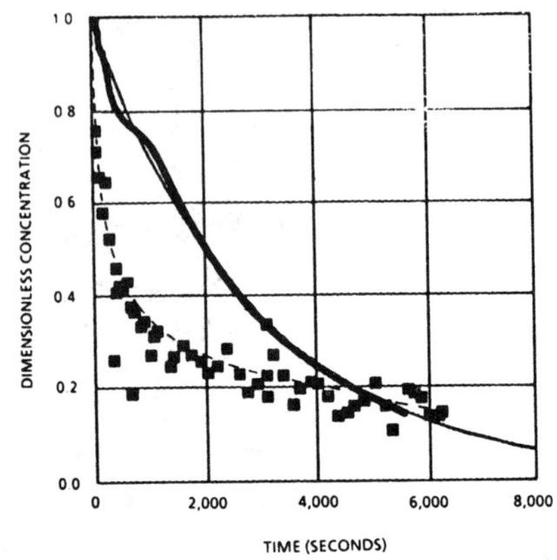

Figure 4B:

Relative change in the level of inclusions at position B1 versus settling time (secs) (- - - - exptl fit, ▬▬▬ predicted, ——— "theory," expon. best fit)

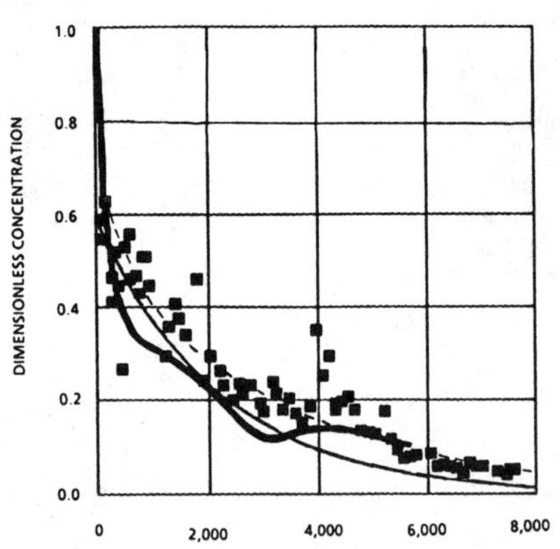

Figure 4C:

Relative change in the level of inclusions at position AB1 versus settling time(secs) (- - - - exptl fit, ▬▬▬ predicted, ——— "theory," expon. best fit)

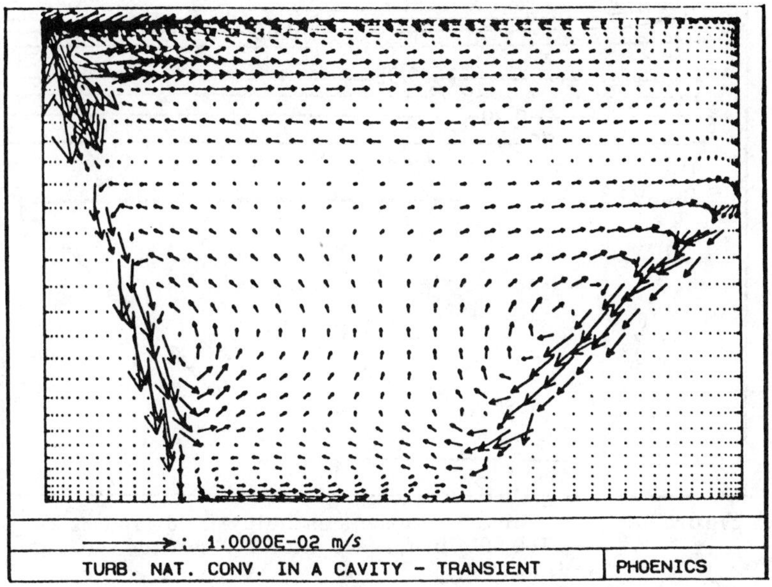

Figure 5A: Predicted transient velocity field after 310 sec. settling time

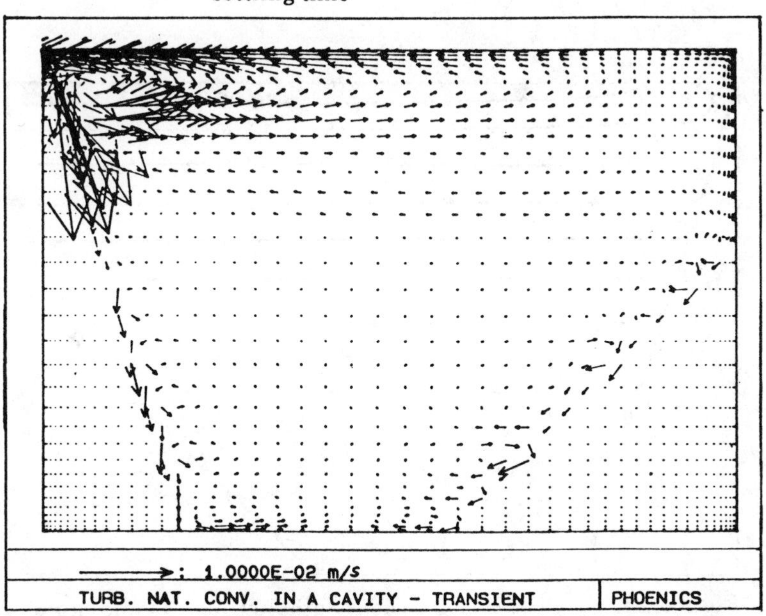

Figure 5B: Predicted transient velocity field after 910 sec. settling time

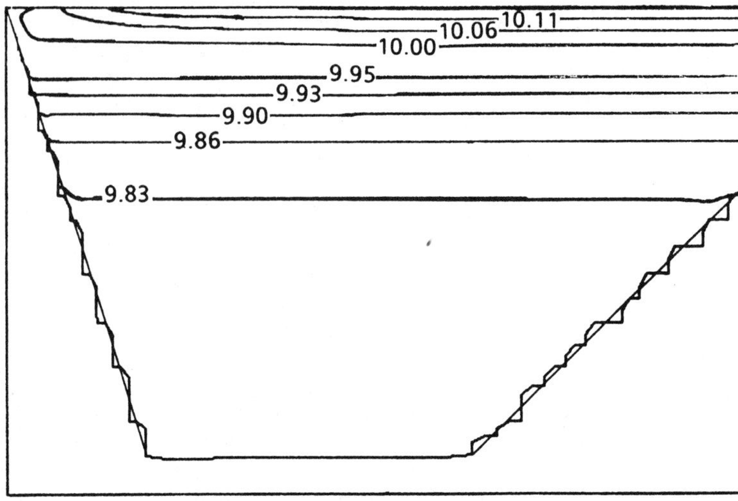

Figure 6A: Predicted transient enthalpy field (Joules/kg x100,000) in furnace after 310 seconds settling time.

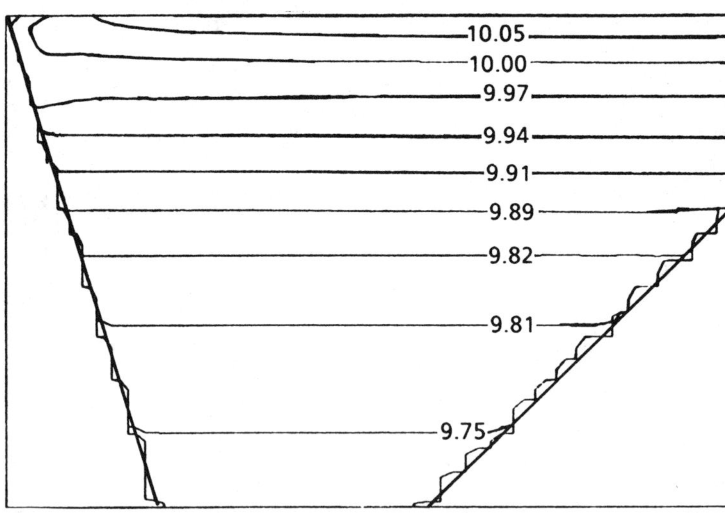

Figure 6B: Predicted transient enthalpy field (Joules/kg x 100,000) in furnace after 910 seconds settling time.

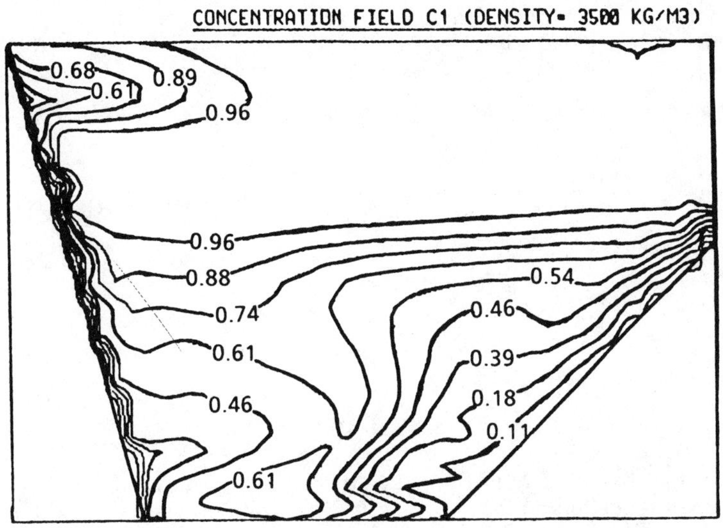

Figure 7A: Predicted transient concentration field in holding furnace after 310 seconds settling time

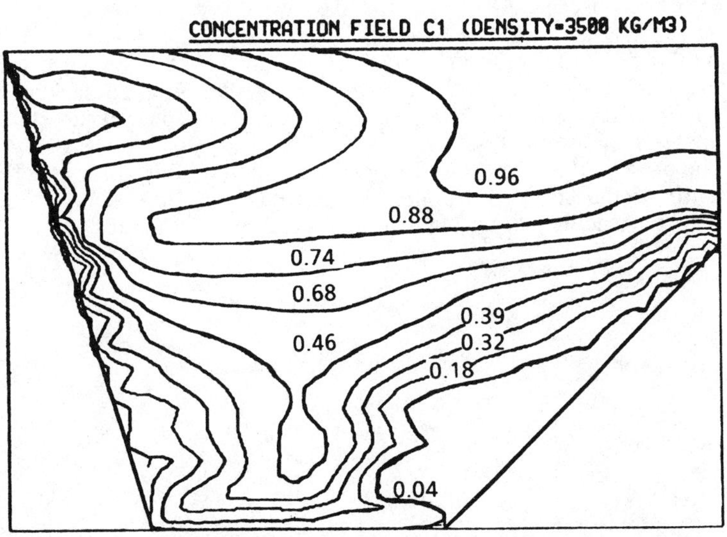

Figure 7B: Predicted transient concentration field after 910 seconds settling time

Modelling of DC casting of aluminium alloys

L. Katgerman, S.C. Flood
Alcan International Ltd. Banbury Laboratories,
Banbury, Oxon OX16 7SP, England

A.H. Langille
Kingston Research and Development Centre, Kingston, Ontario, Canada, K7L 5L9

ABSTRACT

The solidification, heat and fluid flow and stress build-up during Direct-Chill casting (DC) have been modelled mathematically in order to investigate the influence of the casting parameters and thereby formulate recommendations for improvements in casting practice and mould design.

The thermal and fluid flow fields were calculated by the control volume finite difference technique using a modified version of the CFD package PHOENICS. The effect of natural convection in the liquid metal is included through the Boussinesq approximation and turbulent effects are included using the κ-ϵ model.

The resulting thermal distribution formed the basis for a finite element calculation of the stress distribution arising from the thermal strains. The ANSYS package was used to calculate the stresses assuming firstly elastic and then non-linear elastic (temperature dependent material properties) behaviour.
Results will be presented to show the significance of the fluid flow prior to solidification and the nature of the stress distribution during casting. The effect of the casting parameters on the tendency of an ingot to cold crack will be discussed.

THE CASTING PROCESS

DC casting is currently the most common continuous casting process for aluminium alloys (Figure 1). The molten alloy is introduced continuously into the top of a short, bottomless, water-cooled mould; here a solid shell is formed , and this serves to contain the semi-solid and the molten metal in the core of the ingot. The partly solidified ingot passes out the bottom of the mould, and solidification is completed by direct chilling with water (hence "DC" casting).

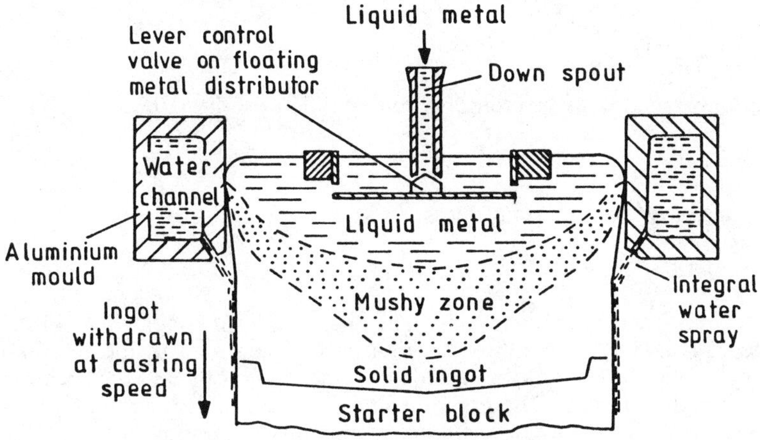

Figure 1- A schematic explanation of D.C. casting

MATHEMATICAL DESCRIPTION

Formulation

A phase averaged formulation is adopted, and this is detailed elsewhere (1). In essence, conservation equations are solved for quantities, Φ, averaged across both the solid and liquid phases:

$$\Phi = f\Phi_s + (1-f)\Phi_l$$

Φ_s and Φ_l are quantities per unit mass in the solid and liquid phases, and f is the mass fraction of solid.
The process is assumed to have obtained a steady state. The commercial CFD package PHOENICS was used to solve equations of the form:

$$\text{div}\,(\rho v\Phi - \Gamma \text{grad}\,\Phi) = S_\Phi$$

for temperature (Φ=T), the velocity components and mass continuity (Φ=1).
The turbulent kinetic energy and its rate of dissipation are also solved for when the κ-ϵ turbulence model (2) is applied to the molten metal. The compositional field is neglected.
ρ, v and Γ are respectively the average density, local average velocity and an exchange

coefficient; S_Φ is the net source of Φ per unit volume. PHOENICS solves the discretised conservation equations by SOR and uses the SIMPLEST algorithm (3) to couple conservatively the pressure and velocity fields.

Both cylindrical and rectangular ingots have been studied. Usually, symmetry enables the cylindrical problems to be reduced to two dimensions, and the rectangular Cartesian ones to only a quarter section of the whole ingot.

The Thermal Module

The Temperature Equation

The phase averaged enthalpy conservation equation can be written in terms of (Φ=T). Then

$$\Gamma = \rho\kappa + \Gamma_t$$

where κ is the phase averaged thermal diffusivity:

$$\kappa = f\kappa_s + (1\text{-}f)\kappa_l$$

and is calculated and inserted explicitly into the diffusion terms of the temperature finite difference equation. Γ_t represents the transport of heat due to the motion of turbulent eddies: it is the product of the turbulent dynamic viscosity and the reciprocal of the turbulent Prandtl number. The precise value for the turbulent Prandtl number is uncertain. It may well be in excess of unity, and maybe substantially so, since, because of the high laminar Prandtl numbers of metals, heat diffuses rapidly out of the turbulent eddies, and, consequently, the contribution of turbulent motion to the transport of heat may be small (4). The sensitivity of the results to the turbulent Prandtl number is not great, however, and uncertainty over its size does not determine the validity of the calculations.

The latent heat of solidification produces a source:

$$S_T = \rho\ L/c\ \text{div}\ (\mathbf{v}_s f)$$

where L is the specific latent heat, c is the phase averaged specific heat capacity

$$c = fc_s + (1\text{-}f)c_l$$

and $\mathbf{v}_s$ id the local velocity of the solid phase. In purely thermal calculations, when the fluid flow field is not calculated, **v** and $\mathbf{v}_s$ are set to the casting velocity throughout the calculation domain. Otherwise, **v** is calculated and then $\mathbf{v}_s$ is deduced from **v**.

The local mass fraction of solid, f, is assumed to be only a function of temperature and is determined either experimentally by differential scanning calorimetry (DSC), or theoretically. For binary alloys in a eutectic system, there are good theoretical reasons for using an expression for f derived from the Scheil equation (5):

$$f_s = \begin{cases} 0 & T > T_l \\ 1 - \left[\dfrac{T_m - T}{T_m - T_l} \right]^{1/_{k-1}} & T_l > T > T_e \\ 1 & T_e > T \end{cases}$$

where T_l, T_e and T_m are respectively the temperature of the liquidus, eutectic reaction and the melting point of the pure solid. k is the distribution coefficient. Experimental measurements of f for more complex alloy systems have been fitted to this mathematical form, with the distribution coefficient being used as a fitting parameter.

The boundary conditions

The heat transfer across the external face(s) of an ingot is specified in terms of heat transfer coefficient, h(z), and cooling temperatures, T_{ex}.

The boundary conditions are of the form

$$\frac{\partial T}{\partial \mathbf{n}} = -h\,(T-T_{ex})$$

where **n** is the unit vector perpendicular to the surface of the ingot. Three different cooling regions exist:

(i) The Mould. The heat transfer within the mould is extremely complicated: an airgap opens up between the metal and the mould soon after solidification starts. Fortunately, only around 10% of the heat loss occurs in the mould and so the mould boundary condition is not critical. Experiments and observations suggest that the airgap forms within 10 mm of the start of the solidification (6) and that, for some alloys, the combination of a high heat transfer coefficient (2-4 kW/m²/K) over 6 mm of solidification, and then a low one (0.1 kW/m²/K) thereafter, gives a good matching of calculated and measured sum profiles. Alternatively, the development of the air-gap can be approximated by blending the high and low coefficients as a function of the solid fraction. For some alloys an expression of the form:

$$h = h_{high}^{\,1-f} . h_{low}^{\,f}$$

has been used; the heat transfer coefficient tends to h_{low} as the solid fraction at the mould wall increases.

(ii) Just below the mould. The DC water does not flow on to the surface of the ingot immediately below the mould. There is a small gap between the bottom of the mould and where the water first impinges; in this region, the cooling is by air only. This gap is between 5 and 20 mm normally.

(iii) Water cooling. The Weckman analysis (7) is used to describe the direct chilling by the cooling water: the overall surface heat flux is the sum of convective and nucleate boiling components. The surface heat flux is calculated as a function of water flow rate and the surface and cooling water temperatures. The combined heat flux per unit area of water-cooled surface, q, is:

$$q = (-1.67\ 10^5 + 704\ T^*)\ Q'^{1/3}\ \Delta T + 20.8(\Delta T_x)^3$$

where T^* is the average of the wall and water temperatures. Q' is the volume flow-rate of cooling water per unit horizontal length of ingot surface, ΔT is the temperature difference between the cooling water and the surface and ΔT_x is the excess temperature of the ingot surface above the boiling point of water. The term involving ΔT_x is set to zero when $\Delta T_x < 0$. The heating of the cooling water is calculated by integration of the heat gained from the ingot as the water falls past the surface.

When solving for temperatures only the top boundary condition is simply $T = T_p$, where T_p is the pouring temperature. The bottom of the ingot is assumed to be sufficiently far from the region where there are significant gradients, and, therefore a far field Neumann condition is placed there:

$$\frac{\partial T}{\partial z} = 0$$

The relaxation of the latent heat source term.

The large magnitude of the L/c (approximately 300K for Aluminium) in the latent heat source causes the temperature solution to be prone to destabilisation from fluctuations in the fraction solid. Therefore the fraction solid is relaxed to make the calculation more stable. Originally, severe linear relaxation was used but now a more sophisticated and less constraining treatment has been developed. After each SOR pass, the solid fraction is updated in each cell according to the procedure:

(1) Calculate the cell specific enthalpy, H, using the current solid fraction and temperature, $f^\ominus$ and $T^\ominus$:
$H = cT^\ominus + (1-f^\ominus)L$

(2) Use the fraction solid-enthalpy graph to obtain a value of fraction solid, f^* consistent with H: $f^* = f(H)$

(3) Linearly relax f^* to get the new solid fraction, f:
$f = \alpha f^* + (1-\alpha)f^\ominus$

Often values as high as $\alpha = 0.9$ can be used in step 3. This technique can also be used for pure metals because f(H) is always continuous.

The Fluid Flow model

The velocity equations

For the phase averaged velocity equations (Φ = axial velocity components) $\Gamma = \mu_{eff}$, where μ_{eff} is the effective dynamic viscosity and is the sum of laminar and turbulent viscosities. The turbulent viscosity is calculated from the κ-ϵ model.

S_Φ is the sum of three sources:

Pressure gradient- There is momentum source arising from the pressure gradient in the liquid.

Buoyancy source- The buoyancy source arises from the variation of density with temperature. It only acts on the component of the velocity in the direction of gravity, and is:

$$S = \rho_{ref}\, \beta\, g\, (T - T_{ref})$$

where ß is the thermal expansivity of the liquid (10^{-4} K^{-1} for Aluminium) and g is the magnitude of the gravitational acceleration.

Mush source- The interaction of the solidifying mush with the velocity field is complicated but is approximated by the inclusion of a source:

$$S = \frac{\mu_l}{K} (\mathbf{v}_s - \mathbf{v})$$

where $\mathbf{v}_s$ is the velocity of the solid, and μ_l is the laminar dynamic viscosity. K is the permeability of the mush, and is dependent on the structure of the semi-solid, the dendrite arm spacing, the direction of the flow and the fraction of solid. Many experimentally based correlations have been reported. In this paper K is approximated as solely a function of fraction solid; three alternative expressions were considered:

(1) $K = \alpha\, (1-f^2)$ (ref.(8))

(2) $K = \beta\, (1-f)^3 / f^2$ (ref.(1))

(3) $K = \begin{cases} Y_1\, (1-f)^2 & \text{for } f > 2/3 \\ Y_2\, f^{2/3}\, (3 + 4/f - 3(8/f - 3)^{1/2}) & \text{for } f < 2/3 \end{cases}$ (ref.(9))

α, ß, Y_1 and Y_2 were assumed to be constants but they are really functions of dendrite armspacing and, albeit weakly, fraction solid.

The turbulent kinetic energy and dissipation rate are both set to zero in the mush and solid.

The velocity of the solid phase, $\mathbf{v}_s$, is present in the latent heat and mush permeability source terms. The solid grains eventually coalesce and then move with the casing velocity v_c. At the other limit, when the grains first start to grow, they move with virtually the same velocity as the liquid v_l. The solid velocity is approximated by a function which matches these two limiting cases and, in between, is dependent on a "consolidation factor" F:

$$v_s = F v_c + (1-F) v_l$$

where $0 < F < 1$, and $F = F(f)$. $F = 1$ when the solid has coalesced and moves with the casting velocity, and $F = 0$ when the grains move with the liquid velocity.

Thermal stress model

Formulation

The thermal stresses due to the temperature distribution in the ingot are considered in the fully solidified part of the ingot. The solid metal is described as an isotropic (non)-linear elastic material in which strain is generated by thermal contraction. Assuming classical small deformation theory the total strain vector can be expressed as:

$$\epsilon = \epsilon^e + \epsilon^T + \epsilon^i$$

in which the superscripts e,T and i stand for elastic, thermal and initial.
The thermal contractions can be expressed by an associate thermal strain vector:

$$\epsilon^T = -\int_T^{T_{ref}} \alpha_T(T)\, dT$$

where $\alpha_T(T)$ is the temperature dependent coefficient of thermal expansion and T_{ref} the stress-free temperature taken as the solidus temperature.
The elastic strain is given by Hooke's law written as:

$$\sigma = \mathbf{D} \cdot \epsilon^e$$

in which **D** is the material stress-strain tensor.

The basic equation in finite element formulation is derived by relating the total strain to the displacement $\epsilon = \mathbf{B.u}$, where **B** is the strain-displacement tensor, and by applying the principle of virtual displacements and variational methods (10):

$$\mathbf{K} \cdot \mathbf{u} = \mathbf{F}^T + \mathbf{F}^a$$

where **K** represents the total stiffness matrix (sum of element stiffness matrices), $\mathbf{F}^T$ and $\mathbf{F}^a$ the applied element thermal load vector and applied element body force load vector respectively. The resulting set of simultaneous equations is solved by the commercial finite element code ANSYS (11) using a 4-node isoparametric element (STIF42) for axisymmetric cases and a 8-node isoparametric element (STIF45) for the three dimensional rectangular ingot.

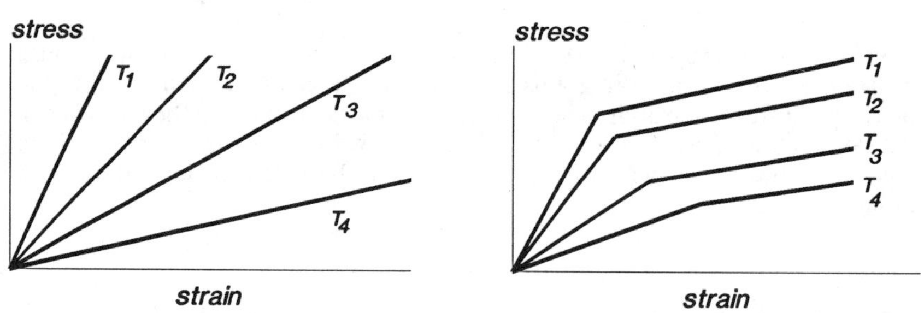

Figure 2- Temperature dependent elastic and non-linear elastic material behaviour

The non-linear elastic treatment is an attempt to allow the stresses to saturate, as they would do due to yielding (Figure 2). This approach does not allow irreversible flow such as would occur due to plastic deformation or creep. The non-linear elastic option is such that unloading occurs along the same path as loading. This behaviour is still conservative (path-independent). The 'plastic' strain for this option should be interpreted as a pseudo plastic strain, since it returns to zero when the material is unloaded (no hysteresis). The material behaviour is described by a piece-wise linear stress-strain curve.

NUMERICAL CONSIDERATIONS

The Grid

The dimensions of an ingot cross-section range from 0.15 to 2 metres and the bottom of the domain is generally of the order of one to two metres from the top (calculations are performed for different positions of the bottom boundary to ensure that its position does not influence the solution). Regular, orthogonal grids are used, with variable cell sizes to resolve accurately and efficiently the trends in the variables. The cells range from 5 mm where rapid changes in variables are anticipated (e.g. in the water-cooled region) to around 75 mm where more gradual changes are expected (e.g. near the bottom of the domain). The grids are typically 40*90 for an axisymmetric calculation and 20*14*64 for the three dimensional rectangular case.

In purely thermal calculations, depending on the aspect ratio of the ingot cross-section, the grid parallel to the rolling face can be made very coarse with only a negligible effect on the accuracy of the solution. When the rolling face is approximately three or more times longer than the ingot thickness, then the heat transfer in an ingot is predominantly in the plane perpendicular to the rolling face, and it is necessary only to resolve the gradients in this plane. The effect of the ends is not great.

A grid sensitivity analysis indicates that the stress distribution can be resolved satisfactorily with a crude mesh in which the centres of the control volumes used in the temperature solution form the nodal points in the finite element mesh. An element is assigned solid properties when at least three quarters of its nodes are below the solidus temperature, otherwise it is assumed to be stress free.

Convergence and Relaxation

The calculated temperature and velocity fields are regarded as acceptably close to the true solution when:

(1) the maximum variation in the cell values between sweeps is small; and

(2) the sums of the absolute values of the residuals are small relative to the dominant sources, namely the latent heat source for the temperature, and the buoyancy and permeability sources for the velocities.

Typically, an axisymmetric thermal only calculation on a 90*40 grid reaches a satisfactory converged solution after 500 sweeps: the maximum temperature variation in each cell is 0.2 K over one hundred sweeps, and the sum of the absolute residuals is less than one percent of the latent heat source. As a rough guide, convergence with a grid of n cells is normally obtained by 5 +n/250 cpu minutes on a Convex XP2 computer. No relaxation is applied to the temperature equation, but a relaxation parameter of $\alpha = 0.9$ is used for the fraction of solid adjustment.

The interaction of the highly buoyant velocity field with the solidifying mush causes the combined temperature and velocity calculations to be inherently unstable. Great care has to be taken to prevent the solution from diverging catastrophically. Generally the solution procedure is first, to solve for temperature only, until a reasonable converged solution is obtained; second to do several pressure and velocity sweeps with the temperature switched off. Finally, all the variables (temperature, velocities, pressure and turbulent kinetic energy and dissipation rate) are solved for.

The three dimensional solutions are more difficult to obtain, and need more severe relaxation and sometimes over 2000 sweeps to obtain a satisfactory degree of convergence.

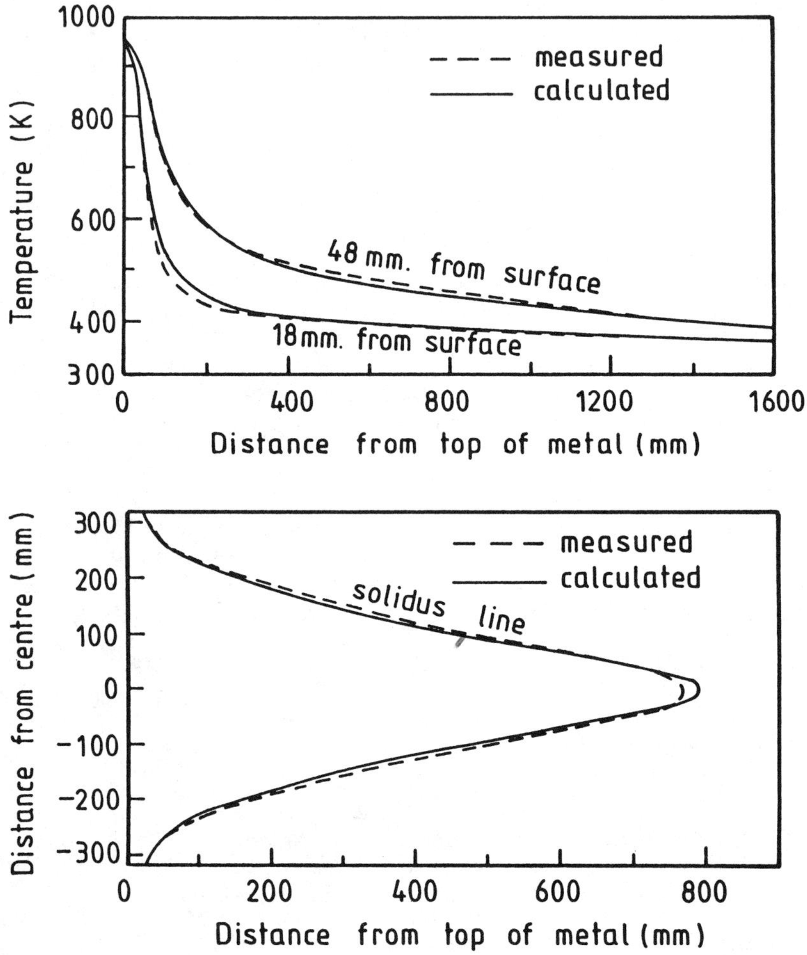

***Figure** 3-Comparison of experimentally measured and calculated (a) temperatures and (b) solidus positions in a rectangular ingot.*

A significant improvement in the rate of convergence is achieved for pure materials (i.e. when all the latent heat is evolved discontinuously at one temperature) when the solid fraction values are interpolated intelligently during the calculation of the phase averaged thermal diffusivity; stability is enhanced when the interpolated solid fraction at the cell face is set to unity when f>0.99 at an adjacent cell centre. When adjacent cells have solid fractions which are 0.01 and 0.99, then the interpolated value for f is set to either zero or one (the choice is arbitrary but should always be consistently one or the other).

The incremental procedure used in the non-linear elastic solution requires an iterative solution. After each iteration the load vector is modified so that the stress calculated in the next iteration approaches the stress which the material can support at that strain. The non-linear convergence criterion states that convergence occurs whenever the ratio of the plastic strain increment to the elastic strain, for all elements, is less than a 1%.

RESULTS AND DISCUSSION

Thermal and Fluid Flow Calculations

The results demonstrate the dominance of the latent heat over the sensible heat. Purely thermal calculations, which do not include the convective transport of heat or solid in the molten metal (other than that due to the casting velocity), agree well with both temperatures measured in the mushy zone and solid and the experimentally determined solidus position (Figures 3(a) and (b)). Naturally these calculations do not reproduce the experimentally rapid dissipation of the superheat due to convection, but the agreement with the measurements in the mushy zone and the solid show that the effect of convection on the temperature field is swamped soon after the start of solidification by the evolution and removal of the latent heat.

Assuming no motion of the solidifying material other than the casting speed, the extent of the mushy zone gives the local solidification time, and this has been used to predict the as-cast cell size. The formula used is:

$$d_s = a\, t_f^b$$

where d_s is the cell size and t_f is the local solidification time. The coefficients a and b can be obtained from regression of experimental data (e.g. (12); b is normally around 1/3.

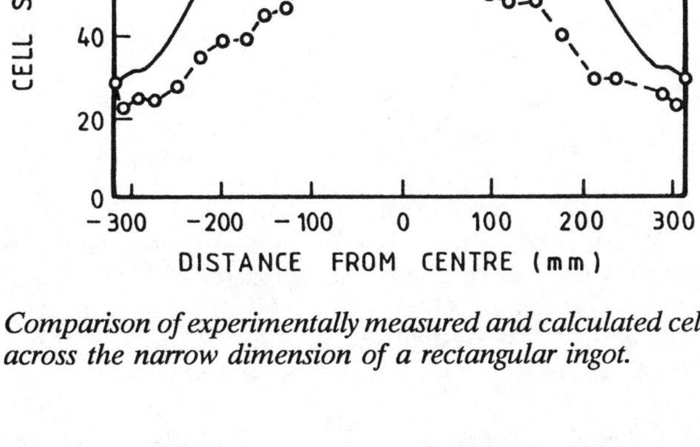

Figure 4- *Comparison of experimentally measured and calculated cell-size distributions across the narrow dimension of a rectangular ingot.*

Figure 4 shows a comparison of predicted and experimentally measured cell sizes across an ingot. The agreement of measurement and prediction is fair; the discrepancy might arise from the convective motion of fluid and equiaxed grains during solidification.

As expected, the inclusion of fluid flow in the calculation increases dramatically the calculated cooling of the molten metal through the convective mixing. When the bulk flow is allowed to penetrate in the mush, the convection of solid extends the mushy zone and, in addition, the remelting of convected solid provides an efficient means of removing sensible heat. The benefit of the fluid flow calculations is that they demonstrate possible interactions of the flow and the mush, and, therefore suggest the degree and nature of the convection of solid and solute. The flow does not influence significantly the temperature field over the major part of the solidification interval but the movement of solid and solute can affect substantially the final metallurgical structure by determining the level of macrosegregation, the coarseness of the dendritic structure and the uniformity of the macrostructure. The significance of the fluid flow calculations is their structural ramifications.

In the absence of a strong flow entering the ingot, natural convection is caused by the thermally induced difference in density between hotter metal at the centre of the ingot and colder metal nearer the solidification front towards the centre (Figure 5(a)). A strong entry velocity tends to oppose this convective motion (Figure 5(b)) and when there is significant penetration of the mush (Figures 5(c)).

The inclusion of the turbulence model increases the rate of transport of heat in the liquid phase and reduces the gradients there. The effect on the solidification process of variations in the turbulent Prandtl number is insignificant, as might be anticipated because of the very low laminar Prandtl number.

The flow patterns and isotherms shown sofar have been only for round ingots. Results have been obtained for rectangular ingot but they are more difficult to present graphically. An interesting feature of the three dimensional rectangular solution is that the flow patterns in the two vertical planes of symmetry can be quite different. Three dimensional flow patterns are far more complicated than two dimensional ones and it is often difficult to appreciate fully the details of the flow.

Thermal Stress Calculations

The stress state of the ingot was characterised by the equivalent (or von Mises) stress calculated as:

$$\sigma_e = 1/\sqrt{2}\,[(\sigma_1 - \sigma_2)^2 + (\sigma_2 - \sigma_3)^2 + (\sigma_3 - \sigma_1)^2]^{1/2}$$

where σ_i are the principal stresses.

It is essential to interpret the relative severity of the calculated stress distribution in the light of the temperature distribution in the ingot and the variations with temperature of the mechanical properties of the metal. For this reason, the calculated stress is represented as the equivalent stress normalised with respect to the temperature dependent yield stress $\sigma_y(T)$ and then this normalised quantity $\sigma_e/\sigma_y(T)$, rather than the absolute stress, is used as an indicator for predicting cold cracking.

The results for the elastic stress calculations show that the casting speed in the range of 40- 60 mm/min. has little influence on the maximum equivalent stress or

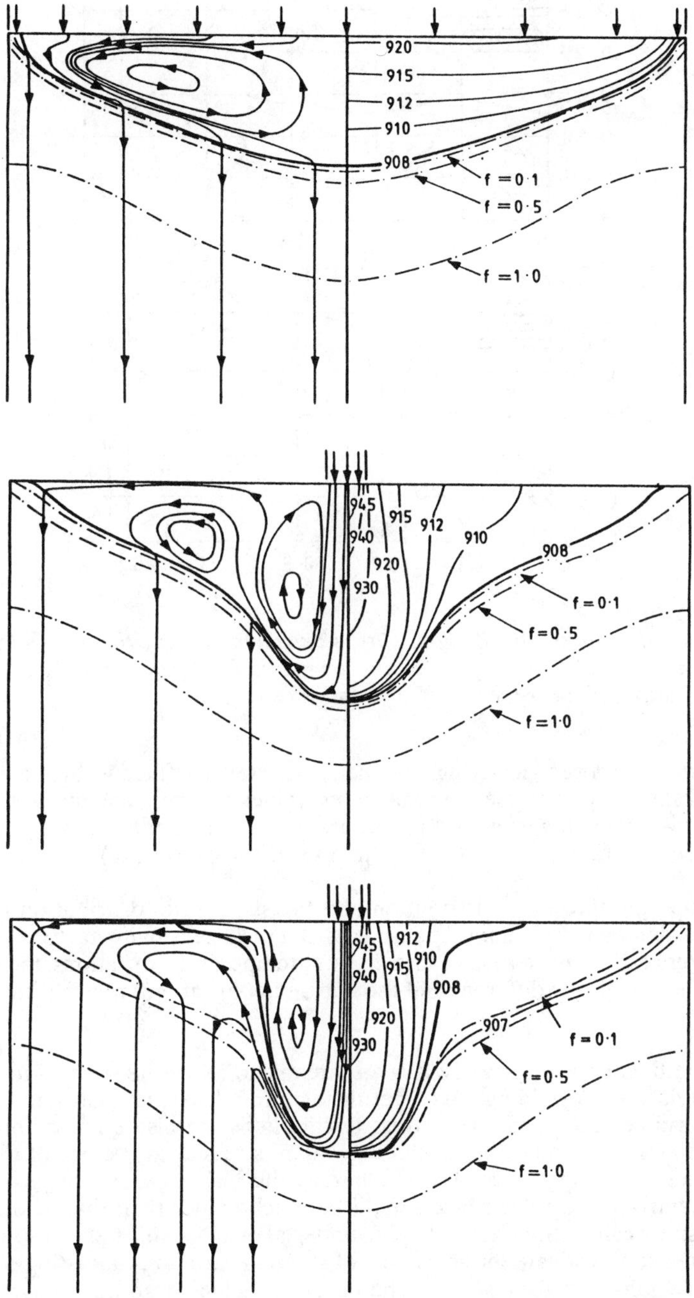

***Figure** 5-Calculated streamlines (left), and isotherms in a round ingot (right). (a) Metal enters over entire surface; (b) Metal enters through a dip tube; (c) as in (b) but with greater permeability in the mush*

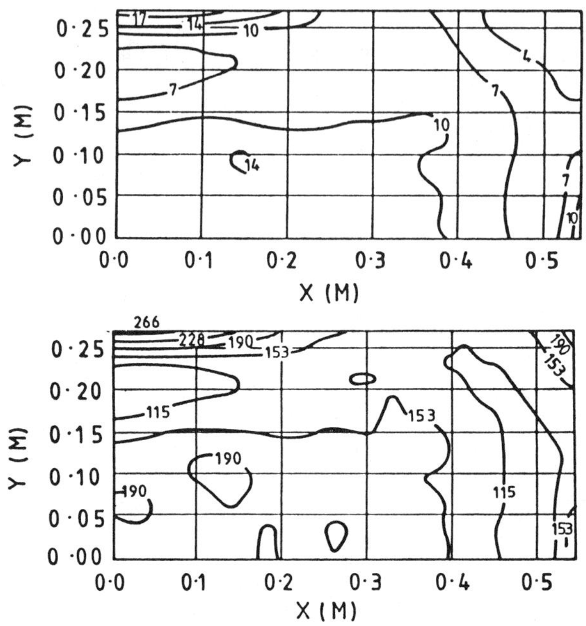

Figure 6- *Elastic stress distribution. (a) Equivalent stress; (b) equivalent stress normalised with respect to the temperature dependent yield stress for a horizontal cross-section of a rectangular ingot.*

$\sigma_e/\sigma_y(T)$ ratio. However, lowering the heat transfer coefficient by an order of magnitude substantially decreases the maximum values of stress and stress ratio (13). The last effect is commonly exploited in DC casting of strong alloys through additions of CO_2 (14), pulsed water cooling (15) or the use of a wipe-off (16).

By comparing the stress distribution and the stress ratio distribution it is clear that the equivalent stress cannot be used as a crack sensitivity factor: when the normalised stress ratio is used to take into account temperature variations, the resulting distribution is topologically different and the maxima occur at different positions within the ingot.

The non-linear elastic treatment is an attempt to allow the stresses to saturate, as they would do due to yielding. As a result the stress levels are substantially lower than in the elastic calculations. The stress distributions are also quite different: The non-linear elastic results indicate that the equivalent stress at the centre of the rolling face and corner of the ingot are similar whereas the elastic results suggest that the stress at the centre of the rolling face should be much higher than that at the corner. However the stress calculations are not being undertaken to establish the absolute stress levels, but rather to investigate the sensitivity of stress to variations in casting conditions and to locate regions of highest stress. The lack of reliable constitutive equations and mechanical properties for as-cast materials means that a more sophisticated and lengthy analysis including elasto-plastic and creep behaviour may not be justified anyway.

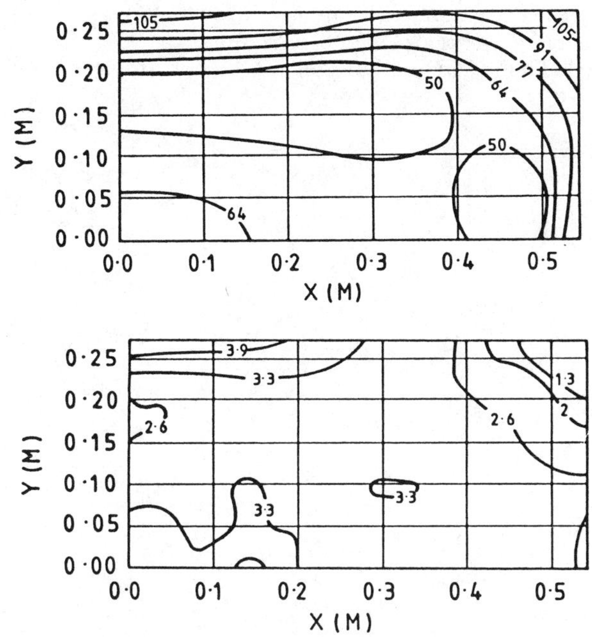

Figure 7- *Non-linear elastic stress distribution. (a) Equivalent stress; (b) equivalent stress normalised with respect to the temperature dependent yield stress for a horizontal cross-section of a rectangular ingot.*

REFERENCES

1. W.D. Bennon and F.P. Incropera, Int.J.Heat Mass Transfer **30**(1987), 2161-2170
2. B.E. Launder and D.B. Spalding, Comp.Meth.Appl.Mech.Eng **3**(1987), 269-289
3. D.B. Spalding, CFDU Report HTS/80/1, Imperial College, London (1980)
4. N. Jischa and H.B. Rieke, "Turbulent Heat Transfer" in "Recent Contributions to Fluid Mechanics", pp. 151-160, ed. W. Haase, Springer Verlag, New York (1982)
5. S.C. Flood and J.D. Hunt, Applied Scientific Research, 27-42 ,1987
6. E.K. Jensen, Light Metals 1984, pp.1159-1175
7. D.C. Weckman and P. Niessen Metall.Trans **13B**(1982), 593-602
8. T.S. Piwonka and M.C. Flemings, Trans. TMS-AIME **236**(1966), 1157-1165
9. R. West, Metall.Trans **16A**(1985), 693
10. O.C. Zienkiewitz, The Finite Element Method, McGraw Hill, London (1977)
11. G.J. Desalvo and J.A. Swanson, ANSYS Engineering Analysis System User's Manual, Houston
12. L. Backerud, E. Krol and J. Tamminen, "Solidification Characteristics of Aluminium Alloys. Volume I: Wrought Alloys", Skan Aluminium/ Universitetsforlaget AS, Oslo (1986)
13. S.C. Flood, L. Katgerman, A.H. Langille and C.M. Read, pp. 553-561, 966-967 in Modelling of Casing and Welding Processes IV, eds. A.F. Giamei and G.J. Abbaschian, The Minerals, Metals and Materials Society (1988)

14. H. Yu, "Ingot casting Method", US Patent 4,166,495 (1979)
15. N.B. Bryson, Can.Met.Quart. **7(1)**(1968), 55- 59
16. A.T. Taylor, E.H. Thompson and J.J. Wegner, Metals Progress **72**(1957), 70- 74

Modelling evolution of microstructure during hot-rolling by plane-strain compression

S.P. Timothy, H.L. Yiu, J.M. Fine and R.A. Ricks
Alcan International Limited, Banbury Laboratories, Banbury, Oxon, OX16 7SP, England

Abstract

The evolution of microstructure during a single-pass of laboratory hot-rolling (47% reduction) of a 5083 aluminium alloy has been investigated at different positions through the thickness of the plate. The thermomechanical histories of the sub-surface and centre regions of the hot-rolled slab were analysed using instrumented rolling blocks and finite element modelling; the pass history was then simulated physically using plane-strain compression testing, by deforming the test specimens to the appropriate equivalent strain at a representative equivalent strain rate and temperature associated with the rolling deformations. Microstructures formed by the plane-strain compression simulations were then compared to the actual hot-rolled microstructures. The main features of the microstructures, particularly the sizes and volume fractions of the recrystallised grains, were reproduced very well. Experimentally-determined temperature profiles, mean strain -rates and strain accumulated at the centre of the slab during rolling were in reasonable agreement with the predictions of the finite element model; the measured strain immediately beneath the surface was however much higher. The implication of these results for the modelling of multi-pass hot-rolling are briefly discussed.

Keywords

Rolling, Plastic Deformation, Finite Element Modelling, Aluminium Alloys, Microstructure.

Introduction

It is possible to simulate satisfactorily the average deformation that a slab of metal experiences during a hot-rolling operation by reproducing its thermomechanical history in a controlled manner in the laboratory. Previous workers (1-7) have analysed single pass and multi-pass hot-rolling using torsion and plane-strain compression deformation to understand the magnitude of the forces involved and how microstructure evolves under these conditions; the "equivalent" strain(8) associated with a given rolling reduction is imparted to the test specimen according to the deformation mode at the appropriate temperature and strain rate, and the microstructure is subsequently examined. In the steel and aluminium industries, for example, useful information on how microstructure evolves during industrial hot-rolling is obtained in this way before resorting to expensive plant trials; microstructural development is important at this stage of the processing route since it often controls properties in the final product.

In general however, the resultant strain within the as-rolled metal is not distributed uniformly through the thickness of the slab(9). If we consider a slab of hot metal entering the roll-bite from the left hand side of Fig. 1, the gradient in deformation is evident from the distortion of an originally square grid pattern located along the centre plane of the rolling block. The increased deformation nearer the surface of the hot-rolled metal relative to the centre results from the redundant shearing close to the rolls, which is superimposed upon the plane-strain compression resulting from the overall reduction in thickness. The strain, strain-rate and temperature of an element of metal passing through the roll-bite during hot-rolling will, in general, depend on its original position in the slab thickness; we may expect a gradation in resultant microstructure from the surface to the centre of the rolled metal since microstructural evolution depends strongly on these parameters(10-13).

The whole basis of being able to reproduce the complex strain history rests on the physical significance of "equivalent" strain(8); the concept of equivalent strain rate follows on automatically. Equivalent strain is an invariant function used to simplify a complex state of strain by combining shear and compressive/tensile strains into a single scalar parameter. The strain path taken by surface and centre elements in Fig. 1 may be characterised by an equivalent strain increment for each sequential step through the roll-bite; the increments may then be summed along the paths taken by the elements to yield a total equivalent strain imparted to each thin slice within the hot-rolled metal as it emerges from the roll-gap. We may then determine the variation of total equivalent strain with through-thickness position, and simulate the rolling deformation of a given region by deforming a relatively large specimen to the same equivalent strain at the relevant equivalent strain-rate and temperature using a servo-hydraulic plane-strain compression apparatus. The strain, strain rate and temperature histories can be derived either experimentally using instrumented rolling blocks (10, 14) or from a mathematical model, e.g. a finite element model(15-17).

The main purpose of the present work was to investigate whether key microstructural features generated at the near-surface and centre of a slab during a single pass of hot rolling could be reproduced by a nominally pure plane-strain compression deformation. The strain, strain-rate and temperature

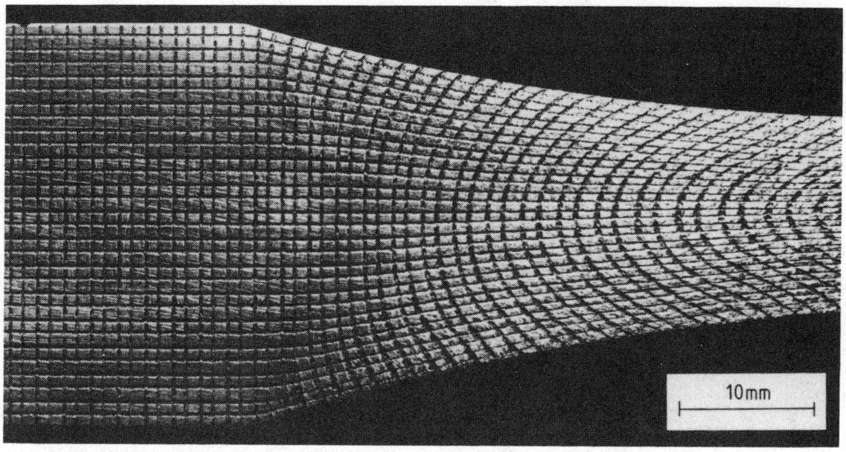

FIG. 1 Pattern of deformation resulting from a 50% reduction during a hot-rolling pass; the grid pattern, originally square, was inscribed along the centre-plane of the rolling block.

profiles which provided the boundary conditions were derived from instrumented rolling blocks, although comparisons were also made between these profiles and those predicted using a finite element model. Preliminary work on hot-rolling of a 5083 aluminium alloy showed that a reduction of ~50% produced a gradient in resultant microstructure through the thickness of as-quenched samples; the metal close the the as-rolled surface had recrystallised to a large degree relative to that at the centre. This alloy and these conditions were then selected as the basis for the simulation studies since the contrast between surface and centre microstructures was quite marked. Previous workers (18-22) have also shown that a high magnesium content in Al-Mg and Al-Mg-Mn alloys promotes recrystallisation during hot-rolling/hot-rolling simulation studies.

Experimental Method

As-cast 5083 aluminium alloy (Al-4.5% Mg - 0.9% Mn) was given an homogenisation treatment, cooled to 460°C, and hot-rolled with a reduction of 47% using a laboratory mill. The rolls were at a nominal temperature of 20°C and rotated at a speed of 10 rpm. Rolling blocks (30 mm thick x 75 mm wide x 150 mm long) were prepared in one of two ways.

One set of rolling blocks was fitted with gridded inserts (10 mm wide) so that the gridded face lay along the central plane of the rolling block; a square grid geometry was chosen, with grid lines scribed at 1.5 mm intervals. These specimens were partially rolled to expose the deformed grid pattern within the roll-bite, for example as shown in Fig. 1. The rolling mill was stopped during the reduction and raised to release the specimen; it was assumed that the strain distribution in the roll-bite depended primarily on the deformation geometry, and to a lesser extent on roll-speed etc.

The total equivalent strain imparted during rolling immediately beneath the surface (~1.5 mm beneath the original surface) and centre of the hot-rolled metal was calculated, assuming no lateral flow, from the distortion of the grid pattern relative to the original(23); the method is similar, in principle, to that described by Jain et al(24). The positions of the nodes in the grid pattern (e.g. Fig. 1) are recorded using a digitising tablet relative to a given set of reference X, Y axes; the region of interest contains undeformed elements as they enter into the roll-bite, deforming elements within the roll-bite and fully deformed elements as they emerge. A lower bound to the accumulated equivalent strain at a given depth is calculated from the deformation of a given grid element as it passes beyond the roll-exit plane, relative to an undeformed element. Finite homogeneous strains e_{xx}, e_{xy}, e_{yy} are calculated from the change in length and direction of three vectors defined within the original element; the total equivalent strain, ϵ_{lower}, is then derived using the appropriate equations for plane-strain plasticity(24). An upper bound to the total accumulated equivalent strain, ϵ_{upper}, is calculated by summing the individual increments of equivalent strain along the strain paths taken by the elements*; the above calculations are performed element by element along the flow lines to the roll-exit plane. The true value of equivalent strain imparted during rolling, ϵ_t, will lie between these values such that $\epsilon_{lower} < \epsilon_t < \epsilon_{upper}$.

A range of possible strain values was thus derived from the deformation of each row of grid elements; for the purpose of the rolling simulation studies however, single values of accumulated equivalent strain from within these ranges, ϵ, were taken to be representative of the deformations at these locations, and used to define the reductions in the plane-strain compression experiments (Table 1).

---

*In theory, the true value of equivalent strain imparted during rolling, ϵ_t, should be equal to ϵ_{upper}, but in practice $\epsilon_t < \epsilon_{upper}$ since errors arise during calculations due to uncertainty in positions of digitised nodal points, out-of-plane plastic flow etc; these points are discussed more fully in ref. 23.

TABLE 1

Equivalent strain, ϵ, mean equivalent strain rate, $\bar{\dot{\epsilon}}$, and initial temperature, T_o, used in plane-strain compression simulation studies.

Experiment	ϵ	$\bar{\dot{\epsilon}}(s^{-1})$	T (°C)
Surface Simulation	1.4	5.4	365
Centre Simulation	0.6	2.3	460

Mean equivalent strain rates immediately beneath the surface of the metal and at the centre were calculated from the total equivalent strain divided by the estimated duration of deformation t, where

$$t = L/V \simeq (R\Delta h)^{1/2}/\omega R \qquad (1)$$

L is the length of the arc of contact, V is the peripheral roll velocity, R is the roll radius, Δh is the reduction in thickness, and ω is roll velocity. In the rolling experiments, t was calculated to be 0.26s.

The second set of rolling blocks were instrumented 1.0 mm below the surface, at the quarter-thickness and mid-thickness positions with thermocouples, whose output signal was recorded during rolling to record their temperature histories; the thermocouple wires were nickel-chrome/nickel-aluminium, encapsulated in 1.0 mm diameter 18/8/1 stainless steel sheaths with magnesium oxide insulation. The specimens were rolled to completion and held for 15s after deformation before being water quenched. Metallographic sections were prepared parallel to the three principal working directions for microstructural examination. The microstructures were characterised by the size and volume fraction of recrystallised grains, and the size of the remaining deformed grains.

The boundary conditions for the plane-strain compression simulation studies were determined by the total equivalent strain, mean equivalent strain rate and a characteristic temperature associated with the deformation of the sub-surface and centre planes. The tests were carried out at a constant strain rate equal to the mean equivalent strain rate above to the pre-determined strain at the appropriate initial temperature (Table 1). Plane-strain compression specimens (10 mm thick x 55 mm wide x 50 mm long) were machined from as-homogenised metal; the specimens were soaked for 10 mins at the previous homogenisation temperature and then transferred to the test furnace and allowed to cool to test temperature; the temperature of the platens was set equal to the furnace temperature. The experiments were carried out on the Servotest servo-hydraulic mechanical test machine (load capacity = 0.45 MN) in the School of Materials, University of Sheffield. The specimens were held for 15s before water quenching after deformation, sectioned parallel to the three principal working axes, and prepared metallographically for microstructural examination as described previously.

Finite Element Model

Finite element (FE) modelling of the hot-rolling was accomplished using MARC, a general, non-linear FE code with automatic contact detection in two dimensions; an updated Lagrangian formulation of MARC was used, incorporating large deformation and large strain options, and coupled thermal-stress analysis. The slab (150 mm long x 30 mm thick) was modelled using plane-strain elements so that plastic flow and heat transfer were constrained in two dimensions, a reasonable assumption for the conditions used here; only one half of the deforming region was modelled as the centre line of slab is a line of symmetry (Fig. 2). The roll was assumed to be rigid at constant initial temperature equal to 20°C. The boundary conditions were defined by those of the laboratory rolling experiments.

Contact between the roll and the slab was controlled by the automatic two dimensional contact algorithm. Shear friction(25) rather than Coulomb friction was assumed in the analysis; the shear stress at the interface τ is related to the local flow stress of the metal in shear, k, by the relation

$$\tau = mk \qquad (2)$$

where m is a constant. Two values of m were used in the analysis, 0.80 and 0.95; these values were considered to be typical of conditions associated with 'sticking' friction during hot-rolling. The effective heat transfer at the roll contact was set equal to $15 kWm^{-2} K^{-1}$, whereas the heat transfer coefficient to the surroundings at the free surfaces was set equal to $0.1 kWm^{-2} K^{-1}$. Again, the values of these coefficients were considered to be representative of conditions during laboratory hot-rolling.

The plastic deformation behaviour of the alloy at hot-working temperatures was described by a standard constitutive equation(4), combining the dependence of equivalent stress σ on equivalent strain rate $\dot{\epsilon}$ and temperature T via the Zener-Hollomon parameter Z(26),

$$Z = \dot{\epsilon} \exp (Q/RT) = A_0 (\sinh \alpha\sigma)^n \qquad (3)$$

with the dependence of σ an equivalent strain ϵ,

$$\sigma = K_0 \epsilon^x \qquad (4)$$

to give

$$\sigma = K_1 \epsilon^{x(T)} \operatorname{arcsinh} (A_1 Z)^{1/n} \qquad (5)$$

The exponent x is actually a function of temperature in equation (5) rather than being a constant as in equation (4); this enabled a closer fit to be made between the calculations of flow stress and the experimental results. K_0, K_1,

A_0, A_1, α and n are all constants; R is the gas constant and Q is the activation energy for the rate controlling mechanism of plastic deformation. The proportion of plastic work dissipated as heat was set equal to 0.95. Other thermal and elastic properties of the alloy were obtained from reference data.

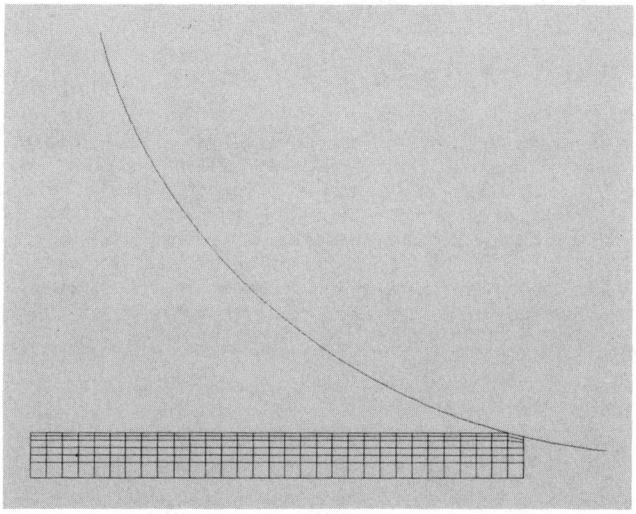

FIG. 2 Finite element mesh for slab and roll at the start of deformation.

Results

Measured temperature profiles derived from the rolling experiments are shown in Fig. 3. The centre-line temperature increases to a maximum value during deformation as the metal is drawn into the roll-bite; the rate of heat generation here resulting from the plastic deformation exceeds the rate of conduction towards the cold-rolls initially. Their cooling effect eventually takes over, and the temperature continues to fall as the metal leaves the roll-gap. The sub-surface temperature passes through a small maximum as deformation commences, but the temperature soon drops rapidly to a minimum value since the chilling effect of the rolls at the contact predominates early on; the temperature rises again to within ~430°C after deformation has ceased since the heat generated in the central regions soon flows towards the surface. The small temperature maximum probably indicates that deformation begins before the metal is in contact with the rolls. The temperature profile at the quarter-thickness position exhibits an intermediate behaviour in between these two extremes. It is interesting to note that the temperature distribution becomes more or less uniform within ~0.5s after leaving the roll-bite, but at a temperature ~30°C lower than before.

Unfortunately however, the absolute values of temperature of the transient regions in Fig. 3, were probably not as accurate as had been hoped for. This becomes apparent when the measured temperature rise T = 40K at the centre of the rolling block (Fig. 3) is compared to that which would result if the metal was deformed adiabatically under equivalent conditions. Assuming that 95% of the plastic work per unit volume is dissipated as heat, then the latter, $\Delta T_{adiabatic}$, is given by,

$$\Delta T_{adiabatic} \leq 0.95\ \sigma_{ss}\ \bar{\epsilon}/\rho.C \qquad (6)$$

where σ_{ss} is the saturation flow stress (equivalent) during hot-working, $\bar{\epsilon}$ is the equivalent plastic strain, and ρ and C are the density and specific heat capacity at 460°C; σ_{ss} = 110 MPa (derived from a plane-strain compression experiment), $\bar{\epsilon}$ = 0.73 (corresponding to a 47% homogeneous reduction, which in fact turns out to be close to the measured strain at the centre), ρ = 2.66 Mgm^{-3}, and C = 1120 $Jkg^{-1}\ K^{-1}$. $\Delta T_{adiabatic} \leq$ 26K which is smaller than the measured ΔT above during rolling. The thermocouple at the centre of the rolling block actually measured an apparent temperature rise that was physically impossible.

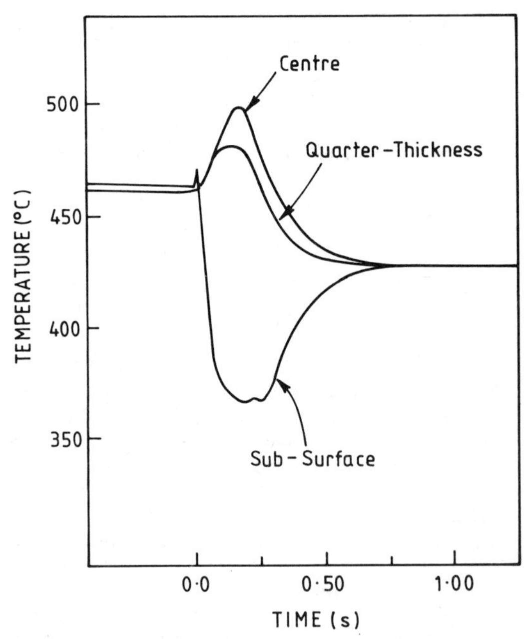

FIG. 3 Measured temperature profiles from centre, quarter-thickness and sub-surface of 5083 roll-block.

Two main sources of possible error in the temperature measurement were identified. First, it was noted that the thermocouple sheaths had undergone plastic deformation during rolling; in section, the thermocouple sheaths had developed aspect ratios between 1.27 - 1.49. Deformation of the filler material around the thermocouple wires may have increased the local temperature, they may have been forced into contact at places along their length other than at the main junction, or the voltage characteristic itself of the thermocouple may have altered. Second, the response time of the thermocouple wires was finite relative to the duration of the rolling deformation (~0.26s); this effect would tend however to give lower apparent temperature changes in the transient, rather than the higher readings observed in practice. The absolute error in the transient temperature profiles in Fig. 3 was therefore not determined, but it is thought that they are representative to within a few degrees of the true temperatures in the metal during rolling.

Total equivalent strains imparted to the surface and centre regions during rolling deformation were calculated from the deformed grid patterns using the methods described previously. A strain of 1.4 was taken to be representative of the sub-surface deformation, whereas a strain of 0.6 was associated with the deformation at the mid-thickness (Table 1). Mean equivalent strain rates were calculated to be $5.4s^{-1}$ and $2.3s^{-1}$,respectively (Tables 1 and 4).

The final microstructures after hot-rolling deformation are shown in Figs. 4 (centre) and 5 (sub-surface). The sizes and volume fractions of the recrystallised grains are given in Table 2, and the sizes of the remaining as-deformed grains are given in Table 3.

The plane-strain compression simulation experiments were carried out according to the conditions in Table 1. An initial temperature of 460°C was chosen to be representative of the centre temperature profile in Fig. 3, whereas 365°C was selected to be the initial temperature of deformation for the sub-surface simulation; the former corresponded to the initial rolling temperature, whereas the latter temperature was associated with the minimum point during deformation. The temperature profiles in Fig. 3 could not be simulated exactly, so the initial conditions were chosen to allow for the temperature rise arising from the plane-strain compression deformation itself. The resultant microstructures are shown in Figs. 4 and 5; the grain sizes and volume fraction of recrystallised grains are given in Tables 2 and 3.

The strain distribution itself was actually non-uniform within the deformed plane-strain compression samples. This effect has been reported in detail previously(27, 28). Specimen geometry was selected to minimise this effect, and metallographic examination was performed close to the centre of the sectioned specimens, away from regions where local strain changes rapidly with position. The local equivalent strain at the centre of the specimen tends to be the same or slightly higher than the nominal value applied during compression(27).

TABLE 2

Comparison of volume fraction and size of recrystallised grains in hot deformed microstructures.

	Rolling Deformation		Plane-Strain Compression	
Experiment	Surface	Centre	Surface Simulation	Centre Simulation
Volume Fraction of Recrystallised Grains (%)	39.2	4.5	37.7	4.6
Mean Recrystallised Grain Size (μm)	16	16	15	16

TABLE 3

Comparison of mean grain intercept length, D, measured parallel to the longitudinal (D_L), transverse (D_T) and short transverse (D_{ST}) directions in the unrecrystallised part of the hot-deformed microstructures.

	Rolling Deformation		Plane-Strain Compression	
Experiment	Surface	Centre	Surface Simulation	Centre Simulation
D_L (μm)	156	156	194	150
D_T (μm)	114	133	137	123
D_{ST} (μm)	61	65	53	67

FIG. 4 Comparison of deformed microstructure from the centre of the actual hot-rolled sample with that formed by plane-strain compression simulation of the centre deformation.

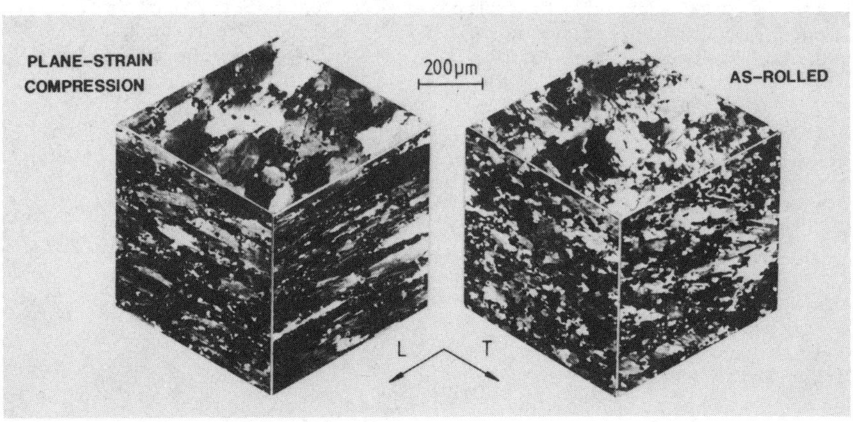

FIG. 5 Comparison of deformed microstructure from the surface of the actual hot-rolled sample with that formed by plane-strain compression simulation of the surface deformation.

The results of the FE modelling (m = 0.80) of the hot-rolling reduction are given in Figs. 6 to 10. Figs. 6 and 7 respectively show the predicted distribution of equivalent strain and temperature through the roll-bite. The variation in accumulated equivalent strain and instantaneous temperature with time at given nodes at the sub-surface, quarter-thickness and centre positions are shown in Figs. 8 and 9 respectively. Equivalent strain reaches a plateau in Fig. 8 corresponding to the final strain as the metal leaves the roll-bite; the total equivalent strain at the sub-surface and centre regions was calculated to be 0.86 and 0.71, respectively (Fig. 10). Increasing the friction coefficient m from 0.80 to 0.95 had little influence on the calculated distribution of temperature and strain-rate during the rolling deformation; its main effect was to increase slightly the total strain imparted to the metal as shown in Fig. 10.

Equivalent strain rates $\dot{\epsilon}^*$ during rolling were calculated from the slopes of the curves in Fig. 8, and compared to mean strain rates estimated from the actual rolling experiments in Table 4. Two strain rates terms are defined from Fig. 8, the maximum strain $\dot{\epsilon}^*_{max}$ in the process zone, and a mean strain rate $\bar{\dot{\epsilon}}^*$ equal to the total equivalent strain divided by the total duration of deformation. The latter relate to $\bar{\dot{\epsilon}}$ derived from the actual rolling experiments.

TABLE 4

Comparison of mean equivalent strain rate during hot-rolling derived from experiment, $\bar{\dot{\epsilon}}$, with corresponding values, $\bar{\dot{\epsilon}}^*$, calculated using the FE model; the predicted maximum strain rate in the roll-bite using the model, $\dot{\epsilon}^*_{max}$, is also given.

Position	$\bar{\dot{\epsilon}}$ (s^{-1})	$\bar{\dot{\epsilon}}^*$ (s^{-1})	$\dot{\epsilon}_{max}^*$ (s^{-1})
Surface	5.4	2.2	5.5
Quarter-Thickness	-	2.1	3.1
Centre	2.3	2.2	3.3

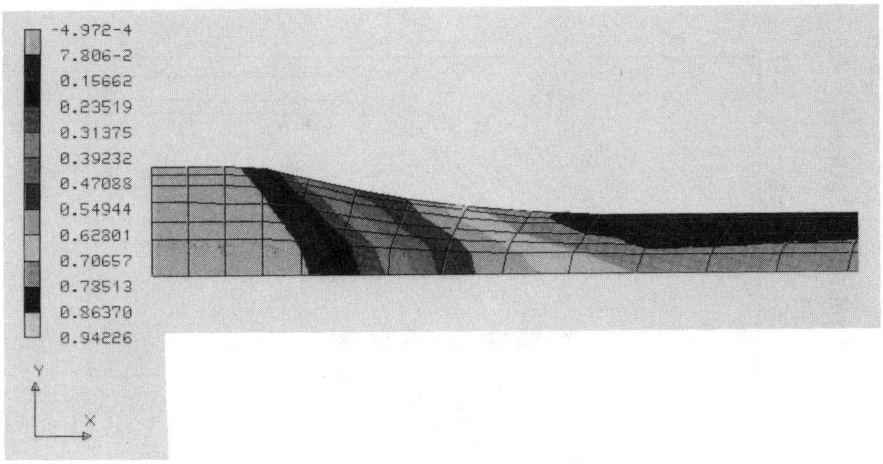

FIG. 6 Distribution of accumulated equivalent strain through one half of the deforming region in the roll-bite as predicted by the finite element model (m = 0.80).

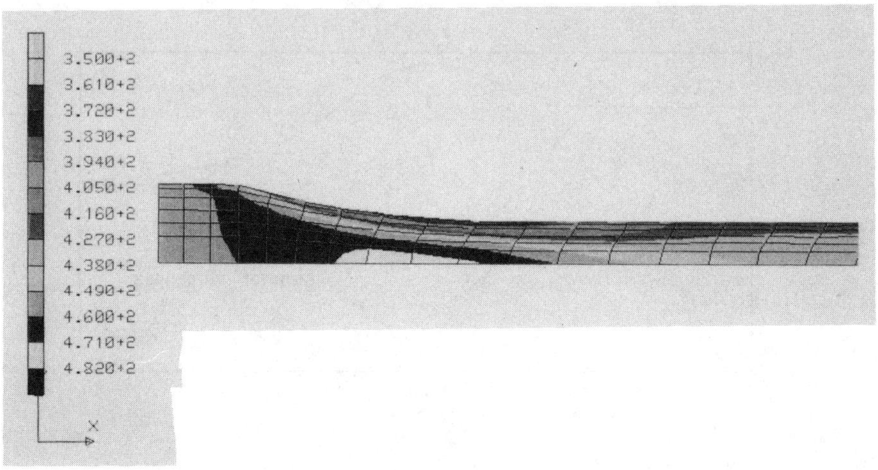

FIG. 7 Distribution of instantaneous temperature through one half of the deforming region as predicted by the finite element model (m = 0.80).

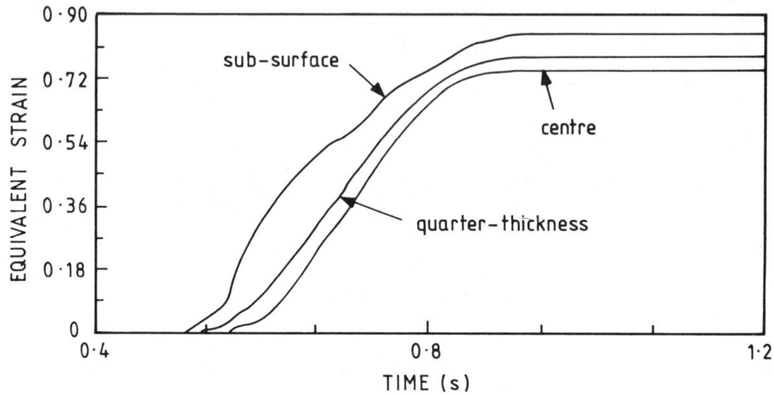

FIG. 8 Predicted equivalent strain histories of centre, quarter-thickness and surface regions having passed through the roll-bite in Fig. 6.

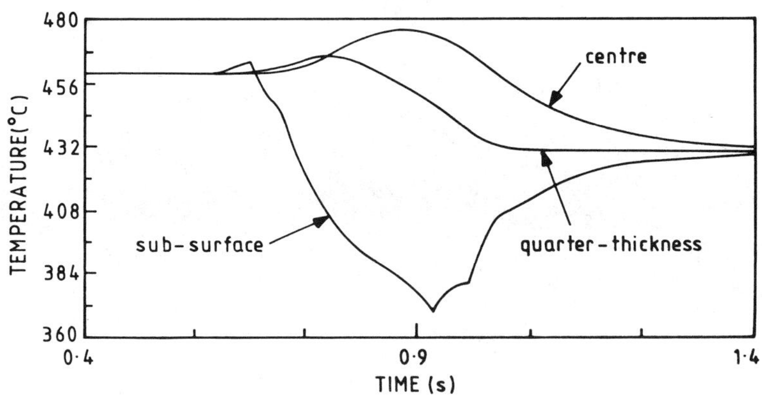

FIG. 9 Predicted temperature histories of centre, quarter-thickness and surface regions having passed through the roll-bite in Fig. 7.

Discussion

It can be seen from Figs. 4 and 5 that the main features of the as-rolled microstructures are reproduced very well in the plane-strain compression simulations; both sets of microstructures have partially recrystallised. The morphology of the deformed and recrystallised grains at the centre of the hot-rolled slab in Fig. 4 are almost identical to those in the simulation; the grain sizes and the volume fraction of recrystallised grains shown in Tables 2 and 3 agree very well. The close similarity arises because the metal at the centre of the slab has undergone essentially plane-strain compression only (see later). The aspect ratio of the deformed grains in the surface simulation in Fig. 5 is much greater (Table 3) than the actual hot-rolled microstructure due to a difference in strain path; both samples have been deformed to the same equivalent strain, but this has resulted from compressive deformation in the former, rather than a combination of compression and shear in the latter. If,however,we choose to characterise the deformed surface microstructures by the size and volume fraction of recrystallised grains only in Table 2, then agreement is very good. Overall, the recrystallised fraction of the microstructure seems to be reproduced accurately during the simulation of the rolling deformation. This result is very encouraging given the critical nature of the experiment. During a hot-rolling operation involving several passes, we may expect this alloy to completely recrystallise one or more times. If the recrystallised microstructure depends primarily on the total level of strain (at a given strain rate, temperature etc) imparted during deformation and to a lesser extent on strain path, then it may be possible to model straightforwardly the types of microstructure produced in given regions of final hot-rolled slab by a physical simulation technique, or by a finite element model incorporating equations describing recovery and recrystallisation kinetics; Sellars et al(11, 12) have already shown satisfactory agreement between predicted and actual microstructures using the latter approach.

Predictions of mean equivalent strain rate (Table 4) and temperature history (Fig. 7) were found to be in reasonable agreement with the experimental results (Table 4 and Fig 3). The correct trends in temperature profile with position in the metal were predicted, with absolute values being at the right level. The mean strain rates at the centre of slab agree well (Table 4), but the discrepancy is much greater at the sub-surface position due mainly to the difference in equivalent strains here (Fig. 10). The strain rate is greatest in the sub-surface region immediately after deformation commences (Table 4); this result agrees with earlier work on the modelling of hot-rolling of aluminium alloys(15).

Fig. 10,however,shows that the rate of change of strain with through-thickness position as calculated by the FE model is lower than the measured gradient. The predicted and measured strains agree well at the centre of the slab where the slab has undergone effectively a homogeneous plane-strain compression reduction. The predicted strains immediately beneath the surface of the slab are significantly lower than the measured value. It is at the roll contact that the boundary conditions are least well known, and it is a region where both strain rate and temperature change rapidly as material passes through the process zone. It is not surprising, therefore, that the two values differ widely here. However, even by increasing the value of the friction coefficient m towards the limit of 1.00 in the computations, the result is only a small increase in predicted strain, not predominantly at the surface

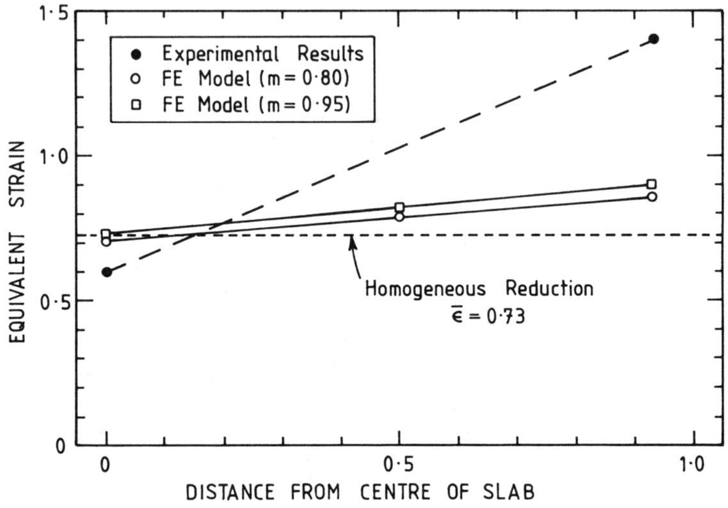

FIG. 10 Comparison of measured and predicted equivalent strain as a function of position through the hot-rolled slab, as it emerges from the roll-bite.

but rather uniformly through the thickness of the slab (Fig. 10). The discrepancy between measured and predicted sub-surface strains still remains large. Gruber(16) has shown that the strain in the sub-surface region relative to the centre increases as the coefficient of friction increases, although the rate at which the deformation pattern changes becomes very small as the coefficient of friction increases from 0.6 to 1.0, as here. It would seem, therefore, that the coefficient of friction by itself may not be the most important factor controlling the sub-surface deformation when high friction forces prevail. The heat transfer coefficient assumed at the contact must also be approximately correct since the temperature profiles in Figs. 3 and 7 are similar. The reasons for the overall poor quantitative agreement between predicted and measured deformations are not well understood at the present time despite the reasonable success in reproducing temperature profiles and mean strain rates; work is on-going to investigate further this problem.

Conclusions

The partially recrystallised microstructures of the sub-surface and centre regions of a hot-rolled slab (47% reduction at 460°C) of a 5083 aluminium alloy have been reproduced satisfactorily by plane-strain compression simulation of their deformation histories. The actual sizes and volume fractions of the recrystallised grains formed at these positions during rolling were reproduced very well in the compression experiments; the evolution of the recrystallised fraction was thus controlled primarily by the total applied strain during deformation rather than on the deformation path. The morphology of the as-deformed grains did, however, depend on the latter; agreement here was better during the simulation of the centre deformation, since the deformation path during the rolling process itself was essentially pure plane-strain compression.

The rolling experiment was analysed using a finite element model; predicted deformation histories at the sub-surface, quarter-thickness and centre positions were compared to measured histories derived using experimental rolling blocks. Agreement between temperature profiles, mean strain rates and strain accumulated at the centre of the slab during rolling were reasonable; the equivalent strain at the centre was found to be close to 0.73, the equivalent strain associated with a homogeneous plane-strain reduction of 47%. The measured strain immediately beneath the surface of the hot-rolled slab was, however, much higher than that predicted using the model.

Acknowledgements

We would like to thank Mr Kevin Smith, Alcan International Limited, Banbury and Mr David Manvell, School of Materials, University of Sheffield, for performing much of the experimental work.

References

1. C Rossard and P Blain, "A Method of Simulation by Torsion for Determining the Influence of Hot-Rolling Conditions on the Structure of Steel", in "Flat Rolled Products III", E W Earhart (ed), Interscience, 1962, pp 3-28.

2. M M Farag, C M Sellars and W J McG Tegart, "Simulation of Hot-Working of Aluminium", in "Deformation under Hot-Working Conditions", Publication No. 108, Iron and Steel Institute, London, 1968, pp 60-67.

3. J J Jonas, C M Sellars and W J McG Tegart, "Strength and Structure under Hot-Working Conditions", Metall Rev (1969) 14, 1-24.

4. C M Sellars and W J McG Tegart "Hot Workability", Int Metall Rev (1972) 17, 1-24.

5. C M Sellars, "Laboratory Methods of Assessing Hot-Working Characteristics", in "Les Traitements Thermomechaniques", 24 ieme Colloque de Metallurgie, Saclay, France, 1981, pp 111-120.

6. H Yada, N Matsuzu and K Nakajima, "Deformation Stress and Structural Change during High Strain Rate Deformation of Carbon Steel", in "Advances in the Physical Metallurgy and Applications of Steels", Book 284, The Metals Society, 1982, pp 325-330.

7. M Ueki, S Horie and T Nakamura, "Simulation of the Hot-Working of 5083 Aluminium Alloy by Means of the Torsion Test", J Mech Work Technol (1985) 11, 365-376.

8. R Hill, "The Mathematical Theory of Plasticity", Oxford University Press, 1950.

9. I Ya Tarnovskii, AA Pozdeyev and V B Lyashkov, "Deformation of Metals During Rolling", translated by M de O Tollemache, Pergammon, 1965.

10. T Sheppard and D S Wright, "Structural and Temperature Variations During Rolling of Aluminium Slabs", Metals Technol (1980) 7, 274-281.

11. J H Beynon, A R S Ponter and C M Sellars, "Metallographic Verification of Computer Modelling of Hot-Rolling", in "Modelling of Metal Forming Processes", J L Chenot and E Onate (eds), Kluwer, 1988, pp 321-328.

12. C M Sellars, "Computer Modelling of Microstructural Evolution During Hot-Rolling", in "Thermec-88", Vol 2, Proc of Int Conf on Physical Metallurgy of Thermomechanical Processing of Steels and Other Metals, Tokyo, Japan 6-10 June 1988, The Iron and Steel Institute of Japan, 1988 pp 448-457.

13. E S Puchi, J Beynon and C M Sellars, "Simulation of Hot-Rolling Operations on Commercial Aluminium Alloys", ibid, pp 572-579.

14. E G Thomsen, C T Yang and S Kobayashi, in "Mechanics of Plastic Deformation During Metal Processing", Macmillan, 1965.

15. K A Woodbury and A J Beaudoin, "Mathematical Modelling of Strip Rolling of Aluminium Alloys", in "Aluminium Alloys: Their Physical and Mechanical Properties" Vol. 1, E A Starke and T H Saunders (eds), Chameleon Press Limited, London, 1986, pp 387-401.

16. H Gruber, "Finite Element Simulation of Hot Flat Rolling of Steel", in "NUMIFORM 86: Numerical Methods in Industrial Forming Processes", K Mattiasson et al (eds), A A Balkeema, Rotterdam, 1986, pp 225-229.

17. J H Beynon, P R Brown, S I Mizban, A R S Ponter and C M Sellars, in "An Eulerian Finite Element Method for the Thermal and Viscoplastic Deformations of Metals in Industrial Hot-Rolling", in "Computational Methods for Predicting Material Processing Defects", M Predeleanu (ed), Elsevier, 1987, pp 19-28.

18. K J Gardner and R Grimes, "Recrystallisation During Hot Deformation of Aluminium Alloys", Metal Science (1979) 13, 216-222.

19. K Lintermanns and H A Kuhn, "Dynamic Recrystallisation During Large Strain Deformation of Commercial Al-Mg-Mn Alloys", as ref 15, pp 529-543.

20. M Ueki, S Horie and T Nakamura, "Hot Deformation and Dynamic Restoration in a Series of Al-Mg Alloys", ibid, pp 419-422.

21. H J McQueen and K Conrod, "Recovery and Recrystallisation in the Hot Working of Aluminium Alloys", in "Microstructural Control in Aluminium Alloys: Deformation, Recovery and Recrystallisation", E Henry Chia and H J McQueen (eds), Metall Soc AIME, 1986, pp 197-217.

22. N Raghunathan and T Sheppard, "Evolution of Structure in Roll Gap When Rolling Aluminium Alloys", Mater Sci Technol (1989) 5, 194-201.

23. S P Timothy and S Rogers, to be published.

24. V K Jain, L E Matson, H L Gegel and R Srinivasan, "Physical Modelling of Metalworking Processes - I: Determination of Large Plastic Strains", J Mater Shaping Technol (1988) 5, 243-248.

25. J A Schey, "Friction Laws in Metal Forming Tribology", in "Advanced Technology of Plasticity 1987", Vol II, Springer Verlag, Berlin, 1987, pp 873-882.

26. C Zener and J H Hollomon, "Effect of Strain Rate upon the Plastic Flow of Steel", J Appl. Phys (1944) 15, 22-32.

27. J H Beynon and C M Sellars, "Strain Distribution Patterns During Plane-Strain Compression", J Test Eval (1985) 13, 28-38.

28. R Colas and C M Sellars, "Strain Distribution and Temperature Increase During Plane-Strain Compression Testing", J Test Eval (1987) 15, 342-349.

Thermomechanical modelling of a work roll of an aluminum hot rolling mill with finite element method

Gy. Juhász, F. Nagy, L. Szabó
HUNGALU, Engineering Development Centre, Hungary
M. Balla
Technical University of Budapest, Hungary

ABSTRACT

This paper contains the finite element analysis of thermal and elastic strain processes taking place in work rolls of an aluminium hot rolling mill during rolling. The investigations are focused on determination of the thermal crown caused by the heat expansion and of thermal stresses arising on the surfaces of the rolls. An axisymmetric and a plane finite element model have been developed The axisymmetric model has been applied to the examination of thermal crown. With the plane model we have determined the transient temperature- and thermal stress distributions arising near the roll surfaces. We have revealed that after a quasi-stationary state has been reached, the temperature of the rolls varies only in a depth of some millimeters but there evolve high temperatures and stresses above yield point in this region. On the basis of our calculations we have pointed out that preheating of the rolls is advantageous both for the forming of thermal crown and for thermal stress distribution as well.

KEYWORDS

Hot rolling of aluminium strips; finite element method; thermoelasticity; thermal crown; thermal stresses.

INTRODUCTION

The instationary thermal and elastic strain processes taking place during the hot rolling of aluminium strips, have a great influence on the lifetime of the work rolls and flatness of the rolled strips. The flatness of strips is directly related to the actual roll gap. The main factors, determining the shape of the roll gap, are: elastic deflection and elastic elongation of the work rolls and the thermal crown. In case of traditional, relatively stiff rolls (work roll diameter/roll face width ≈ 0.45), the most significant one among these factors is usually the thermal crown. As an effect of the thermal shock load elevated temperature values arise in the contact zone. Consequently significant compressive stresses develop leading to local plastic strains in the thin surface layer of the rolls. In the cooling zone, these plastic strains are followed by residual strains which cause tensile stresses. This periodic fluctuation of the thermal load results in thermal fatigue and small surface cracks which can damage the rolls exposed to bending moment.

The danger of roll surface crack development can be diminished by the right selection of rolling technology and application of intensive cooling. Besides these, the thermal crown can also be influenced by appropriate formation of the ground crown. By the use of roll bending or special expandable rolls, the roll gap can be varied during rolling as well. However, to be able to intervene effectively during rolling and to select the correct technology, we have to understand and determine the processes taking place in the work rolls.

Except a thin surface layer, the strain of the major part of the work roll remains within the elastic range, consequently the processes can be described by basic equations of thermoelasticity. Many researchers have already dealt with the investigation of thermal processes only and developed analytical solutions expanded in series by means of one-, two-, or three-dimensional models [1, 2, 3]. These solutions, however, generally refer to merely simple loads and shapes, so they do not give a chance to consider many essential effects. More exact analyses can be made by the use of numerical methods, among which the finite element method has become the most widespread one. By using commercial finite element program systems, thermal stress and displacement fields can generally be obtained only if the temperature field is known. We have developed a finite element program system which calculates the mentioned fields simultaneously. We have applied this program to model thermal and elastic strain processes taking place in the work rolls of a four-high hot rolling mill. The investigations have been focused on the determination of the thermal crown and thermal stresses originating on the surfaces of the rolls.

The first part of the paper gives a brief survey of basic equations of thermoelasticity and their finite element formulation. The second part contains a qualitative analysis of the processes taking place in the work rolls and the description of the applied plane and axisymmetric finite element models. This is followed by the introduction and detailed evaluation of the gained results and the conclusions of the examinations.

Basic Equations of Thermoelasticity and Their Finite Element Formulation

In case of homogenious, isotropic continuum, the governing equations of the classical linear theory of thermoelasticity for displacement and temperature fields consist of the coupled partial differential equations as follows [4]:

$$\mu u_{i,jj} + (\lambda + \mu) u_{j,ji} + \beta T,_i + \varrho f_i - \varrho \ddot{u}_i = 0 \ , \tag{1}$$

$$k T,_{jj} - \varrho c \dot{T} + r + \beta T_0 \dot{u}_{j,j} = 0 \ , \tag{2}$$

in which λ and μ are the Lamé's constants; β is the thermal modulus; ϱ is the mass density; u_i and f_i are the Cartesian components of displacement and body force vector, respectively; T is the absolute temperature; k is the thermal conductivity; c is the specific heat at constant strain; r is the heat source and T_0 is the reference temperature of the natural, stress-free state. Superposed dots ($\dot{}$) and commas $(\ ),_i$ denote material time differentiation and partial differentiation with respect to Cartesian co-ordinates x_i (i=1,2,3).

When formulating a given problem, the following initial and boundary conditions have to be joined to the equation of motion and the heat conduction equation:

- initial conditions:

$$u_i(\underline{r},t=0) = u_{i0}(\underline{r}) \ , \quad \dot{u}_i(\underline{r},t=0) = \dot{u}_{i0}(\underline{r}) \ , \quad T(\underline{r},t=0) = T_0(\underline{r}) \ , \tag{3}$$

- kinematic boundary condition:

$$u_i\Big|_{A_u} = \hat{u}_i \ , \tag{4}$$

- dynamic boundary condition:

$$\sigma_{ij}n_j\Big|_{A_p} = \hat{p}_i \ , \tag{5}$$

- prescribed temperature condition:

$$T\Big|_{A_1} = \hat{T} \ , \tag{6}$$

- heat flow and convective condition:

$$q_i n_i\Big|_{A_2} = \hat{q}_n + \alpha(T_W - T_\infty) \ . \tag{7}$$

In the equations $\underline{r}$ is the space vector; t is the time; A_u, A_p, A_1, and A_2 are parts of the surface of the examined domain; n_i, σ_{ij}, p_i and q_i are the Cartesian components of the surface normal unit vector, the stress tensor, the distributed load and the heat flux vector, respectively; q_n is the heat flux in normal direction; α is the coefficient of convective heat transfer; T_W is the surface temperature and T_∞ is the reference temperature.

For the finite element solution of the outlined thermoelastic problems, field equations (1) and (2) have to be discretized. Assuming a continuous time dependence of the field variables, we first discretize the equations in space. As there is no variation principle for the displacement and temperature variables, we deduce the element matrices and load vector components by means of a special version of weighted residuals, the Galerkin method. Applying the method for one finite element, we get the matrix differential equation system as follows [5]:

$$\begin{bmatrix} \underline{\underline{M}}^{dd} & \underline{\underline{0}} \\ \underline{\underline{0}} & \underline{\underline{0}} \end{bmatrix} \begin{bmatrix} \ddot{\underline{d}}^e \\ \ddot{\underline{\theta}}^e \end{bmatrix} + \begin{bmatrix} \underline{\underline{0}} & \underline{\underline{0}} \\ \underline{\underline{C}}^{\theta d} & \underline{\underline{C}}^{\theta\theta} \end{bmatrix} \begin{bmatrix} \dot{\underline{d}}^e \\ \dot{\underline{\theta}}^e \end{bmatrix} + \begin{bmatrix} \underline{\underline{K}}^{dd} & \underline{\underline{K}}^{d\theta} \\ \underline{\underline{0}} & \underline{\underline{K}}^{\theta\theta} \end{bmatrix} \begin{bmatrix} \underline{d}^e \\ \underline{\theta}^e \end{bmatrix} = \begin{bmatrix} \underline{\Phi}^d \\ \underline{\Phi}^\theta \end{bmatrix} , \tag{8}$$

where $\underline{d}^e$ is the vector of nodal displacements of the element and $\underline{\theta}^e$ is the vector of nodal temperatures of the element. In case of plane problems, the further quantities in (8) are:

- elements of the generalized stiffness/conduction matrix:

$$K_{ij}^{dd} = \int_{\Omega^e} \left[\begin{array}{c|c} D_1\frac{\partial N_i}{\partial x}\frac{\partial N_j}{\partial x} + D_3\frac{\partial N_i}{\partial y}\frac{\partial N_j}{\partial y} & D_2\frac{\partial N_i}{\partial x}\frac{\partial N_j}{\partial y} + D_3\frac{\partial N_i}{\partial y}\frac{\partial N_j}{\partial x} \\ \hline D_2\frac{\partial N_i}{\partial y}\frac{\partial N_j}{\partial x} + D_3\frac{\partial N_i}{\partial x}\frac{\partial N_j}{\partial y} & D_1\frac{\partial N_i}{\partial y}\frac{\partial N_j}{\partial y} + D_3\frac{\partial N_i}{\partial x}\frac{\partial N_j}{\partial x} \end{array} \right] d\Omega \ , \tag{9/a}$$

$$K_{ij}^{d\Theta} = \int_{\Omega^e} \beta \begin{bmatrix} \frac{\partial N_i}{\partial x} N_j \\ \hline \frac{\partial N_i}{\partial y} N_j \end{bmatrix} d\Omega \, , \quad K_{ij}^{\Theta\Theta} = \int_{\Omega^e} k(\frac{\partial N_i}{\partial x}\frac{\partial N_j}{\partial x} + \frac{\partial N_i}{\partial y}\frac{\partial N_j}{\partial y}) d\Omega + \int_{S_2^e} \alpha \, N_i N_j \, dS_2 \, ,$$

– elements of the generalized capacity/damping matrix:

$$C_{ij}^{\Theta d} = \int_{\Omega^e} -\beta T_0 \begin{bmatrix} N_i \frac{\partial N_j}{\partial x} & \vdots & N_i \frac{\partial N_j}{\partial y} \end{bmatrix} d\Omega \, , \qquad C_{ij}^{\Theta\Theta} = \int_{\Omega^e} \varrho c \, N_i N_j \, d\Omega \, , \tag{9/b}$$

– elements of the mass matrix:

$$M_{ij}^{dd} = \int_{\Omega^e} \varrho \begin{bmatrix} N_i N_j & 0 \\ 0 & N_i N_j \end{bmatrix} d\Omega \, , \tag{9/c}$$

– elements of the mechanical load vector:

$$\Phi_i^d = \int_{S_p^e} N_i \begin{bmatrix} \hat{p}_x \\ \hat{p}_y \end{bmatrix} dS_p + \int_{\Omega^e} \varrho \, N_i \begin{bmatrix} f_x \\ f_y \end{bmatrix} d\Omega \, , \tag{9/d}$$

– elements of the thermal load vector:

$$\Phi_i^\Theta = \int_{S_2^e} \alpha \Theta_\infty \, N_i \, dS_2 + \int_{\Omega^e} r \, N_i \, d\Omega - \int_{S_2^e} \hat{q}_n \, N_i \, dS_2 \, , \tag{9/e}$$

where D_1, D_2, D_3 parameters are formed from elastic material properties; N_i are shape functions applied to the interpolation of displacement and temperature fields; Ω^e is the domain and S^e is the boundary of the element. Subscripts i and j extend from 1 to the number of nodes per actual element. The global differential equation system of the whole structure can be obtained from (8) by means of the usual element assembly process of the finite element method:

$$\underline{\underline{M}} \, \ddot{\underline{\delta}} + \underline{\underline{C}} \, \dot{\underline{\delta}} + \underline{\underline{K}} \, \underline{\delta} = \underline{\Phi} \, , \tag{10}$$

where $\underline{\delta}$ is a vector formed from the nodal unknowns of the structure. Equation (10) can be discretized in time by the use of quadratic time finite elements or three time-level difference schemes. Among these, the Newmark scheme [6] seems to be the most effective one because of its simplicity and its favourable stability and accuracy. Using the scheme, the solution of the matrix differential equation system (10) can be reduced to the solution of a linear system of equations in each time step. The kinematic boundary conditions and the prescribed temperature conditions must be taken into account during the solution of this so-called effective equation system.

FINITE ELEMENT MODELLING OF PROCESSES TAKING PLACE IN WORK ROLLS

If material properties and loadings of the work rolls are known, temperature, displacement and thermal stress fields can be determined by the use of the finite element program system built on the basic equations introduced in the previous section. However, in order to get sufficiently exact solutions, the whole volume of the work roll has to be divided into an extremely large number of finite elements and very small time steps must be selected. These can extraordinarily increase the calculation time especially when using small personal computers. In order to diminish running time, it is expedient to use some modelling tricks to develop finite element models containing a relatively low number of elements, but giving reasonable solutions for the problem, in spite of their simplicity.

As it was mentioned in the introduction, the present work has got two main targets: determination of roll gap which influences the flatness of rolled strips and determination of temperature and thermal stress peak values that may damage the work rolls. When we developed the finite element model of the problem, we set out from the following assumptions:

- In case of a given ground crown, the roll gap fundamentally evolves as a superposition of three effects. These effects are: elastic deflection of the work rolls, elastic elongation of the work rolls and the thermal crown arising from the heat expansion of the work rolls.
- The time dependent variation of the thermal state of work rolls can basically be divided into two parts: a quick cyclic fluctuation as an effect of the rotation of the rolls and an essentially slower variation. The actual thermal state of the rolls can be approached by the superposition of these processes.

The slower process results from the fact that the heat capacity of the rolls is extremely large, thus the inner layers react to the changes taking place on the surfaces with a great delay. The short time or fast changes of the thermal loads of the rolls do not considerably modify the state of the core. As an effect of long lasting continuous rolling, the temperature of the core starts to increase and after a certain amount of time the temperature of the major part of the roll becomes stationary in time. In spite of the quick processes taking place on the surface, this state can be considered as quasistatic. The developed quasistatic state is only altered if a permanent modification occurs in the thermal load of the work rolls (e.g. change of the width of the rolled strip).

The faster process is the result of the rotation of the rolls. In the contact zone a part of the cylinder jacket picks up a significant amount of heat from the strip, then loses it (partly by convective dissipation towards cooling fluid and air, partly by thermal conduction towards back-up rolls) when the roll goes on rotating. While the quasistatic state is beeing reached, one part of the picked up heat warms up the core of the roll. However, after the quasi-static state is established the whole quantity of the heat picked up in the contact zone is lost in the process of one rotation. Subsequently, this fast process alters the state of only a thin surface layer and superimposes on the above-mentioned slow process as a transient one.

On the basis of the above mentioned assumptions we have elaborated two finite

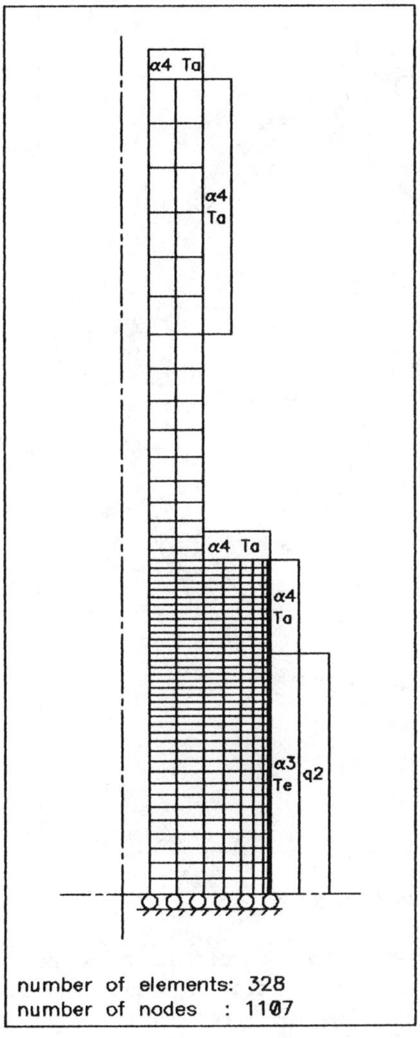

Fig. 1. Axisymmetric model.

element models: an axisymmetric and a plane one. With the axisymmetric model (see Fig. 1.) we have examined the thermal crown, both in transient and quasistatic states. Because of the axial and midplane symmetry we have meshed only a quarter of the work roll with eight noded, quadratic isoparametric finite elements and applied rolling supports along the straight line which represents the midplane (see Fig. 1.). We fined the mesh near the surface because of high temperature gradients may occur in case of transient calculations. We have also applied a finer mesh at the edges of the rolled strip. The axial symmetry of the model means that it is not geometry only that must be axisymmetric but the loads as well. Consequently, the heat flux from the strip and the coefficients of convective heat transfer belonging to different kinds of cooling zones have been uniformly distributed along the periphery of the roll. Of course, this distribution of thermal loads 'blurs' the processes taking place on the surface of the roll during one rotation, but it does not have any effect on the state of inner regions. As the thermal crown is primarily the result of heat expansion of the roll core, it can be assumed that the model gives the magnitude of the thermal crown with adequate accuracy but it shows only average values of the surface temperature and thermal stresses.

The transient thermal processes, which arise as effects of work roll rotation, have been examined in the midplane of the roll with the help of the plane model. The periphery of the roll has been divided into 120 segments. The contact zone and the cooling zones have been specified on these segments according to their real lengths of arc (see Fig. 2.). The cooling effect of the air along the arc sections between the emulsion coolings has been neglected. The rotation of the work roll has been simulated by shifting the loads from segment to segment. The time step has been defined as 120-th part of the period time and this way each load has shifted exactly to the next segment during one time step. On the basis of the above mentioned considerations, we have assumed that after the core has been heated up (in quasistatic case) the temperature does not change during one rotation even in the depth of 50 mm. Consequently, we have meshed

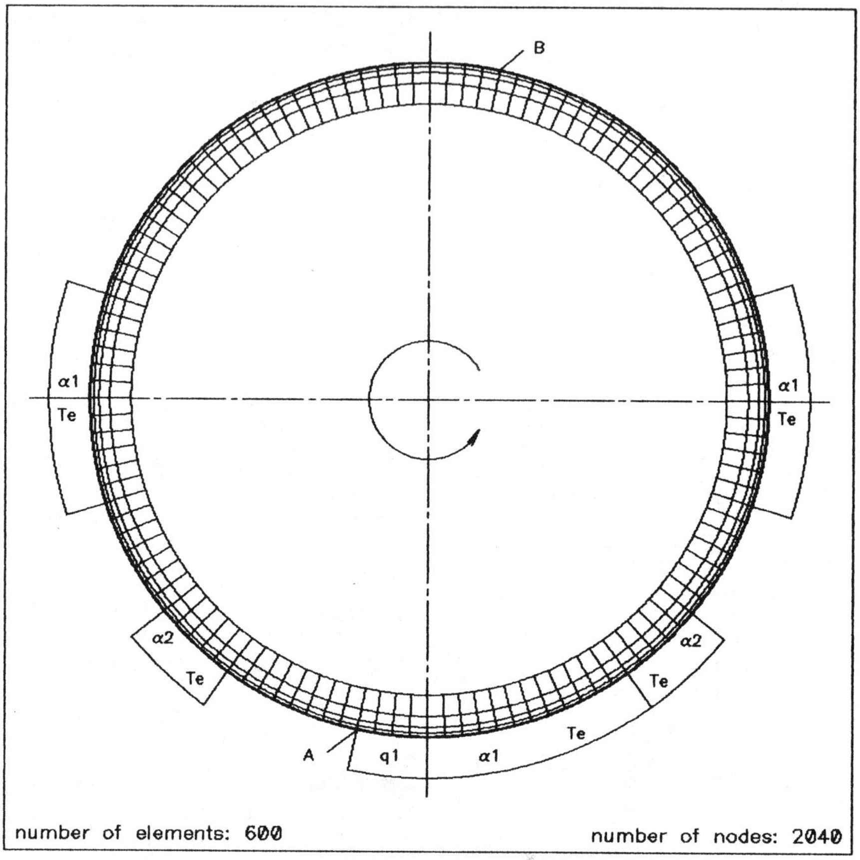

Fig. 2. Plane model.

only this outer domain with finite elements which have resulted in a notable diminution of calculation time. From the viewpoint of heat conduction the examined annular part has to behave the same way as the whole roll does, so we have prescribed that the inner cylinder jacket of this ring should be isolated. The initial temperature has been chosen in accordance with the temperature of the quasistatic state. The thermal stresses have been calculated from the already known temperature field in some time steps. For the calculations we have meshed the whole cross section of the roll and presumed plane strain state. We have prescribed the calculated temperatures in the nodes of the outer ring and the quasistatic temperature in the inner nodes. As the roll can really move free in axial direction and this motion has been restricted with the assumption of plane strain state, we have substracted the thermal stress belonging to the core temperature from the axial stress in order to get adequate results.

By means of joint application of the above introduced models, thermal processes can essentially be examined in the whole domain of the work rolls. Note that elastic deflection and elastic elongation of the work rolls caused by mechanical loads can only be determined with three-dimensional models.

Presentation of Calculated Results

With the help of the finite element models introduced in the previous section, we have analysed thermal processes in the work roll of a four-high hot rolling mill on an IBM AT 386 personal computer. The schematic representation of the work roll and the cooling system as well as important geometry data can be seen in Fig. 3. The physical characteristics and technological parameters are summarized in Table 1. The values of convective heat transfer coefficients given in the table have been determined by laboratory tests. The heat transferred from the rolled strip to the work roll has been modelled by a uniformly distributed heat flux along the contact zone. The value of this heat flux has been calculated on the basis of one of our previous papers [7].

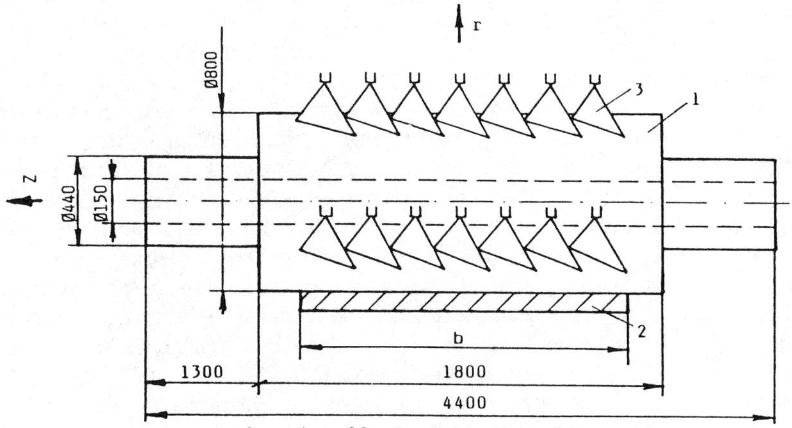

1 work roll, 2 strip, 3 cooling system

Fig. 3. Scheme of the work roll and the cooling system.

TABLE 1.

Work Roll Parameters and Rolling Conditions

thermal conductivity	: 32 W/m°C	strip widths	: 1000/1300/1600 mm
density	: 7400 kg/m^3	rolling time of one strip	: 426 sec
specific heat	: 480 J/kg°C		
linear thermal expansion	: 1.25e-5 1/°C	heat fluxes q1:	11300 kW/m²
Young's modulus	: 2.1e5 Mpa	q2:	377 kW/m²
Poisson's ratio	: 0.3	coefficients of convective heat	
alloy	: Al 99.5	transfer α1:	34000 W/m² °C
average roll speed	: 2 m/s	α2:	38000 W/m² °C
entry temp. of strips	: 495 °C	α3:	10700 W/m² °C
entry thickness of strips	: 400 mm	α4:	50 W/m² °C
final thickness of strips	: 6 mm	emulsion temperature Te:	55 °C
number of passes	: 11	air temperature Ta:	20 °C

First we have used the axisymmetric model to determine the quasistatic state of the roll developing as an effect of long lasting continuous rolling under unaltered technological conditions. In accordance with the considerations detailed in the previous section, the heat flux and convective heat conduction coefficients have been distributed along the pheriphery of the roll according to their lengths of arc (see Fig. 2). We have also given these distributed values (with their markings used in Fig. 1) in Table 1. From the investigation of heat equlibrium of a thin axial section near the midplane, we have got 90 °C average temperature using the above defined distributed thermal loads. This temperature corresponds to the quasistatic temperature of this domain. We have prescribed convective heat conduction towards the air on the non-cooled surface sections as well as on the end face of the roll. We have also prescribed the temperature of the lubricating oil at the bearing support. Three calculations were carried out for three different strip widths (1000/1300/1600 mm). In each case, the width of cooling zones was the same as that of the rolled strip. Out of the calculation results, we have shown only the temperature, the radial displacement and the Von Mises stress of the work roll surface as functions of axial co-ordinate (see Figs 4-6). Fig. 4 indicates that the average surface temperature agrees with the quasistatic temperature where the roll contacts the strip, but there is a steep decrease of temperature on the surface section beyond the strip. As a result of high temperature changes, there arise significant thermal stresses at the edge of the strip as compared to other surface areas. Fig. 6 shows that the maximum values of stresses essentially do not depend on the width of the strip. The curves in Figs 4 and 6 give an approximate picture of axial direction distributions, but it should be taken into consideration that these illustrate only average values for a single rotation upon which there have to be superimposed transient distributions calculated by the plane model. Fig. 5 shows the radial displacement of surface points. According to our previous assumptions (which have later

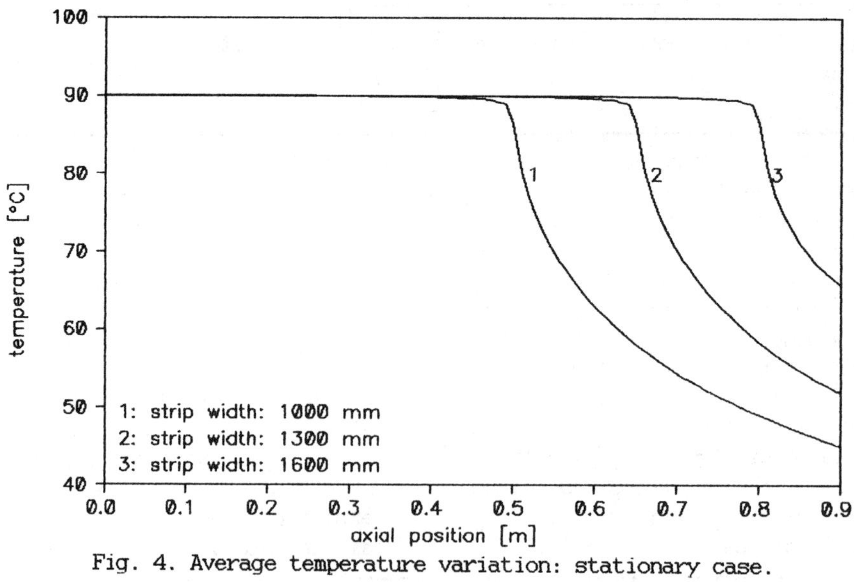

Fig. 4. Average temperature variation: stationary case.

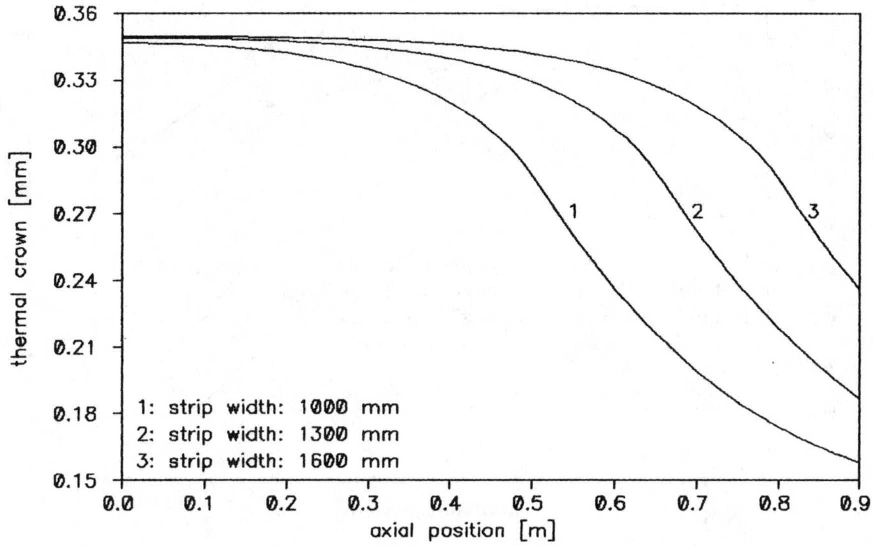

Fig. 5. Radial displacement variation: stationary case.

been proved through calculations with the plane model) the temperature fluctuation caused by the roll rotation extends only to a thin surface layer, thus it has no significant effect on the thermal crown. Accordingly, the curves in Fig. 5 give the wanted thermal crowns for various strip widths with adequate accuracy. In the figure it is clearly shown that the change of the strip width does not basically influence the magnitude of the thermal crown and the type of its shape.

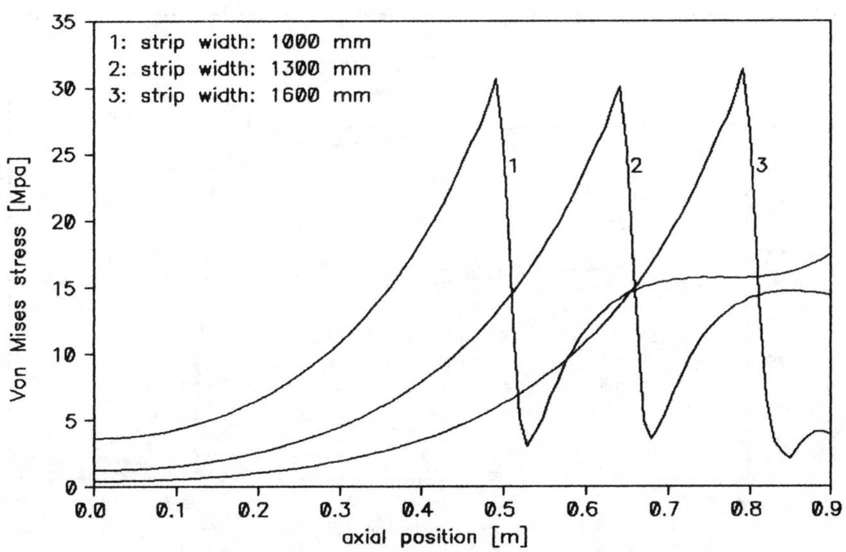

Fig. 6. Average thermal stress variation: stationary case.

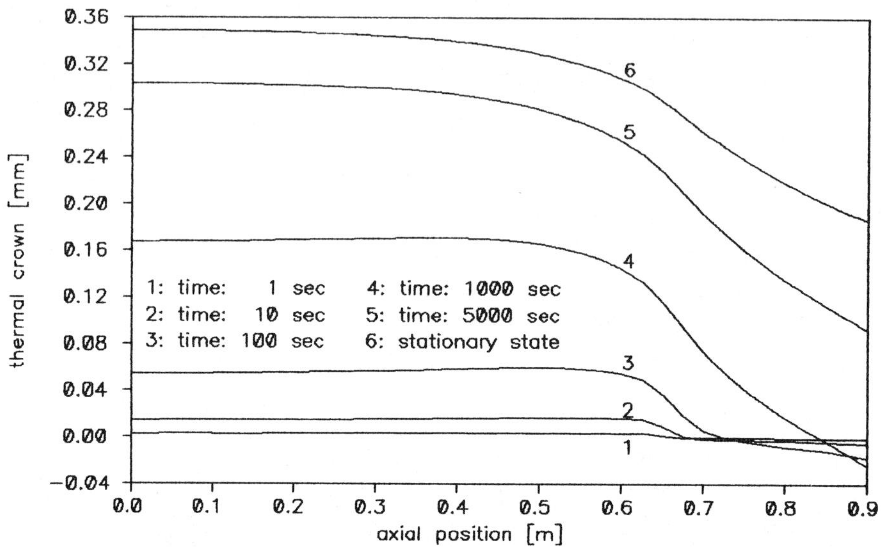

Fig. 7. Radial displacement variation: instationary case.

With the help of the axisymmetric model, we have also examined the gradual heating up of the roll of initially 20 °C temperature under continuous rolling of 1300 mm size strips. In order to be able to exactly calculate the quick initial processes in the surface layer of the roll, we have chosen the first time steps for some tenths of seconds, then increased to some hundreds of seconds during the calculations, according to the decelerating rate of the processes. Figs 7 and 8 show the temporal development of stationary distributions of the 1300 mm strip as shown in Figs 5 and 6. Fig. 7 proves that the development

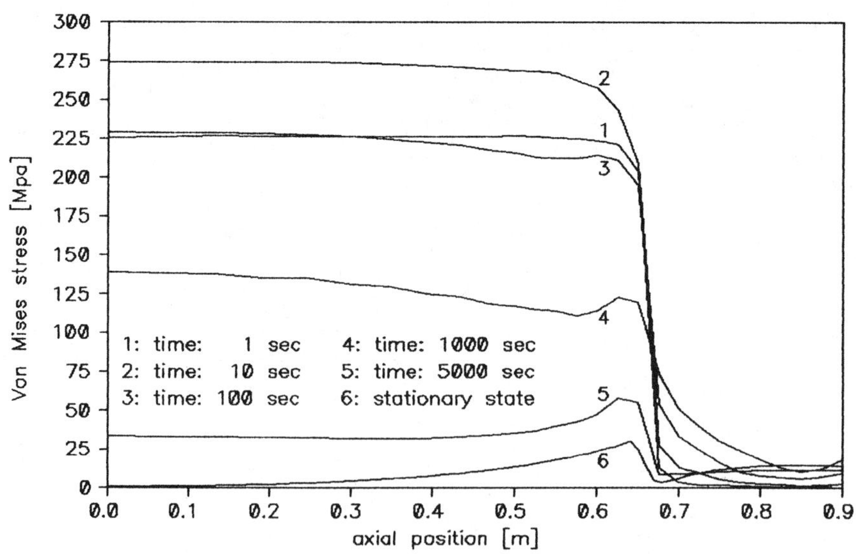

Fig. 8. Average thermal stress variation: instationary case.

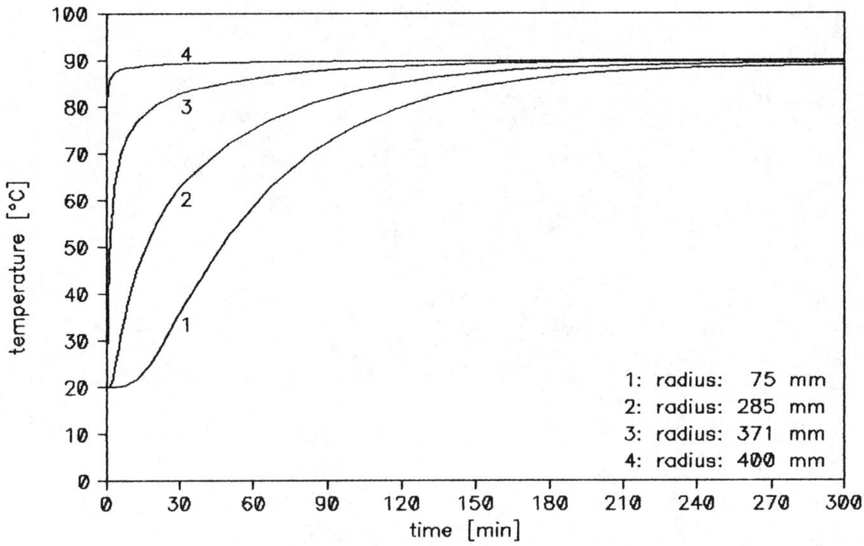

Fig. 9. Instationary average temperature variation.

of the thermal crown is a rather slow process, since it depends on the rate of heating up of the core. It can be seen from the figure that the crown measured between the middle and the edge of the strip reaches its maximum value about 15–20 minutes after the beginning of rolling and not in the quasistatic state. After some minutes of rolling the thermal stresses on the roll surface also rise to maximums which are approximately ten times greater than their quasi-static values (see Fig. 8). In the first minutes the stresses are uniformly high along the places where the roll contacts the strip, while the typical

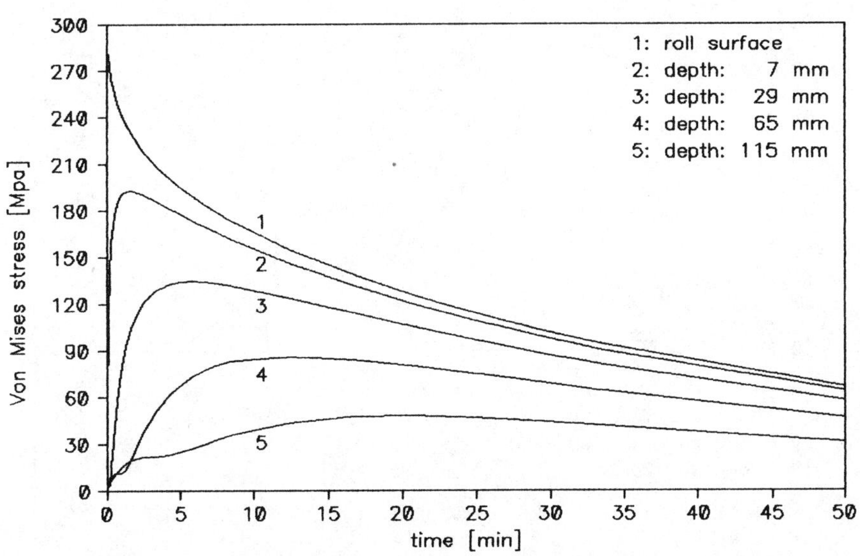

Fig. 10. Instationary average thermal stress variation.

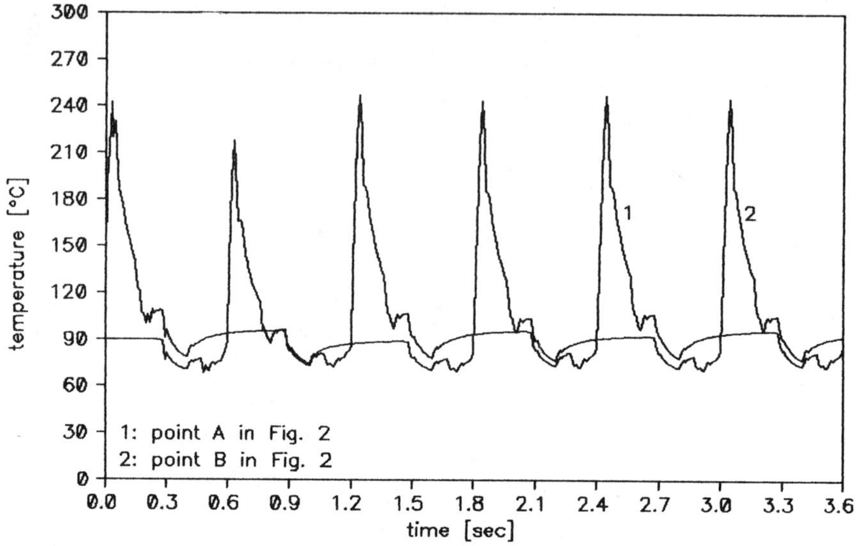

Fig. 11. Surface temperature variation.

quasistatic distribution takes shape only later. Figs 9 and 10 show the temperature- and thermal stress variations along the midplane in different depths. Under the given rolling conditions the work roll is entirely heated up in approximately three hours (this means the rolling time of 25 strips) as shown in Fig. 9. Fig. 10 indicates that the highest thermal stresses arise on the surface of the roll and in the starting phase of heating up, as stated above.

The results of the calculations with the transient plane model can be seen in Figs 11–14. Fig. 11 shows temperature variations during the time of three

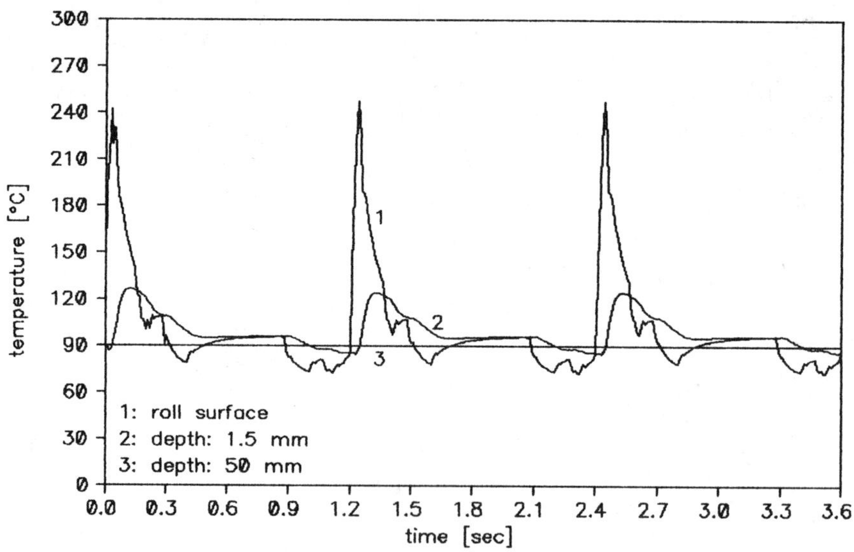

Fig. 12. Instationary temperature variation.

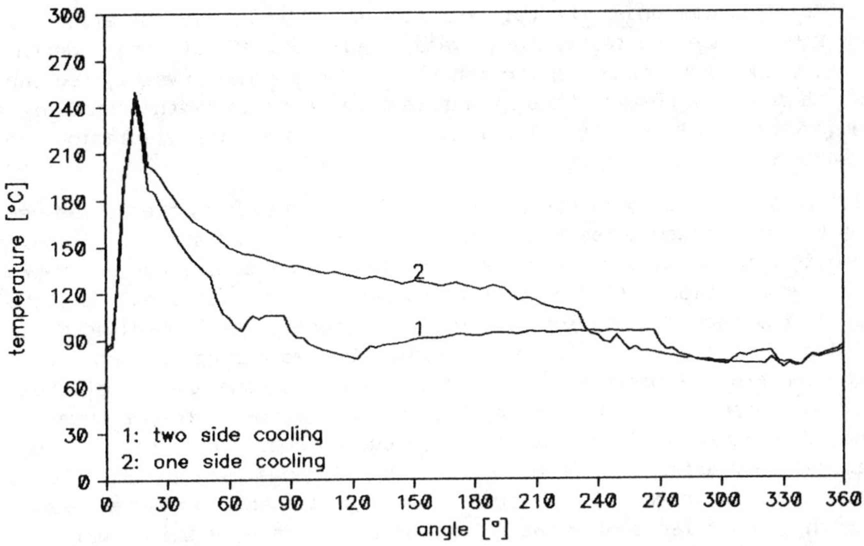

Fig. 13. Comparison of surface temperatures for two different coolings.

rotations in points marked as A and B in Fig. 2. Except the initial transient term the shapes of curves 1 and 2 correspond to each other almost completely and typical phases are repeated periodically even from the second rotation on. This proves that we have modelled the rotation of the roll with adequate accuracy by dividing the whole pheriphery into 120 segments and shifting the loads on them step by step. Fig. 12 indicates the temperature time variation in three points along the radius belonging to point A. The surface temperature significantly rises in the contact zone and exceeds the value of 240 °C.

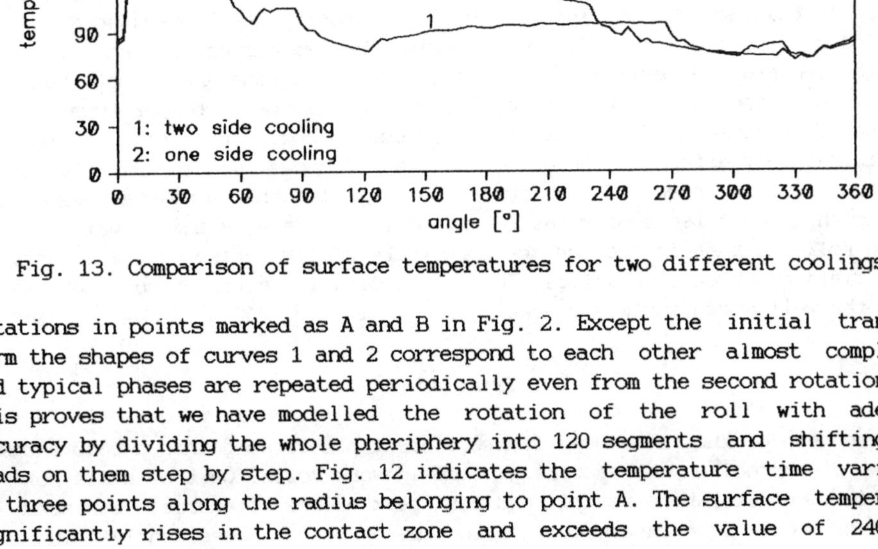

Fig. 14. Thermal stress variation.

In a depth of 1.5 mm under the roll surface, there is hardly any effect of the contact zone, since the temperature rises merely 30 °C in this depth. The temperature does not change on the inner surface of the investigated annular part (in a depth of 50 mm), thus it was reasonable to apply the annular model instead of the total cross section model and to prescribe isolation on its inner surface.

Figs 13 and 14 show temperature and stress distributions on the cross section area of the roll after three rotations. Two temperature distributions belonging to two different cooling systems can be compared in Fig. 13. Curve 1 represents temperature variation referring to the rather uniform cooling, sketched in Fig. 2. In the second case, the same cooling capacity has been divided on the entering half part of the roll. By comparing the two curves, it can be stated that the temperature maximums arising in the contact zone are almost the same, but in the second case there develops a higher average temperature on the surface. Thus it seems to be more advantageous to begin the cooling of the work roll immediately after the contact zone. Fig. 14 shows stress variations for the second cooling case. The figure also verify the previous assumptions stating that the effects of rotating extend to only a very thin surface layer of the roll and that in the contact zone there really arise thermal stresses above yield point on the surface. These periodically alternating peak values cause the well known surface cracks.

Conclusion

With the help of the introduced finite element models we gained a comprehensive picture of thermal processes taking place in work rolls. On the basis of the investigations it can be established that peak values of thermal stresses are caused by sudden temperature changes in time or place. Consequently, in order to diminish peak values, we have to try to get as uniform temperature distribution as possible which, of course, would favourably affect the thermal crown as well. Thus the temperature of the entire roll has to be kept near the quasi-static temperature during the whole time of rolling. This can be solved: 1. by preheating the roll, 2. by heating the roll beyond the strip during rolling, 3. by controlling the temperature of the lubricating oil. With the described finite element models, the effects of these factors can be analysed and an effective intervention can be planned.

References

1. C.F. Peck, J.M. Bonetti and F.T. Mavis, Iron Steel Eng. 31, 6, p. 45 (1954)
2. P.G. Stevens, K.P. Ivens and P. Harper, Iron Steel Inst. 201, 1, p. 1 (1971)
3. C. Troeder, A. Spielvogel and J-W. Xu, Steel Research 56, 7, p. 379 (1985)
4. J.L. Nowinski, Theory of Thermoelasticity with Applications, p. 135 Sijthoff & Noordhoff Int. Publ., The Netherlands (1978)
5. M. Balla, Periodica Polytechnica 33, 1, p. 59 (1989)
6. N.M. Newmark, Proc. A.S.C.E. 8, p. 67 (1959)
7. F. Nagy, Gy. Juhász, F. Szabó and L. Szabó, 4th Seminar on Metal Forming, p. 49, Győr, Hungary, (1988) (in German)

Aluminum smelting and the protection of the environment

Reduction of fluoride emissions and effluents from Alcan's Kitimat smelter

G.J. Gurnon and R.L. Smart
Alcan Smelters and Chemicals Ltd., P.O. Box 1800,
Kitimat, British Columbia, Canada V8C 2H2

Abstract

Reductions in fluoride-containing gaseous emissions and liquid effluents at a vertical-stud Soderberg anode aluminum smelter are discussed.

These improvements have been achieved by the evolution and modernization of facilities for cell gas collection and treatment as well as by improvement in process operation and control.

The reduction in gaseous fluoride emissions has had a significant impact on the condition of nearby forest growth.

Keywords

Air pollution, water pollution, V.S. Soderberg cells, fluoride collection, fluoride emissions, gas cleaning in smelter, fluoride effect on forests.

Introduction

Kitimat Works is a vertical-stud Soderberg anode aluminum smelter built between 1954 and 1966.

The first two potlines of 342 cells were based on an Elektrokemisk design, of which very few similar ones were built elsewhere. Later potlines, comprising 570 cells, were built after a Pechiney design, comparable to most of the western-world V.S.S. capacity built since the 1950's. Basic operating data are summarized in Table 1.

TABLE 1 - Basic Operating Data

Annual Rated Capacity (MT)	268,000
Number of Potlines	7
Number of Electrolytic Cells	912
Amperage (KA)	106 to 127

Kitimat is located 400 air miles north-west of Vancouver at the north end of a 60 mile long ice-free channel running to the Pacific Ocean. It is situated in a long, steep-sided north-south trough which contains Douglas Channel and the Kitimat River (Figure 1).

FIG. 1 - Kitimat Smelter - Looking North

Gas Collection and Scrubbing

Several contaminants are emitted during the production of aluminum from Soderberg electrolytic cells. Among these are carbon monoxide, sulphur dioxide, tars (measured as benzene soluble materials) and gaseous and particulate fluorides. This discussion will deal only with fluorides. In the case of Kitimat's V.S.S. cells, most of the contaminants are collected in a gas channel around the lower perimeter of the anode casing and are treated in scrubbing systems. Those that are not collected pass into the building air and are vented to the atmosphere by the building ventilation system. Government "permits" impose standards on the air quality inside the potrooms and on the emissions to the environment. Workers' exposure to total fluorides must not exceed 2.5 mg/m3 for an 8-hour workday and air emissions of fluoride (expressed as HF) must not exceed 1.90 kg/tonne of aluminum produced. Workers' exposure has never been a problem, however in order to meet the emission standards, a high gas collection efficiency must be maintained at the cell. In addition, scrubbing efficiencies of the collected gases must be kept at a high level.

A sketch of a typical gas collection-scrubber system is shown in Figure 2. There are eight similar arrangements in total at Kitimat. A brief description and history of the evolution of the major system components follows:

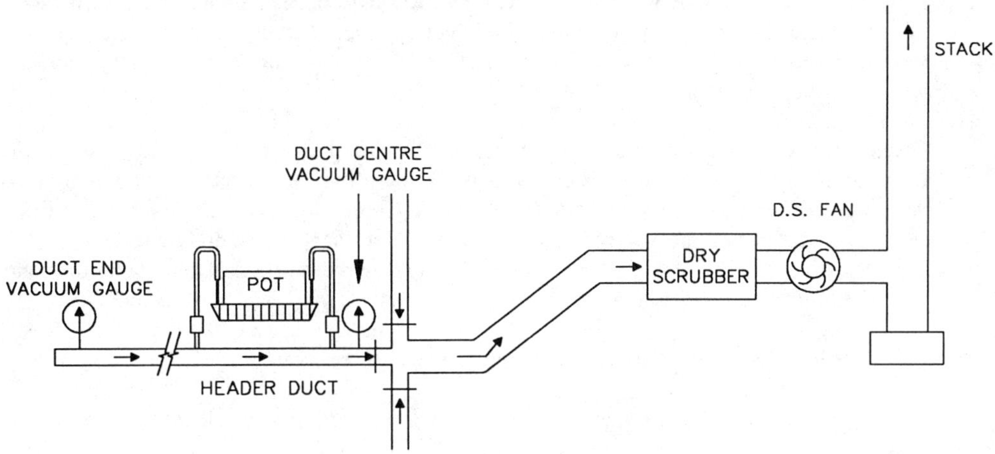

FIG. 2 - Gas Collection-Scrubber System

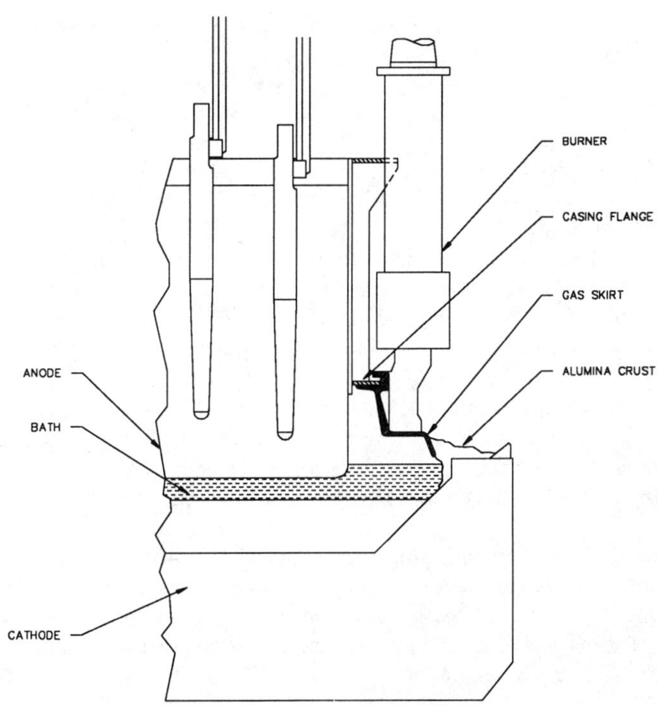

FIG. 3 - Cell Gas Collection Arrangement

1) Gas Skirts/Pot Flange

The set of 20 cast iron gas skirts are supported from the lower anode casing flange in a ring around the pot (see Figure 3). They are the front line gas collection equipment, forming a gas channel between the bath, alumina crust, and the lower portion of the anode. It is important that gaps between skirts, and between skirts and the casing flange, be kept to a minimum to maintain the small underskirt vacuum provided by the exhaust system, thereby preventing the escape of gases to the potroom. This vacuum is also lost if the alumina crust does not completely seal the skirts all around the pot.

The first gas skirts installed in the 1950's were bolted both to the casing flange and to each other. This was unsatisfactory, primarily because leaks developed at both locations allowing gas to escape. Replacement of skirts and flange repairs were also very disagreeable jobs. Develop ments over the intervening 30-plus years has led to many changes, the highlights being:

- Skirts now hook over the flange, eliminating built-in holes in both flanges and skirts, and the need to replace bolts on operating pots.

- Overlap joints have replaced the bolted connections between skirts.

- Skirts are now made of ductile cast iron, rather than ordinary cast iron.

- Anode casings have been strengthened to minimize distortion and the mild steel lower casing flange has been changed to stainless steel.

The overall effect of these improvements has been a much tighter gas seal and at least a two-fold extension of skirt life.

2) Gas Burners

The pot gases collected by the gas skirts are channeled to two locations around the pot where the burners are connected to the skirts. The burner is simply a chamber with holes for air entry. Its functions are two-fold:

1) transmit the vacuum provided by the exhaust system to the gas collection channel behind the skirts, i.e. hole area must be properly sized;
2) admit air not only for good combustion of pot gases, but also to provide for sufficiently high duct velocities to prevent solids settling out and blocking pipes.

If air is admitted in the proper quantities and locations, combustion is spontaneous in most cases. Carbon monoxide is the principal fuel, however some tars are also burned. An unlit burner does not create a serious situation in the short term as the fluorides are still collected. However, over a longer period of time tarry solids will plug either the burner or the ducts.

Initially, pots were equipped with one burner per pot, fabricated from mild steel, and designed to act as hoods, transmitting no vacuum to the skirt area. Today's pots have two symmetrically located burners, use stainless steel or cast iron in critical areas, and are designed to transmit the necessary vacuum to skirts. The result is much less gas leakage around the pot, longer burner life, and improved combustion leading to less plugging.

3) Duct Systems

Burner gases are collected in one of several header ducts connected to the scrubbers. There are anywhere from 7 to 35 pots connected to each header with about 120 pots on each scrubber system. The function of these steel ducts is to provide each pot with a uniform and equal vacuum, thereby permitting the burners and skirts to operate properly.

The original duct systems failed badly in this respect with pots furthest from the scrubbers being severely under-exhausted. This was largely due to the original designs not taking into account the nature of the gases (i.e., sticky solids) which resulted in build-up and higher pressure drops than expected. This was compounded by the inaccessible location of many of the ducts which made cleaning and maintenance very difficult. Over the years all these ducts have been replaced, (some more than once) at a cost of several million dollars. The shortcomings of the previous systems have been largely overcome with both cleaning and maintenance requirements drastically reduced.

4) Scrubbers

All the potroom gas collection systems were originally equipped with wet-scrubbers (Figure 4). The design of these was simple, consisting of a fan pushing the gases through a wood tower filled with chevron grids and a once-through fresh water spray system. The operation and maintenance was anything but simple. Problems and shortcomings encountered were:

- Fans required weekly cleaning, shutting the entire system down. Vacuum to the collection system dropped off prior to cleaning.

- The scrubber had to be by-passed to the atmosphere during periods of R&M. These were numerous, especially in the early days when maintenance of spray banks and nozzles was high.

- Overspray of the acidic liquor out the top of the tower created extensive corrosion of surrounding structures.

- Scrubbing efficiency of collected solids was very poor, at an estimated 15%. Fortunately, the scrubbing efficiency of gaseous fluoride (85% of the total fluoride) was in the high 90's.

- The scrubber liquor was discharged to Douglas Channel, not only polluting the water, but also throwing away valuable fluorides.

In the early seventies, dry scrubbers of Kitimat design (Figure 5) were installed on all eight systems with wet scrubbers being relegated to a back-up role. Fluoride scrubbing in these units is based on the ability of relatively high surface area aluminas (>50 m2/gm) to efficiently chemisorb HF. At Kitimat, 100% of a potroom's alumina requirement is fed into a gas duct where it is entrained in the gas stream and carried to a cyclone. After about three seconds nominal contact time, 85% of the alumina, containing 90% of the HF, is removed from the gas stream.

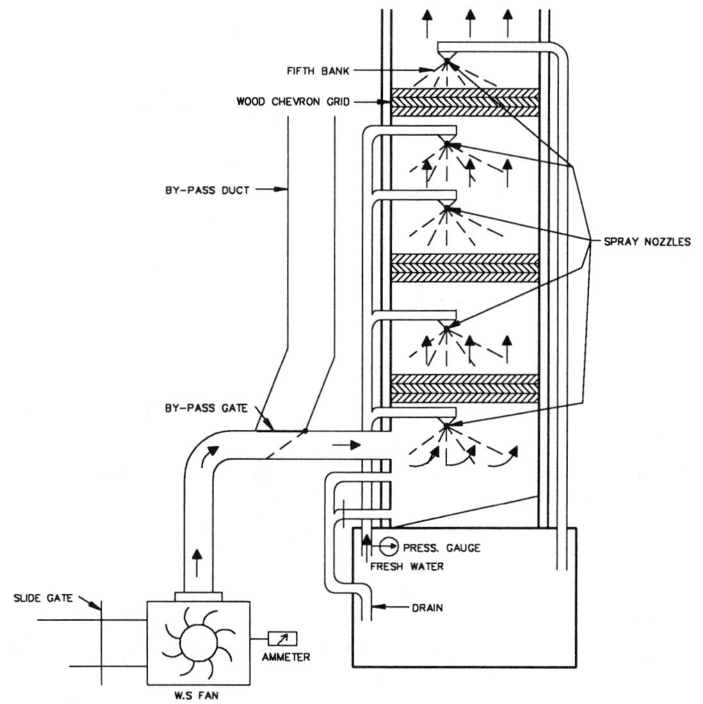

FIG. 4 - Wet Scrubber

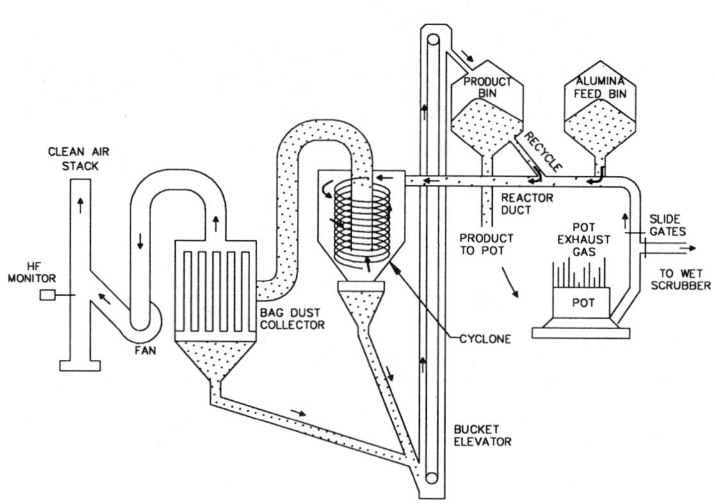

FIG. 5 - Dry Scrubber

The remaining alumina is removed in a baghouse, which scrubs most of the remaining HF and all the pot solids. The combined cyclone-baghouse catch is then returned directly to the pots. Overall fluoride scrubbing efficiences are about 99.6%. These fluorides, mainly in the form of aluminum fluoride, are valued at about $13 million per year.

One major drawback to the use of dry scrubbers is that several trace metallic elements are contained in the captured pot solids, mainly iron, resulting in a decrease of 0.05% in the purity of the aluminum produced. Some plants, because of their product mix, need to install electrostatic precipitators to capture these solids before they enter dry scrubbers.

5) Potroom Process Control

The final significant improvement, though certainly not the least, was the re-emphasis in the early 1970's on improving control of the electrolytic process. If major pot operating problems are being experienced, it is impossible to achieve good gas collection. This will not be gone into in detail here, but suffice it to say that many of the equipment improvements would have gone for naught had we not made the progress we did in this area.

Fluoride Recoveries - Emissions

The cumulative effect on fluoride gas recoveries of all these improvements is shown on Figure 6. As can be seen, recoveries have improved from around 20% in the 1950's to an impressive 94% today. Total plant emissions have dropped from a peak rate of 6600 MTPA in 1970 to today's figure of about 650 MTPA with further improvements expected.

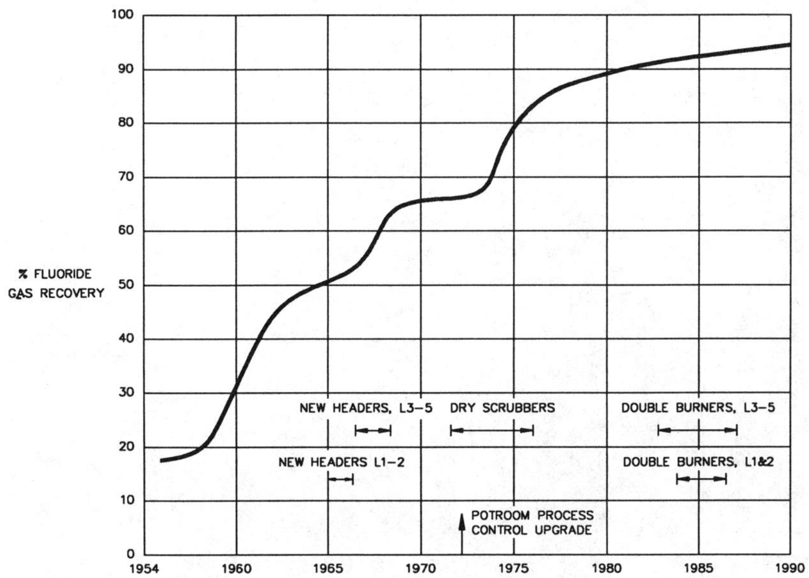

FIG. 6 - Kitimat Fluoride Gas Recovery

Impact of Fluoride Emissions on the Surrounding Forest

Two basic factors influence the climate in this area. The first is the general atmospheric circulation patterns, and the second is the local topography. North-south winds prevail and overnight ground-based inversions occur about 50% of the time. Temperatures are moderated by the ocean, normally within a relatively small 20°C range, however, extremes of -23°C in December and 41°C in July have been recorded. Precipitation in the form of rain and snow occurs about one-half the days in a year and averages 2800 mm per year (110").

The condition of the forest system at Kitimat prior to the coming of Alcan in the early 1950's was that of an uneven-aged, over-mature forest. It had passed from full maturity into a condition of over-maturity when individual trees died of old age and were replaced by naturally regenerated young trees. The average ages of hemlock and fir exceeded 300 years, with many older than 425 years. The forest contained dead and dying trees, and "spike" tops were common. This is a common condition in coastal B.C. forests where the damp climate precludes forest fires.

Samples of foliage in the forest in the Kitimat valley were analyzed for fluoride content as early as 1953, before the smelter began production. The fluoride in foliage increased over the years as production increased. The most devastating influence on the forest was an insect epidemic that began in 1960, spanning nine years. The area of attack coincided with the emission plume from the smelter, however the relationship between the infestation and the emissions remains a mystery.

The reduction in fluoride emissions from the potlines and the concomitant reduction in fluoride in hemlock is shown in Figure 7. The uptake of fluoride in the forest has resulted in growth loss over the years.

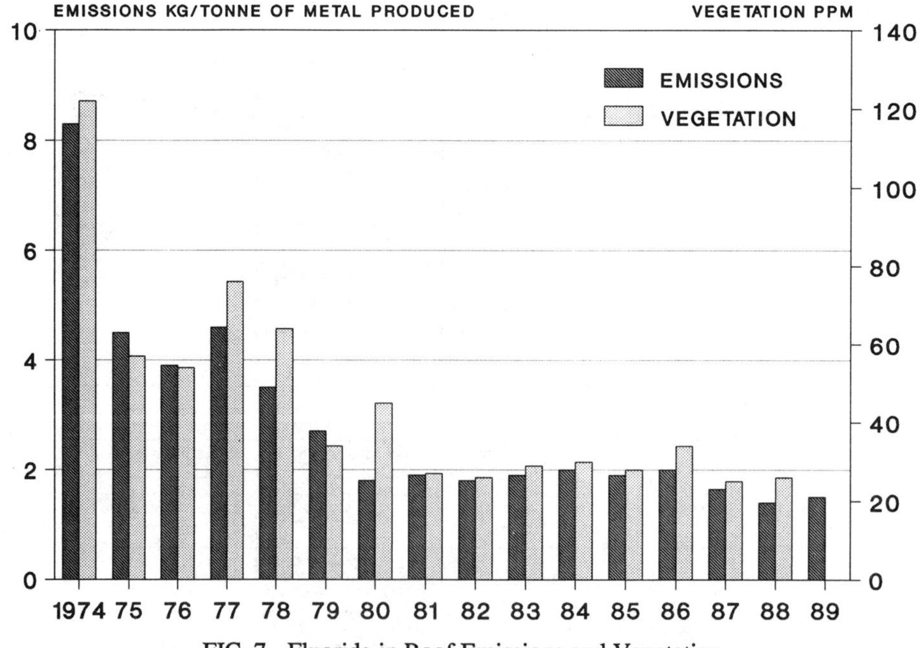

FIG. 7 - Fluoride in Roof Emissions and Vegetation

In 1973, Reid, Collins and Associates Ltd., a consulting company specializing in forest management, was hired to determine the impact of fluoride emissions on the forest. Initially, the study concentrated on the mature forest. The progress of logging and removal of dead trees killed in the insect epidemic provided the beginning for a substantial second growth. The study was expanded to include the second growth in 1983. Other aspects of the study included interalia lichens, small mammals, moose browse, and insect densities. Visual inspections have also been carried out biennially to assess the physical damage to foliage since 1970. These subjective analyses were carried out by Len Weinstein, a renowned expert in plant physiology, of the Boyce Thomson Institute, Ithaca, New York. The observer must be able to distinguish fluoride damage from other similar markings caused by insects or disease. The details of the physical surveys will not be discussed; the results are consistent with those found in the growth studies.

Methods

The details of the mature and second growth studies are extensive so only a brief description of the major components will be given here.

Mature Growth - 1974, 1979, 1984

The boundaries of the study area were selected using historical fluoride data, topographic features, wind patterns, lichen distribution, etc. Within these boundaries 64 permanent sample plots were established using a grid system. Sixteen of these were control plots located approximately thirty kilometers north-northwest of the smelter.

Each tree in each plot over a certain size was tagged and measured for height and diameter. The health of each tree was also recorded, noting defects and abnormalities. Four increment cores were taken from two dominant trees of each species in each plot. The core samples were analyzed by X-ray and a densitometer. In 1974, 161 cores were analyzed which represent 51,520 annual tree ring growth measurements.

Results

Emissions from the smelter have reduced the growth rate of the adjacent forest by about 3764 cubic meters in the five-year period 1979-1983. The previous five-year period 1974-1979 showed a growth reduction of 4635 cubic meters. Growth in the outer strata (defined later under "Second Growth") actually increased by 2725 cubic meters. This has led to the proposition that low levels of fluoride may stimulate tree growth. The emissions have not affected the secondary values of the forest such as watershed protection and wildlife habitat.

Where second growth replaced over-mature stands, the regeneration rate was equivalent to that in the control plots.

Insect activity was considered normal over the total period of the study and has remained stable to this day.

Second Growth - 1982, 1984, 1986

The vegetation sampling program conducted by Alcan in September of each year provided the data necessary to divide that part of the forest located in the fume path into three strata of foliage fluoride concentration. The plume path was designated as that area where fluoride in hemlock exceeded 25 ppm.

The control plots were located in the area where fluoride was less than 15 ppm. The area where concentration was between 15 and 25 was designated as a buffer.

Sample plots were selected in each of the three strata in the plume path with due consideration for species mixture, stocking density, ground condition, climate, etc. A set of paired sites was chosen from the control and each stratum that had near identical site characteristics. Twenty 100 meter long sample lines were established in each stratum. Trees of each species were selected by height and abundance criteria within one meter of the sample line. Eleven hundred and fifty trees were tagged after being measured for height and diameter. Height and diameter ratios by stratum were used to express comparative growth.

Results

It has been concluded that at the present level of fluoride emissions, the impact on second growth is so low it is barely measurable. A comparison of Figures 8 and 9 shows a large decrease in the size of the high level stratum between 1974 and 1988. In 1974 approximately 200 square kilometers fell within the high stratum, i.e. fluoride concentration over 100 ppm. By 1988 this shrank to just over one square kilometer.

Table 2 shows the growth rates between 1982 and 1984 for four species in the fume path, compared to the same species in the control plots. Cedar and spruce show reduced growth rate, balsam a faster growth rate, and hemlock about the same. The results of the 1988 survey have not been reported yet, but are expected to confirm these data, or indicate a continuing improvement. An emission limit of 1.9 Kg/Tonne has shown to be adequate and appropriate for protection of vegetation.

Concluding Remarks

The foregoing presents a brief history of the evolution of the gas collection-treatment system and the subsequent reduction in fluoride emissions and impact on the surrounding forest. All the improvements made now allow us to operate the smelter without marked effects on the environment. As this plant has a long economic life, well into the next century, we will strive constantly to pursue improvements through technology, operating methods, analytical methods and maintenance.

TABLE 2

Growth Rates by Species - 1982 to 1985
Cm/yr

Hemlock	Control	38.75
	Fume Path	38.00
Cedar	Control	47.25
	Fume Path	38.25
Balsam	Control	28.75
	Fume Path	30.00
Spruce	Control	41.75
	Fume Path	31.00
All Species	Control	38.25
	Fume Path	37.50

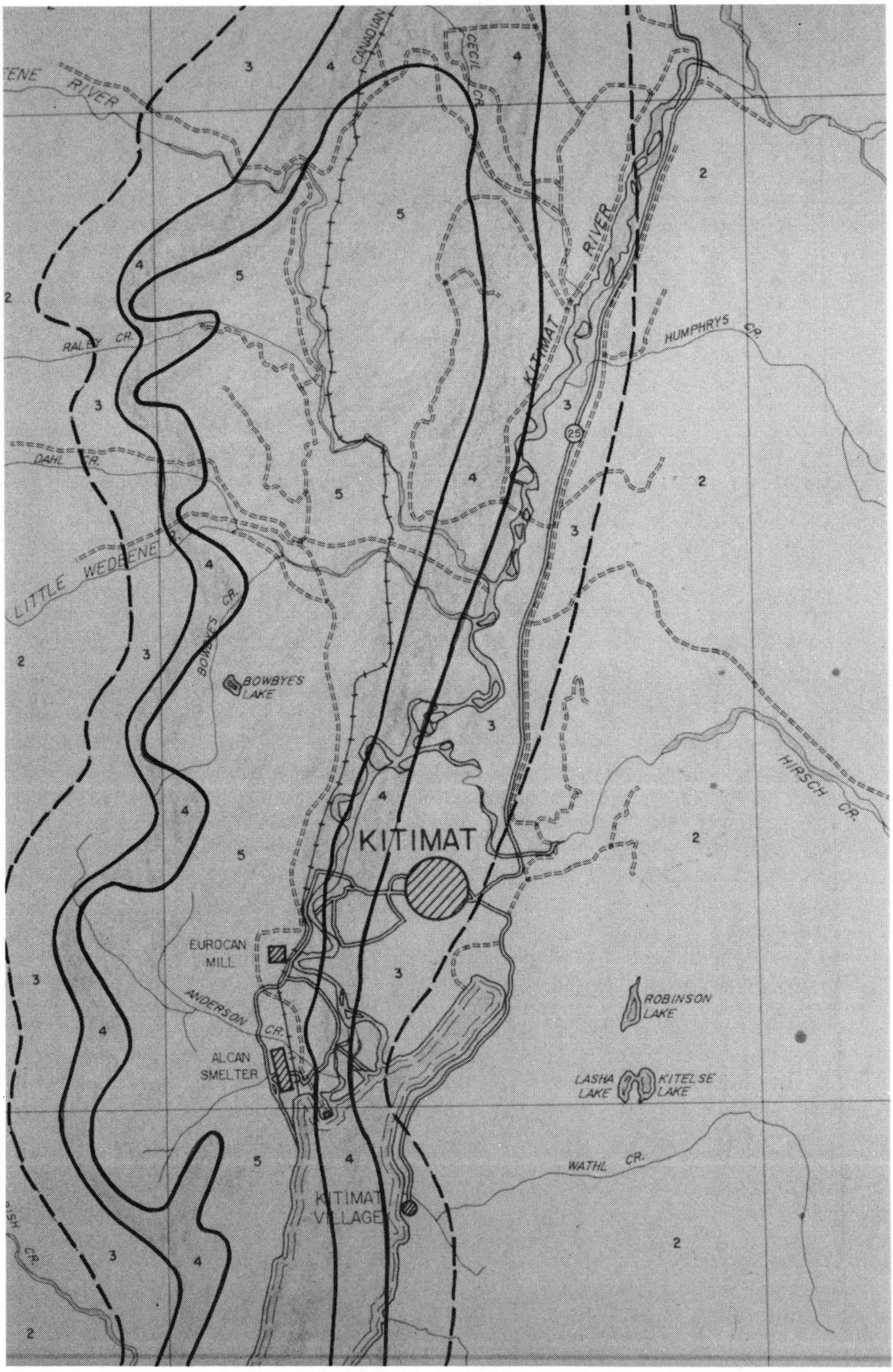

FIG. 8 - Fluoride Content of Hemlock Foliage - 1974

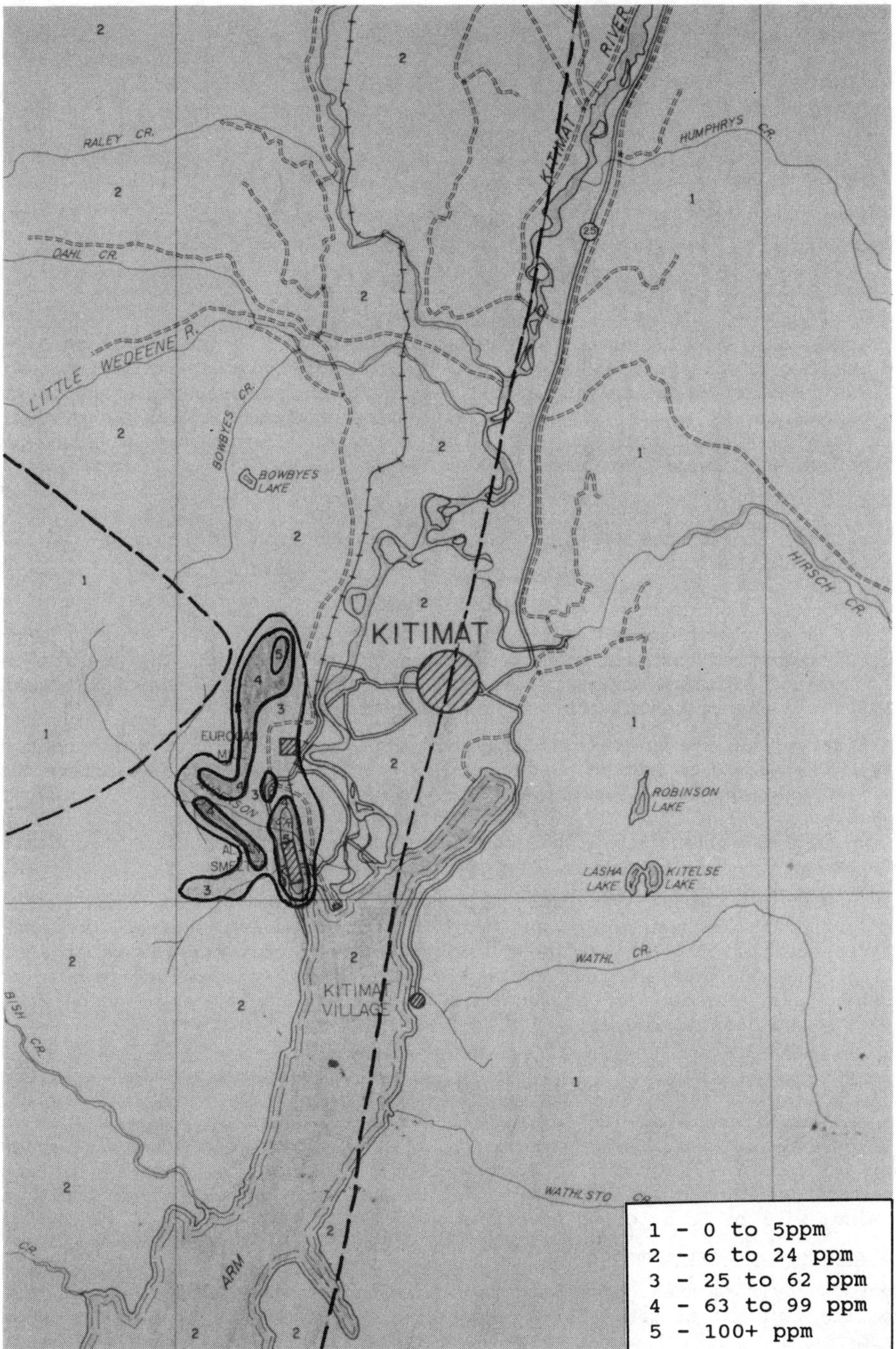

FIG. 9 - Fluoride Content of Hemlock Foliage - 1988

Environmental protection and aluminum smelting a successful integration

M. Lalonde
Alcan Smelter and Chemicals Ltd., Montreal, Quebec, Canada

Abstract

On December 1, 1989, Alcan Smelters and Chemicals Ltd. began operation of its new smelter at Laterrière, Quebec. In designing and constructing this $750 million facility, much time and effort were devoted to addressing environmental concerns such as water recycling, air emission treatment and groundwater protection. This publication gives an overview of the main environmental protection aspects of the project.

Keywords

Laterrière, environment, environmental protection, aluminum smelter, transmission line, railway.

Introduction

Alcan's smelter replacement program, now under way in Quebec, is the largest capital investment project ever undertaken by the company. The new smelters will replace existing capacity that is less efficient in regard to working conditions, environmental control and power consumption.

The first milestone in this replacement program is Laterrière Works near the city of Chicoutimi in Quebec's Saguenay--Lac-Saint-Jean region. Laterrière Works now in operation since December 1, 1989, is scheduled to be in full production by February 1991. This $750 million smelter will provide around 200,000 tonnes of primary aluminum, replacing an equivalent capacity at Alcan's Jonquière Complex.

Aluminum will be produced by two cell lines in four buildings. Each building has 108 prebaked-type cells (or "pots"). Two casting centres, a general maintenance facility, a storage and distribution system for raw materials, a bath storage and crushing facility and four air scrubbing centres are also part of the works. Anodes are produced at Grande-Baie Works, whose anode baking capacity has been increased.

Electrical power is supplied via a new 20-kilometre power line connecting Jonquière Complex with Laterrière Works.

Raw materials and aluminum ingots will be transported by railway. A 14-kilometres line will link the Works to the existing Alcan Roberval Saguenay railway system (Ville de la Baie sector). While the railway is under construction -scheduled for completion in December 1990 - raw materials and aluminum will be transported by truck.

Protecting the Environment

Numerous environmental concerns directly influenced the design of the project from its earliest stages. Three exhaustive environmental impact studies were performed and the railway involved public hearings. The capital investment related to the quality of the environment inside and outside the works amounts to approximately 20% of the total project cost.

The large number of measures taken in designing, constructing and operating the works, the transmission line and the railway will minimize their impact on the environment.

Power Line

Electrical power is provided via a 161kV transmission line that runs the approximately 20 kilometres from Jonquière Complex to Laterrière Works (figure #1).

Three potential routes were studied. After identifying sensitive areas such as recreational areas on the Chicoutimi River, agricultural lands and the urban areas of Jonquière and Chicoutimi, a route was selected that best satisfies the overall environmental concerns of the region.

Much of the selected route runs through forested land. More than 90% of the route runs along property lines; this is especially important in agricultural areas where every step has been taken to avoid parcelling farm land.

For more than 23% of its length the route runs alongside a planned Hydro-Québec service corridor, integrating the project into future power infrastructures. Moreover, in certain key areas, the visual impact and land requirements have been minimized by using AVA (Avec Vue Améliorée - aesthetically improved) transmission towers.

Railway

A detailed environmental impact study was also performed for the railway line. A multi-faceted examination of environmentally sensitive areas and of the various path options led to the selection of the optimal path (figure #2). One of the many advantages of this work was that the selected line avoids passing near private residences.

The route is some 14 km long. To avoid interfering with farming activities, it runs alongside all agricultural areas, maintaining the land's integrity. The line takes advantage of the forest of the Laurentian Shield for more than 80% of its length to reduce its visual and audible impact.

The natural milieu has been protected by reducing deforestation to a minimum; by preserving streams and other watercourses; and by protecting flora in sensitive areas. Extraordinary measures have been taken to ensure that contamination risks are minimized. For example, for 50 metres on both sides of streams and other watercourses, untreated railway ties are used. In addition, chemical maintenance of vegetation is forbidden in these areas and a vegetation program is planned for slopes with a high risk of erosion.

During construction of the railway, special measures are being used to protect the environment. For example, machinery and materials are not allowed to sit in standing or flowing water.

Finally, a number of operating restrictions will apply once the railway is up and running. The total number of trips will be limited and the line will not operate at night or on Sundays. Taken together, the measures employed during the design, construction and operation of the railway will minimize its environmental impact.

Aluminum Smelter

Located near Chicoutimi, the Laterrière municipal area (4000 inhabitants) is primarily a residential, agricultural and recreational-tourist environment. Bounded on the north by farm land and on the south by forest, the smelter is located near the du Moulin river, upstream from the nearby village of Laterrière.

The area around the smelter and its current use must then be protected to the greatest possible extent. From the very beginning, environmental protection has been an integral part of the smelter's design and environmental concerns about water, air and ground quality have all been addressed.

Water

A major concern has been to minimize the amount of water used. For this reason, recirculation systems have been installed on most cooling systems, allowing over 70% of the works' daily water consumption (around 1380 m3/d at peak periods) to be recycled. Additional measures have been taken to further reduce consumption; for example, air-cooled air-conditioning systems are used where feasible. The combined measures have reduced the smelter's total discharge to a dry-season maximum of 570 m3/d, equivalent to less than 0.5% of the minimum flow of the du Moulin River.

Another major concern was to reduce water contaminants at source. As no chemical processes will be used at the smelter, no process water will be produced. Water will be used exclusively for sanitary purposes (around 200 m3/d) and cooling (around 1180 m3/d). Mostly indirect cooling will be used in the different smelter areas in order to minimize water contamination. Only the DC casting centre will use water for direct cooling of aluminum and this water will be extensively treated before release (figure #3). Moreover, to discourage the use of water for cleaning floors and equipment, etc., floor drains have not been installed in most buildings.

A potential source of water contamination is the accidental spilling of oil from, for example, power transformers or diesel fuel tanks. All such equipment has therefore been equipped with recovery tanks so that potential contaminants can be contained before they reach the river.

For the direct casting centres (at 620 m3/d, the main users of water) a number of measures have been taken. For example, biodegradable vegetable and animal oils are used in priority over mineral and synthetic oils. The entire water recirculation and treatment system was designed to work as a closed circuit and only a single treated purge amounting to under 140 m3/d (constantly monitored), will be discharged into the smelter's holding pond.

In the garage, the only washing station for the smelter's vehicles is equipped with an oily-water treatment system. Washing water flows into a settling tank and then in an oil separator where it is treated before being sent to the sanitary treatment plant. This plant uses three processes to treat incoming water: biological oxidation, followed by filtering through sand, and finally disinfection using ultra-violet light (the use of chlorine was rejected). The capacity of the water purification plant is 175 m3/d. The purified water is discharged to the smelter's holding tank.

Numerous checkpoints have been implemented to provide control of the water treatment systems. The water quantities used are measured at the main points of use. Quantities are also measured at the discharge points of the main treatment systems.

Water quality is also closely monitored. Sampling stations have been included at several locations throughout the plant. The stations are used for the taking of samples and the continuous analysis of the water for several elements. The cost of the chemical analyses used to monitor water quality exceeds $100,000 a year.

Air

Despite major improvements in Soderberg potrooms, the HS Soderberg facilities generate and remain unable to completely eliminate atmospheric emissions of tar particles, mainly polycyclic aromatic hydrocarbons (PAHs). For Alcan in Quebec, replacement of the HS Soderberg potrooms with smelters using prebake technology is the best solution over the long term, as it practically eliminates all environmental emissions of PAHs as well as exposure for employees.

All sources of atmospheric dust emissions have been isolated and connected to dust collectors. For example, alumina and fluoride are unloaded in a completely closed circuit. More often than not, the collected dust is reused. Each dust collector is equipped with leak detectors to facilitate preventive maintenance.

Potroom gases are scrubbed using four Flakt Ross Inc. dry scrubbers. Each scrubber processes around 150 m3/s of gas and handles some 11.5 tonnes/hour of alumina. To provide optimum control over the effectiveness of the scrubbers, all gas outlets are monitored on a constant basis with a hydrogen fluoride monitor linked to the processing control room.

Several other measures have been implemented to reduce the works' impact on air quality. For example, using natural gas will minimize sulphur dioxide emissions.

In addition to the continuous monitoring of emissions from potroom vents and the regular sampling of all emission sources, air quality will be sampled on a continual basis at four outside stations around the works. Thus all the necessary tools are in place to ensure that emissions are kept to a minimum and closely monitored.

<u>Soil and Groundwater</u>

All water for sanitary and industrial use will be taken from the water table under the works. Given the sandy and therefore highly permeable nature of the soil, a number of special precautions have been taken to protect soil and groundwater quality.

To prevent contamination of groundwater by accidental overflows and spills, tanks that contain liquids have been constructed over or inside leakproof containers. The works' power transformers, which contain no PCBs, have also been constructed over leakproof capture tanks.

To minimize leakage into the ground, special measures were taken in constructing the works' sewer system and holding tanks,making sure they were leakproof. Other measures include the use of untreated wood for railway ties and metal power and light poles throughout the works. These measures fully eliminate the risk of chemical substances contained in processed wood from leaching into the ground. In addition, the surface of the entire site has been covered with clay to provide supplementary protection.

Groundwater is monitored with seven piezometers installed before construction began. Groundwater quality is tested on a regular basis and the water table level is closely monitored.

Byproducts from the works will be separated to facilitate recovery and recycling. This applies to aluminum dross, paper, metal, raw materials like alumina, and bath. Materials that cannot be reused or recycled will be classified for safe disposal according to AS&C's policies and legal requirements in Quebec and Canada.

Finally, to help integrate the works into its physical surroundings, a large-scale landscaping plan has been developed. The plan covers the restoring of several quarry sites located near the works, the building of footpaths, the protection of bodies of water located near the works and the planting of numerous trees. The total cost of the landscaping will be of approximately $5 million.

Conclusion

Like Grande-Baie, Laterrière Works will be a model on which future smelters will be based. The automation and mechanization of operations will translate into improved working and environmental conditions. Alcan has paid special attention to the quality of surrounding and environmental protection where some 20% of the capital invested in the smelter is going toward protecting the environment.

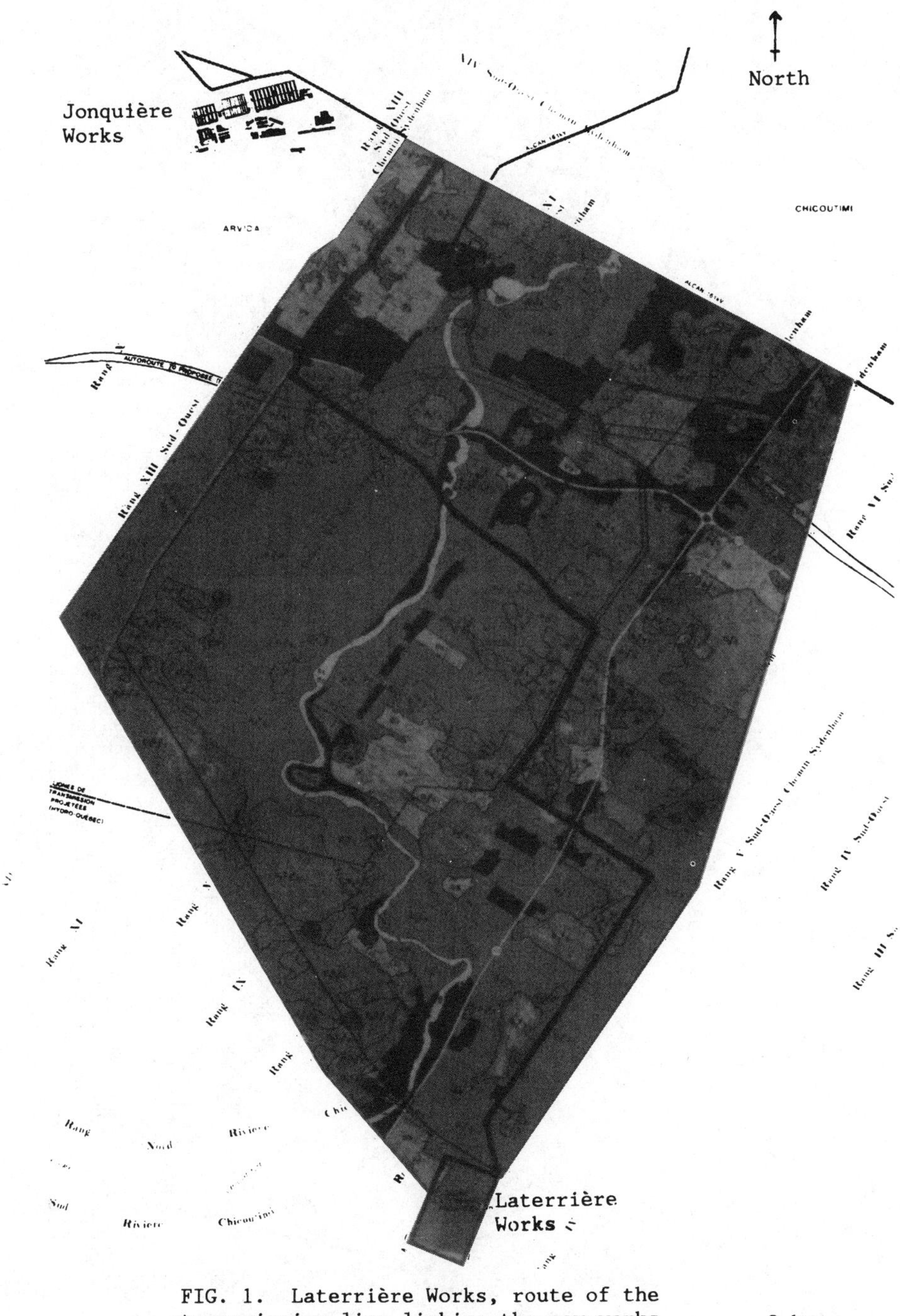

FIG. 1. Laterrière Works, route of the transmission line linking the new works with existing installations at Jonquière.

FIG. 2. Laterrière Works, route of the railway linking the new works with existing installations.

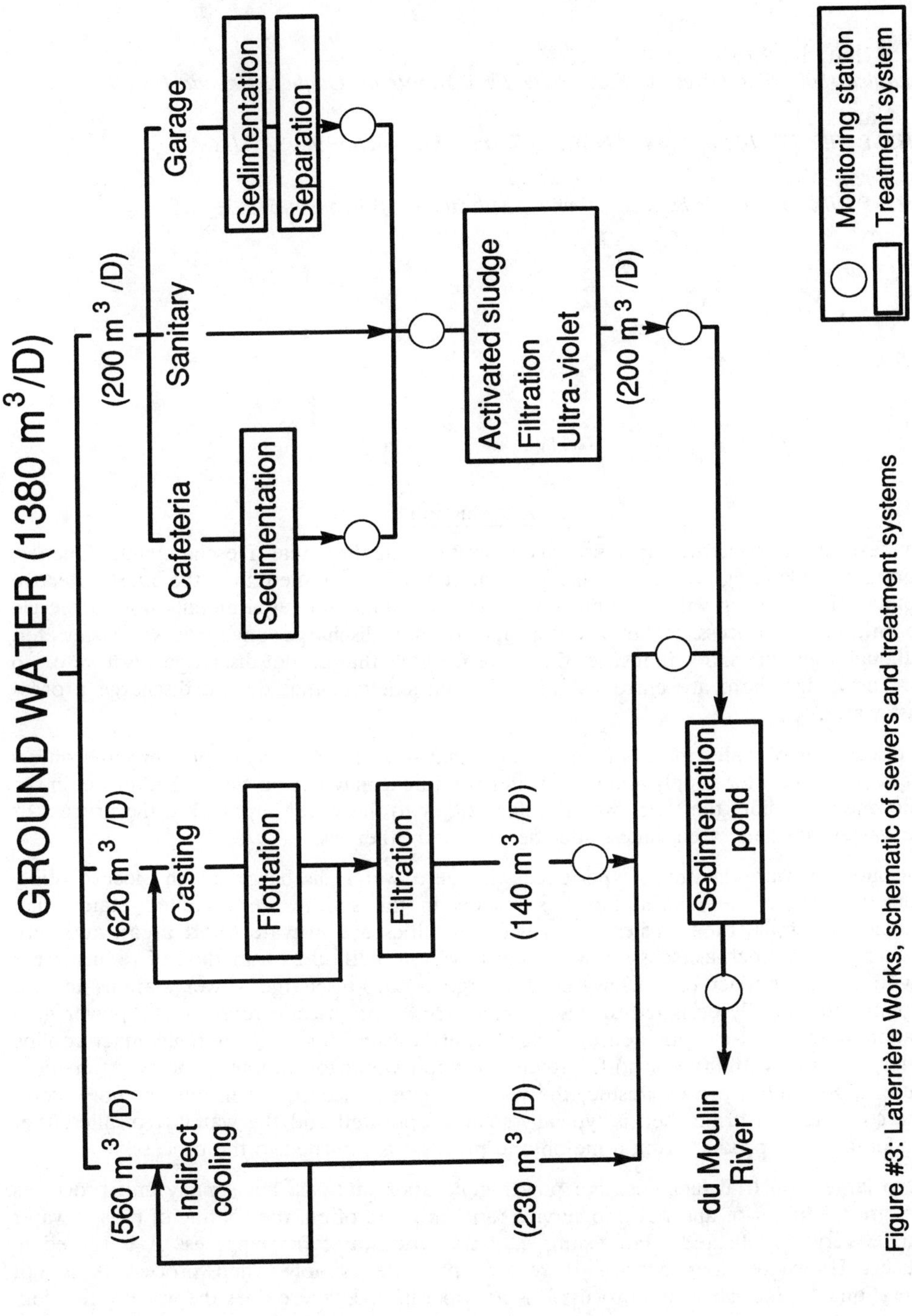

Figure #3: Laterrière Works, schematic of sewers and treatment systems

Zero process water discharge and storm water recycling at the Lauralco aluminum smelter

P. Aylen, D. Leslie
Lauralco, 800 Réne-Lévesque East, Suite 2950, Montreal, Quebec, Canada H3B 1Z1
T. Card
CH2M Hill, 777 108th Avenue N.E., Bellevue, Washington 98009, U.S.A.
D. Drake
CH2M Hill, 310, W. Wisconsin Avenue, Milwaukee, Wisconsin 53203, U.S.A.

Introduction

Alumax, Inc. is constructing a primary aluminum smelter near Deschambault, Quebec. During the planning phases of this project, it was determined that the most effective method of complying with current and future environmental requirements was to recycle stormwater as process makeup water and to not discharge any process wastewater. Although there are many industries that have facilities that do not discharge wastewater, to our knowledge there are currently no aluminum smelters that do not discharge process wastewater.

In discussions of water management systems that have a high degree of reuse, the differences between water supply and wastewater treatment may not be clear. Therefore, in the following discussion, the term "water" refers either to the water supplied to the process or the wastewater generated, unless specifically noted otherwise.

The industry with the most experience with zero water discharge is the electric utility industry. There are several facilities, mostly in the southwestern United States, that completely recycle their wastewater. These facilities are in water-short areas, and zero discharge is as much a necessary water conservation method as a method of reducing the environmental impact of wastewater discharge. Zero-discharge power stations in arid regions are typically designed to have a high degree of cascade reuse in the power generation system. In cascade reuse, a high-quality wastewater, such as noncontact cooling water, is reused with little or no treatment as supply water for another process. After these reuse opportunities are exhausted, the wastewater is concentrated in one or more steps. The concentrated brine then is typically solar evaporated and the water (typically either condensate or a product from a membrane process) is returned to the process.

Since large industrial facilities, like power generation stations, have many unit processes that are taken out of and put into service randomly and often, the routing of reused water can be very complicated. For example, if the wastewater from process A is reused in process B, where does process B receive its water supply when process A is not operational? Likewise, when process B is not operational, where does the wastewater from process A go? This problem is normally solved by storing water of different qualities in large ponds. This provides a buffer so that water can be stored and retrieved at different rates. Typically, water of at least three different qualities is stored on site: high-quality wastewater (which can be reused directly by some processes and is most often used as cooling tower makeup water), low-quality wastewater (which must be treated before reuse), and concentrated wastewater (which must be evaporated).

There are some zero-discharge facilities that operate in geographic areas that do not have a net natural evaporation: more rain falls than evaporates. These facilities must rely on mechanical drying devices instead of solar evaporation for the final separation of water from the solids.

In addition to electric power generation, there are other industries that have operated zero water discharge facilities. These include U.S. Steel in Ogden, Utah; Northwest Alloys in Addy, Washington; and Kubota Tractor in Osaka, Japan. Although the same principles apply to these facilities as to electric power generation facilities, these industries require more complicated systems because of the more complex wastewater that is produced.

Water and Wastewater Characterization

When a zero water discharge facility is being planned, it is essential to know the detailed water requirements and the quality and quantity of wastewater produced from each unit process. In addition, the chemistry of the water supply must be known so that the quality of the water can be predicted as it is concentrated and reused during the process.

Water Sources

At the Lauralco site, the makeup water for the process will primarily come from the surface runoff collected on the site. In addition, water will be supplied from a nearby municipal water system for domestic consumption and process makeup if there is insufficient surface runoff. Deschambault receives a substantial amount of precipitation: more than 1 meter per year. The runoff from this amount of precipitation is much more water than is required for the process uses at the smelter. In order to control the rate of discharge of this surface runoff, stormwater will be stored in large ponds. The ponds are large enough that the stormwater will be able to provide the required process water the majority of the time, even during dry periods.

Water Uses

The process water that is used at aluminum smelters falls into the following categories:

(1) Contact cooling water
(2) Noncontact cooling water
(3) Equipment washing

Contact cooling water is required for casting and anode cooling and is the largest category of water use in a smelter. Uses of noncontact cooling water include air compressor cooling, hydraulic unit cooling, bearing cooling, furnace door and frame cooling, and air conditioning. Very little water is required for equipment washing.

Many of the processes that require noncontact cooling can be designed with self-contained, closed-loop cooling systems. An example of this is hydraulic power units that have air-cooled heat exchangers mounted directly on the units. For zero-discharge systems, it is highly desirable to reduce water use as much as possible by using equipment that requires very little or no water.

Wastewater Generators

The largest source of wastewater is the blowdown from the cooling system used for casting. During the casting operation the aluminum is direct-contact cooled from its molten state at 600°C to 650°C to a temperature of 50°C or below so the casting is safe to handle. The water cools the casting by conduction and evaporation when it contacts the aluminum. The water is circulated through a cooling tower to control its temperature as it approaches the casting. As part of the casting operation, a variety of mold release oils can be applied to the cast part. A portion of these oils end up in the cooling water that is circulated through the system. The concentration of this oil must be controlled in order to maintain the quality of the cast product. The critical water requirements for the casting operation are approach temperature and oil concentration.

The blowdown from this system will contain casting oil, dirt, any cooling water quality control chemicals, and hydraulic fluid if there are leaks in the casting machine. The casting oil can be either vegetable or mineral based. Depending on what oil is used, the oil may be nonemulsified, mechanically emulsified, or chemically emulsified.

Another major wastewater generator is the direct cooling of anodes immediately after fabrication. After the anode is molded it is submerged in a tank full of cooling water. This cooling water will pick up contaminants from the carbon and the binding material. All noncontact cooling water can be directly reused in any of the above processes.

The wastewater from equipment washing will contain industrial oil and grease as well as dirt and soap. The soap will emulsify at least a portion of the oil.

Required Wastewater Quality

During the planning of the Lauralco facility, the criterion for the quality of wastewater discharged was that it not degrade the water quality of the St. Lawrence and tributary waters. This criterion was so stringent that it was decided it would be more practical to treat contaminated wastewater for reuse than to provide the high degree of treatment required for discharge. In addition, the criteria for degradation may change, and there could be wastewater quality criteria established that would not be economically achievable.

Water Management Alternatives

With these process requirements and discharge criterion, the following process alternatives were considered.

Water Supply Management

The primary water supply will be from the site surface runoff. In order to ensure that no solids are deposited in the water distribution system and to prevent plugging in any of the process equipment, the suspended solids in the supply water should be minimized. In addition, there is a possibility that algae will grow in the stormwater retention ponds. Therefore, it was concluded that all process water should be filtered before it entered the water distribution system. The filtration systems considered included pressure filters, continuous backwash filters, and pulsed bed filters. Pressure filters can be costly but may eliminate a pumping stage. They also have more difficulty than the other two systems in handling high solids loadings. Pulsed bed filters generally work the best on algae.

Cascade Wastewater Reuse

Cascade wastewater reuse alternatives consisted of reusing noncontact cooling water in the contact cooling systems and reusing the wastewater from anode manufacturing in the anode baking air pollution control system. For the latter alternative, the wastewater would have to be filtered prior to use in the air pollution control system. The ability to reuse some of the noncontact cooling water is limited by geography: there are some locations in the plant that are too remote for economic reuse of noncontact cooling water.

Wastewater Pretreatment

The wastewater from casting requires pretreatment to remove oils and suspended solids, prior to either reuse or further wastewater treatment. The oil level in the circulating cooling water must be no more than 100 mg/l, and the oil level in the blowdown that goes to final wastewater treatment should be as low as economically possible. The oil removal alternatives considered included free-oil skimming, induced-air flotation, dissolved-air flotation, inorganic chemical emulsion breaking, organic chemical emulsion breaking, coalescing filters, and oil adsorbents.

Dissolved- and induced-air flotation can reduce oil concentration from around 500 mg/l to as low as 15 mg/l; normal performance produces concentration of about 15 mg/l to 50 mg/l. Dissolved-air flotation costs about twice as much as induced-air flotation, but dissolved-air flotation can do a better job of removing emulsified oil by polymer addition. The air inducing brushes on induced-air flotation systems can tear polymers.

Coalescing filters can reduce oil concentration to as low as 5 mg/l. They can handle heavy oil loads, but they generate a lot of backwash water. Adsorbing units can reduce concentration to less than 1 mg/l, but they are very expensive to operate if any significant level of oil is the in the influent.

Oil and suspended solids would also have to be removed from the equipment washing and maintenance areas. It is likely that the oil from these streams would be incompatible with the oil systems used in casting. Therefore, it was decided that the oil systems should be separate.

Wastewater Concentration

The technologies that are available for wastewater concentration include evaporation and membrane processes such as reverse osmosis (RO) and electrodialysis (ED).

Membrane processes can concentrate wastewater from about two times to as much as 10 times under ideal circumstances, and they are less expensive than evaporation. Energy consumption for membrane processes ranges from 0.3 to 0.6 kWh per cubic meter of water processed. The most efficient evaporators require an energy input of 3 to 6 kWh per cubic meter of water processed. Capital costs for membrane processes are also about an order of magnitude less than evaporators. There are, however, disadvantages to membrane processes. First, they have a limited ability to concentrate. In most instances, they can achieve only a 50 percent reduction in volume when concentrating wastewater. Second, they are very susceptible to fouling and other operational problems. This second disadvantage makes it absolutely imperative that membrane processes be pilot tested.

The performance of reverse osmosis and electrodialysis is somewhat similar. The major difference is that noncharged species are not removed in electrodialysis. Silica is the primary noncharged inorganic component of natural water systems.

There are many different configurations of evaporators used for wastewater treatment. The most common evaporator technology for large-scale systems is seeded-slurry vapor compression. This technology requires the maintenance of a seeded slurry in the evaporator that will preferentially precipitate minerals so that the minerals do not precipitate on the evaporator's heat transfer surfaces. The thermal energy for evaporation is provided by a compressor that compresses the vapor so it will condense on the exterior of the evaporator tubes. The heat lost in condensation evaporates the feed water. This technology can concentrate wastewater to a slurry of between 10 percent and 35 percent solids.

One of the major design considerations at Lauralco was the minimization of the use of electricity for other than aluminum reduction. Therefore, in addition to the traditional electric drive vapor compressor, natural gas and gas turbine drivers were also considered.

In addition to vapor compression evaporators, multiple-effect evaporators were evaluated as well. This technology can have a lower capital cost but higher energy costs than vapor compression evaporators. However, natural gas or other fuel can supply the energy. If the operational scenario is to have multiple evaporators for redundancy, one possible configuration is to have a base-loaded vapor compression evaporator and a peaking multiple effect evaporator. In addition to seeded slurry, forced-circulation systems and wiped-film technologies were evaluated.

After water supply quality was reviewed, it was determined that silica scaling would be a problem in the evaporator. The alternative methods for handling the silica problem included using electrodialysis as a preconcentration step. Electrodialysis does not concentrate silica. The excess silica could be discharged with the electrodialysis product water. Silica is almost never a water quality problem, so discharge of the product stream should not be a problem. Another alternative would be to adsorb silica in a softening process. Magnesium hydroxide floc very effectively adsorbs silica, especially at higher temperatures. Unfortunately, there is a softener sludge disposal problem. The sludge is very innocuous, but disposal of any sludge is a problem these days. The third alternative would be to select an evaporator configuration that would tolerate silica scaling. There are several techniques to do this, but there is always a risk of major operational problems.

Wastewater Disposal

The residual after wastewater treatment still has to be disposed. It was decided that these residuals should be a dry powder or crystals with no free water. The technology of choice for manufacturing dry powders is a spray dryer, in which an atomizing nozzle discharges into a hot air stream. The dried product is captured in a bag house. One of the major concerns with this technology was corrosion due to potentially high fluoride concentrations.

Crystallizers produce a slurry that can be dewatered with conventional dewatering equipment such as a centrifuge or belt press. The product normally is a cake of hydrated crystals.

In addition to these technologies, there was also consideration of evaporating the wastewater in the potline air pollution control system. About 30,000 cubic meters per minute of 50°C to 100°C air is passed through the air pollution control system. There would be more than enough energy to evaporate the wastewater in that system. There were concerns, however, of corrosion, contamination of the alumina, and risking operational problems with a key component of the aluminum smelting process.

Description of Proposed Facilities

Water Supply and Stormwater Treatment

The surface runoff will be stored in large ponds that have a 1- to 3-month storage capacity. Water that is either reused or discharged from these ponds will be pH-adjusted and filtered in a continuously backwashed sand filtration system. The process water required to operate the smelter will be pumped into the process water distribution system, and the excess surface runoff will be discharged to a surface stream.

Wastewater Treatment and Reuse

The cooling water from the casting operations will be side-stream treated to control oil concentration. The side-stream treatment will consist of dissolved-air flotation followed by a coalescing upflow filter. This system will be equipped for either an inorganic or organic chemical emulsion break. Waste oil will be stored for off-site disposal. A secondary side stream off this primary side stream will use an adsorption unit to reduce oil concentration to less than 1 mg/l. After the absorption unit, the wastewater will go to the primary evaporator, a vapor compression unit that will operate below the scale threshold. The condensate from this unit will be recycled back into the casting cooling system. The concentrated wastewater will then be further evaporated in a single-effect, forced-circulation evaporator. The solids from this evaporator will be centrifuged and stored for offsite disposal. Expected volume of the solids will be less than 1 cubic meter per day. The condensate from the second-stage evaporator will be recycled back to the casting cooling system. All unit processes will have redundant units, each with 100 percent of full stream capacity.

The hot wells and cold wells of the cooling system will be oversized, both for temperature control and for buffer storage of the cascade reuse water. The system will be designed so that both evaporators can be off line with the cast house at full aluminum casting production for 72 hours and the cooling water will still maintain acceptable quality. When the facility expands, membrane processes can be pilot-tested to determine if it will be cost-effective to preconcentrate the wastewater to the evaporators.

The wastewater from the anode manufacturing facilities will be filtered and then evaporated in the bake oven air pollution control system.

Wastewater from equipment washing and maintenance will be stored and batch processed in a separate oil removal system identical to the casting cooling oil removal system.

Summary and Conclusions

By careful planning and design, Lauralco will be able to eliminate process wastewater discharge at about the same cost as constructing a conventional wastewater treatment and discharge system. The primary advantage of this zero process wastewater discharge system is that it offers the maximum environmental protection to the St. Lawrence and its tributaries, while at the same time it protects Lauralco from costly modifications to the aluminum smelting process if environmental requirements for wastewater discharge are increased.

The disadvantage of this type of system is that it requires a higher level of sophistication both to design and to operate. The normal consequence of failure of a conventional discharge wastewater system is a warning or fine. If a zero-discharge system fails, it is possible that the operation of the entire smelter could be interrupted. This possibility can be minimized by the design of redundant systems and intermediate storage. This will decouple processes so that if a unit process fails, there will be time to correct the problem before the entire facility is stopped.

Zero wastewater discharge has been implemented successfully in many industries and in many countries. In many parts of the world, it has become the only feasible solution to wastewater discharge. As environmental regulations become increasingly stringent, more industries will elect to control their own fate by internally handling all wastewater they produce.

The construction of a greenfield aluminum smelter in a greening environment

Y. Lallemant
Aluminium Pechiney, Paris, France

This paper describes the various steps of the construction of the new Pechiney smelter in Dunkirk (North of France).

The decision to build a 215 000 mt/yr aluminium smelter was made in December 1988.

It is expected that the first pot will be started in November 1991.

The Dunkirk project is designed with a global concept of environmental protection of air water soil and natural resources.

A summary of the French legislation is given in the paper. The preparation and the presentation of an Impact Statement is an important part of the regulatory requirements.

The plant is located in an industrial area. Poor environmental experience of the neighborhood with existing industries has generated suspicion regarding new industrial development. In addition, the surroundings are mainly agricultural. Regaining credibility from the community was part of the challenge as the subjective assessment of risk is for most people more real than the objective statistical assessment of a particular risk.

Note from the editors: the complete paper was not available for publication. Copies are available from the author, Aluminium Pechiney, CEDEX 68, 92048 Paris La Défense, France.

Light metal matrix composites

Sedimentation during liquid processing of metal matrix composites

S. Lafreniere* and G.A. Irons
Department of Materials Science and Engineering, McMaster University, Hamilton, Ontario, Canada

Abstract

During the incorporation of ceramic particles into metallic alloy melts for the production of metal matrix composites, the particles tend to float or sink, depending on their density. In order to study the rate of hindered settling in these systems, a novel electrical resistance technique has been developed. A current is passed between two electrodes, and the potential drop over a fixed distance is measured with two other electrodes. Experiments were carried out in an aluminum-silicon foundry alloy (A356) containing up to 30 volume percent 82 μm silicon carbide particles. The settling rates were much slower than Stokes Law predicts. The particles' behaviour was compared with hindered settling in aqueous systems. It was found that the particles settle in a non-flocculating manner to approximately 55% after passing through a region of intermediate density. The implications for fabrication and remelting of metal matrix composite material are discussed.

Keywords

Metal matrix composites, segregation, sedimentation, hindered settling, fluid mechanics.

Introduction

Metal-matrix composites are a relatively new class of engineering materials which have been found to produce moderate increases in strength and stiffness of aluminum alloys at room and elevated temperature (1). While there has been considerable research on the mechanical properties of such materials, there has been comparatively little on the fundamentals of material processing.

There are three general methods to make metal matrix composites:

(1) Sintering of metal and ceramic powders,

(2) Infiltration of liquid metal into pre-forms of ceramic fibres, and,

(3) Mixing ceramic particles into liquid aluminium.

Currently with: Dofasco, Inc. Hamilton, Ontario, Canada

The third method is potentially the cheapest in situations where only moderate improvements over the un-reinforced alloy are required. Alcan has commercialized a process based on this third method (2).

There are three major fundamental problems associated with mixing ceramic particles into liquid metals:

(1) most particles are not easily wetted,

(2) the particles either sink or float in the melt, depending on the ratio of particle-to-liquid density, and,

(3) the viscosities of the mixtures are very high, and are also shear rate dependent (thixotropic).

Work is underway in this laboratory to address the first problem (3), while the second problem is the subject of this paper.

In the present system (A356 Aluminium-silicon alloy, and SiC), the SiC carbide sinks (ρ_{Al} = 2700 kg/m^3, ρ_{SiC} = 3300 kg/m^3). To achieve a uniform distribution of particles, solidification must occur before appreciable settling has occurred. If settling does occur, the thermal properties of the mixture (thermal conductivity, heat capacity and density) will depend on the local solid volume fraction which in turn will influence the solidification rate. Thus, a quantitative understanding of the settling rates must be achieved for calculation of the macroscopic solidification rate, and to compare particle motion due to gravity with that due to solidification front pushing.

Accordingly, it was decided to develop a technique to measure the in-situ volume fraction of particles, so that settling rates could be determined. The solid volume fractions in commercial applications range up to 30% which is in the regime of hindered settling. This regime has been studied considerably in aqueous, but not metallic systems.

Literature Review

Hindered settling or batch sedimentation has been studied extensively in the chemical engineering field. One of the most useful approaches to the problem was refined by Wallis (4), based upon one-dimensional flow of liquid and solid in the vertical direction. The particle slip velocity is given by:

$$V_s = \frac{V_{ps}}{\theta_p} - \frac{V_{cs}}{1-\theta_p} \tag{1}$$

which is only a function of θ_p and fluid properties. A characteristic velocity for the dispersed phase, V_{cp}, can be obtained directly from Equation 1:

$$\begin{aligned} V_{cp} &= V_s\theta_p(1-\theta_p) \\ &= V_{ps}(1-\theta_p) - V_{cs}\theta_p \end{aligned} \tag{2}$$

Empirically, it is often found that the slip velocity is a function of θ_p in the form:

$$V_s = V_{p\infty}(1-\theta_p)^n \tag{3}$$

where $V_{p\infty}$ is the single particle settling velocity (Stokes velocity for small particles). Consequently, the characteristic velocity is of the form:

$$V_{cp} = V_{p\infty}\theta_p(1-\theta_p)^{n+1} \quad (4)$$

This characteristic velocity is plotted as a function of θ_p in Figure 1. If an abrupt discontinuity or shock occurs between θ_{p1} and θ_{p2} the shock velocity is:

$$V_d = \frac{V_{cp1} - V_{cp2}}{\theta_{p1} - \theta_{p2}} \quad (5)$$

which graphically is the slope of a straight line joining the points on V_{cd} for the two solid fractions, θ_{p1} and θ_{p2}. Depending on the shape of the curve, particular shocks are not permitted if a straight line cannot be drawn directly between the two volume fractions.

Typical settling behaviour is shown in Figures 2 (a) and (b). Initially, the column contains a uniform mixture, B. A clear liquid region, A, appears by a shock from the initial volume fraction to zero volume fraction. A dense sediment appears at the bottom, D, which has the final volume fraction. Depending on the initial θ_p and shape of the characteristic velocity curve, there may or may not be an intermediate region, C, of variable density between regions B and D. Region B eventually disappears, and C, if present, is compressed into D. In aqueous systems, such behaviour can be visually observed, but for the metal matrix system, a probe had to be developed to detect the regimes.

Apparatus and Materials

The sedimentation experiments were carried out in an alumina crucible placed inside a resistance-heated furnace. The melt surface was exposed to air. The SiC particles were mixed into the melt with a graphite marine-type propeller driven by a variable AC motor. The mixer was run at a constant speed of 525 rpm.

The resistivity measurements were taken automatically with a commercially-available micro-ohmmeter (TECRAD DMO 350). The DMO 350 is a portable, microprocessor-controlled, high-accuracy, micro-ohmmeter which was designed for the measurement of resistance in solid alloys, such as aluminum.

For use in liquid aluminum, a 4-point resistivity probe was developed (Figure 3). It was constructed from stainless steel, and coated with an alumina paste. The two outer electrodes were used to carry a very brief (30 ms) DC electrical pulse. The micro-resistance across the inner two electrodes was measured simultaneously.

The micro-ohmmeter was connected to mini-computer through a RS 232 port. The mini-computer controlled the micro-ohmmeter, stored the measurements, and performed data analysis.

The green silicon carbide particles used in the experiments (Norton Advanced Ceramics of Canada) had a mean particle size of 82 µm with a standard deviation of 19 µm. A commercial foundry alloy, A356 (Alcan), which is often used for metal matrix composites, was used. The major alloying elements are: 7.3%Si and 0.33%Mg.

Procedure

Initially, 27.2 kg of the alloy was melted to produce a melt depth of 210 mm. The melt was maintained at 660°C. The probe was then lowered into the melt, beside the stirrer, and fixed at a depth of 50 mm from the melt surface. Subsequently, 5 vol%

SiC particles was added to the melt, and mixed at a speed of 525 rpm for a duration of 10 minutes.

At the end of the mixing time, the mixer was stopped and resistance measurement was started at the same instant. The resistance was measured every two seconds, until sedimentation was complete. The measurement procedure was repeated three times to test the reproducibility of the results.

The above procedure of measurement was carried out for three different probe locations: 50, 100 and 150 mm below the melt surface. The SiC concentration was increased from 5 vol% to 10 vol%, and so on in steps of 5 vol% to a final concentration of 30 vol%.

Results

The output of the probe as a function of time for three different immersion depths is shown in Figure 4 for an initial nominal volume fraction of 10%. In each case the resistance drops as the shock crosses the probe location. The time for the transition takes longer for deeper immersions.

At higher solid fraction, the resistance increased at the deeper probe locations due to the growth of the compaction zone. This is illustrated in Figure 5 for 30% initial nominal volume fraction of SiC. For the 50 and 100 mm immersions, the volume fractions declined, while for 150 mm immersion it increased as the bed rose to that location.

The drop in the resistance during settling as a function of the initial nominal SiC volume fraction is shown in Figure 6. Below 20% volume fraction, the probe response is quite linear, and it becomes more sensitive, and less reproducible at higher volume fractions. This change in resistance is quite close to the equivalent electrical circuit in which the resistances of the melt and particles are in parallel.

Samples were sucked into glass tubes from the vicinity of the probes. The samples were metallographically prepared, and analyzed quantitatively for the area fraction of SiC. There was good agreement between this area fraction, and the nominal volume fraction. Thus, the probe response is proportional to volume fraction below 15% SiC.

From the data in Figures 4 and 5, and those for other volume fractions, Figures 7 to 10 were constructed in the same format as Figure 2b. For the two lowest volume fractions, 5 and 10% in Figure 7, the only transition corresponds to the clarification of the aluminum, that is from the initial state B, to pure aluminum, A.

At 15% SiC, the volume fraction first increases at the 150 mm depth, corresponding to region C, and then becomes clarified (Figure 8). Similar behaviour occurs at 20% in Figure 9, however, the transition to pure aluminum is not as abrupt. The transition from B to A is denoted as A_1.

As Figure 5 shows, the melt does not become clarified, but reaches a final denser packing which is region D in Figure 10. Extrapolation of boundaries A-A' and C-D to the point of convergence at 130 mm depth at 680 seconds, should correspond to the final state of the melt. The volume fraction in region D can be estimated in two independent ways:

(1) The change in resistance from the initial 30% to the final state was 1.1 μΩ. Using Figure 6, this corresponds to 55% SiC.

(2) Applying conservation of silicon carbide volume:

$$\theta_{po} h_o = \theta_{pf} h_f \tag{6}$$

where the o subscript refers to the initial mixture, and f refers to the final mixture. For a final mixture height of 170 mm (300 total depth - 130 mm immersion), this represents an average volume fraction of 53% SiC.

The agreement between these two methods is reasonable, and corresponds to the random close packing of particles observed in packed beds.

The shock velocity between regions A and B, V_{AB}, plotted as a function of $(1-\theta_p)$, according to Equation 3 in Figure 11, is linear up to 15% SiC. The slope yields a value of n equal to 5.6 and a value for the shock velocity at 0% SiC of 3.35 mm/s. As it should be, this velocity is in reasonable agreement with the Stokesian settling velocity for 88 µm SiC particles of 2.7 mm/s.

The characteristic velocity from Equation 4 is plotted in Figure 12 which summarizes the settling behaviour of these particles in aluminum. The front velocities between regions A and B are the slopes of the lines between the origin, and point on the curve at the initial volume fraction. It can also be seen that a direct shock to the final volume fraction (53-55% SiC) is only possible for volume fractions below 5% or above 45% SiC. In all other cases, a region of variable density, C, will form at intermediate times.

Discussion

Clearly, the settling of the silicon carbide particles is much slower than predicted simply by Stokes Law for single particles. Equation 3 provides a framework for considering this behaviour. From Figure 11, based on Equation 3, n was determined to be 5.6. The value of n is a measure of the degree of flocculation in the system; values over 10 are characteristic of significant flocculation or network formation of the solid in aqueous systems. Large n values lower the characteristic velocity, particularly at high volume fractions due to the structure formation. The facts that:

(1) the observed value of n was only 5.6, and

(2) a random close-packed volume fraction was achieved

indicate that flocculation or network formation was not a significant problem. These factors determine the maximum volume fraction of second phase particles which can be incorporated into melts. Flocculation or network formation may preclude high volume fractions if the particles are poorly wetted, or if fibres or whiskers are employed.

Particles used in commercial metal-matrix composites are approximately 15 µm, some 5.5 times smaller than in the present study. Therefore, one would expect the settling velocities to be reduced by 30 according to Stokes Law and Equations 3 and 4. The time scales for clarification would be increased by the same factor. While solidification would be complete before clarification in most small castings, it is still important to consider the amount of settling or motion that may occur in relation to the velocity of the solidification front and dendrites.

During the re-melting of metal matrix composite stock, there may be significant time between initial and complete melting. In such situations, even small particles can segregate. The present analysis can be used to calculate the changes in volume fraction over increments of time during melting.

Finally, it is of interest to note that curves such as Figure 12 govern the co-current or counter-current flow of particles and liquid. Therefore, it can be applied to the filtration of aluminum through a bed of particles. The liquid throughout is the ordinate at $\theta_p = 1$ of a line passing through the origin and V_{cp} at the bed volume fraction. While 88 μm silicon carbide would not be used for this purpose, coarser particles have been to achieve high-throughput metal filtration.

Conclusions

1. A four point conductivity probe was developed to measure the in-situ volume fraction of ceramic particles in liquid aluminum.
2. The probe was applied to the measurement of the sedimentation rate of 88 μm silicon carbide in A356 aluminum alloy from 5 to 30% initial silicon carbide volume fraction. It was found that the particles settled to a final density of 53-55% SiC, after passing through a region of intermediate density.
3. Settling was much slower than according to Stokes Law. The characteristic velocity of this system was determined to follow Equation 4 with $V_{d\infty}$ of 3.4 mm/s close to Stokes velocity and n of 5.6. This latter value is characteristic of those encountered in non-flocculating aqueous systems.
4. The analysis can be extended to predict the sedimentation behaviour in other metal matrix composite systems.

Acknowledgements

The authors wish to thank Dr. G.S. Hanumanth and Mr. O. Kelly for their help with the experimental work. The donations of silicon carbide by Norton Advanced Ceramics of Canada, and aluminum by Alcan are also gratefully acknowledged.

List of Symbols

h	height (m)
V_{AB}	discontinuity velocity between regions A and B (m/s)
V_{cs}	superficial velocity of the continuous phase (m/s)
V_{cp}	characteristic velocity of the particles (m/s)
V_d	velocity of discontinuity (m/s)
V_p	velocity of the particles (m/s)
V_{ps}	superficial velocity of the particles (m/s)
$V_{p\infty}$	single particle settling velocity (m/s)
V_s	slip velocity (m/s)
θ	volume fraction (–)
ρ	density (kg/m^3)

References

1. Processing of Ceramic and Metal Matrix Composites, ed. H. Mostaghaci, CIM Conference of Metallurgists, August 20-24, 1989, Pergamon Press, New York.
2. A.D. McLeod, C. Gabryel, D.J. Lloyd and P. Morris, Processing of Ceramic and Metal Matrix Composites, ed. H. Mostaghaci, CIM Conference of Metallurgists, August 20-24, 1989, pp. 228-235, Pergamon Press, New York.
3. G.S. Hanumanth and G.A. Irons, to be published in International Conference on Fabrication of Particulate Reinforced Metal Composites, September 16-19, 1990, Montreal.
4. G.B. Wallis, Interactions between Fluids and Particles, Instn. Chem. Engrs., London, pp. 9-16.

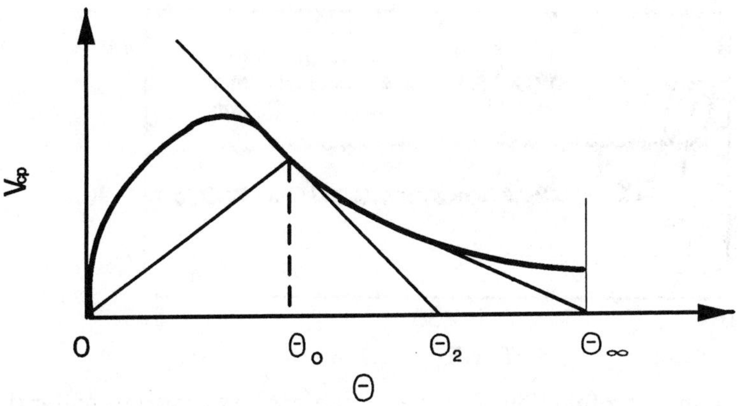

Fig. 1: The characteristic velocity as a function of particle fraction showing how shocks or discontinuities can propagate in the system.

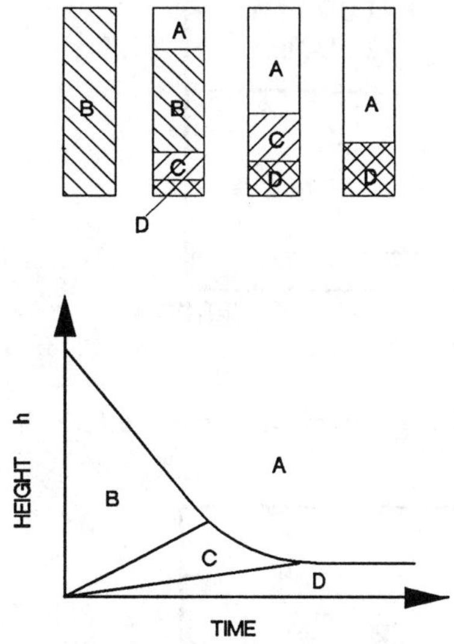

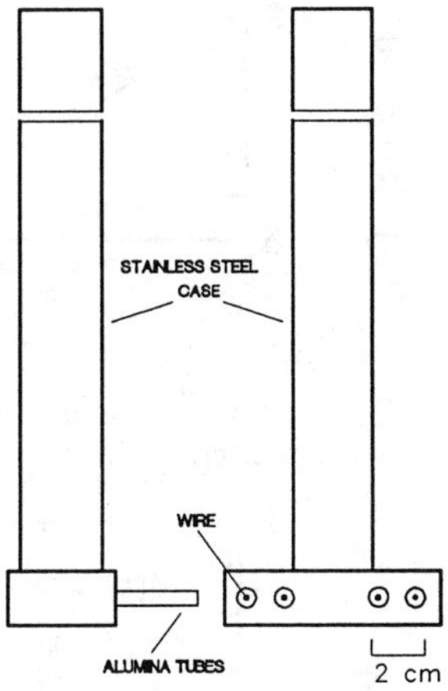

Fig. 2: (a) Typical history of batch sedimentation from the original density B, to a clarified region, A, and a region of maximum final density, D, after passing through an intermediate region, C.

(b) The position of the fronts with time corresponding to (a).

Fig. 3: Diagram of the 4-point probe, to scale. The two outer electrodes carry the applied current, while the inner two are used for resistance measurement.

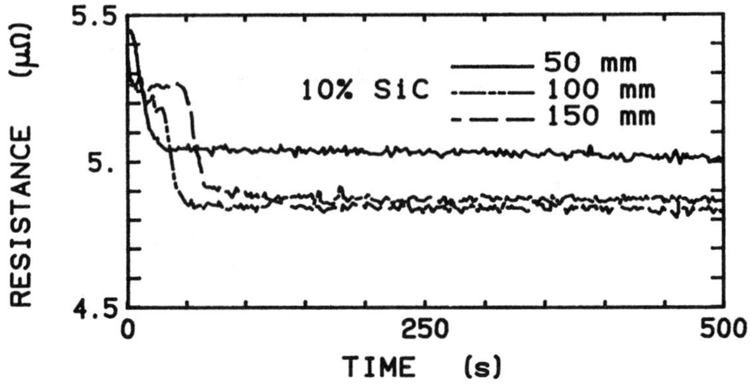

Fig. 4: The resistance as a function of time after the stirrer was turned off for the 10% silicon carbide MMC.

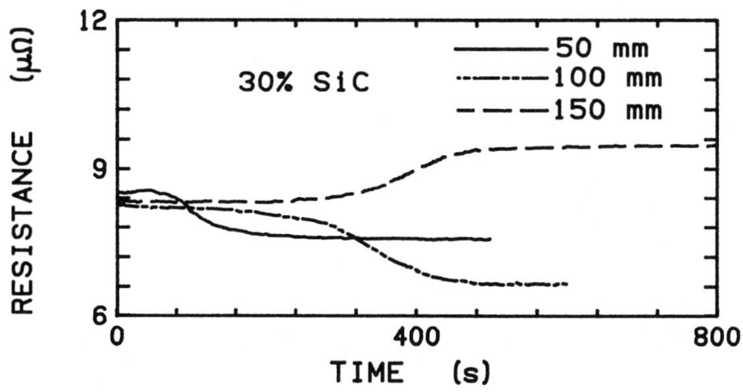

Fig. 5: The resistance as a function of time after the stirrer was turned off for the 30% silicon carbide MMC.

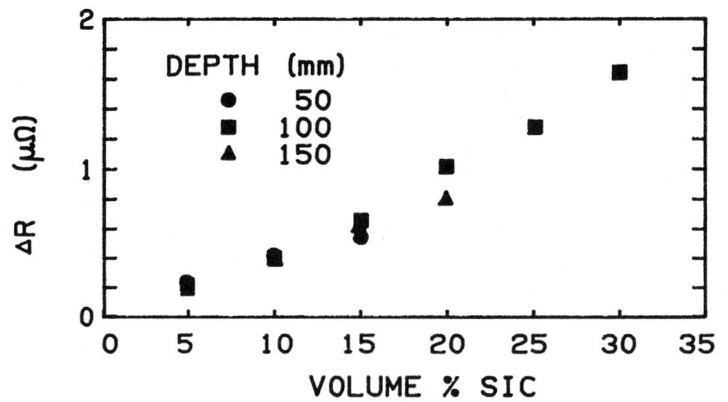

Fig. 6: The change in resistance during settling for the different initial volume fractions of particles at different depths of probe immersion.

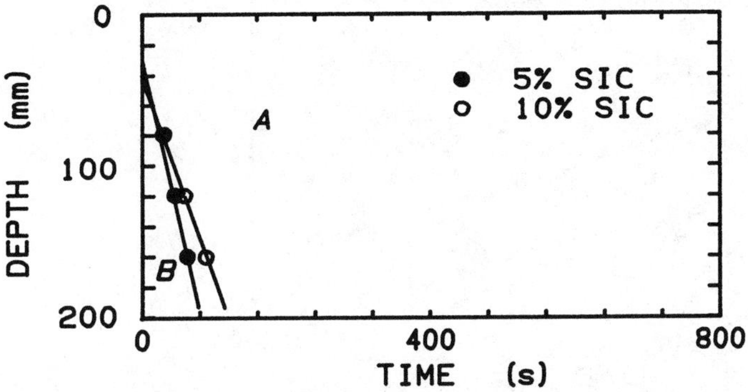

Fig. 7: The time to pass from the initial volume fraction region B, to a clarified region, A, for the 5 and 10% volume fractions at the different heights.

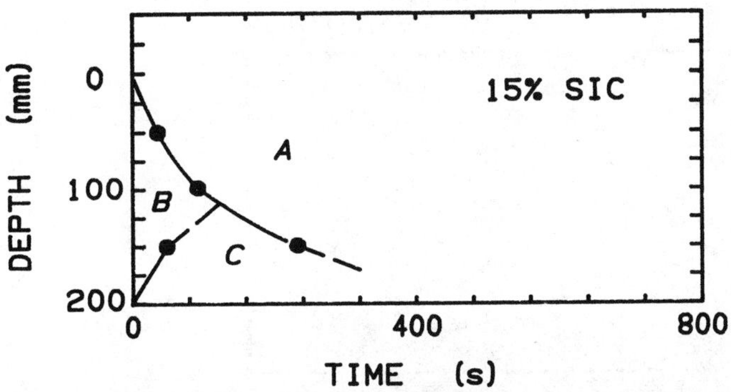

Fig. 8: The time to pass from the initial volume fraction region B, to a clarified region A, for the 15 volume fractions at the different heights. A region of higher volume fraction, C, was also observed.

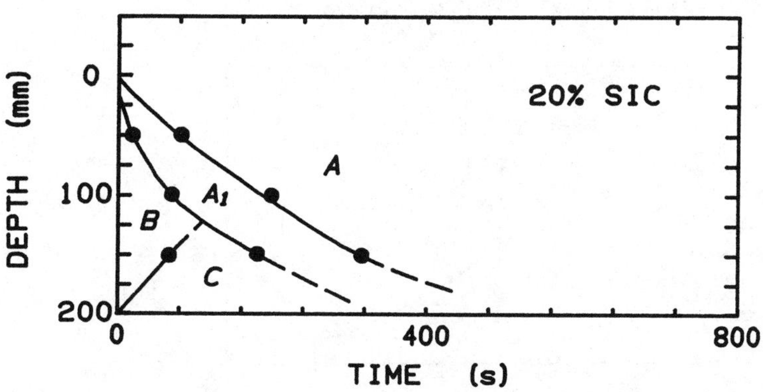

Fig. 9: The time to pass from the initial volume fraction region B, to a clarified region A, for the 20 volume fractions at the different heights. A transition region, A_1, of intermediate density was noted. A region of higher volume fraction, C, was also observed.

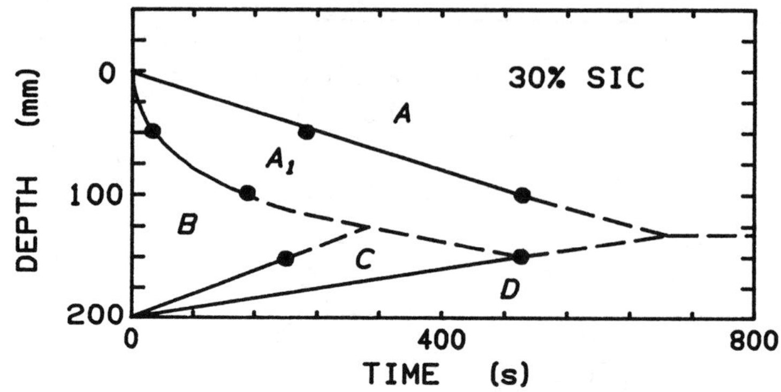

Fig. 10: The time to pass from the initial volume fraction region B, to a clarified region A, for the 30 volume fractions at the different heights. A transition region, A_1, of intermediate density was noted. A region of higher volume fraction, C, was also observed before a final densification to region D.

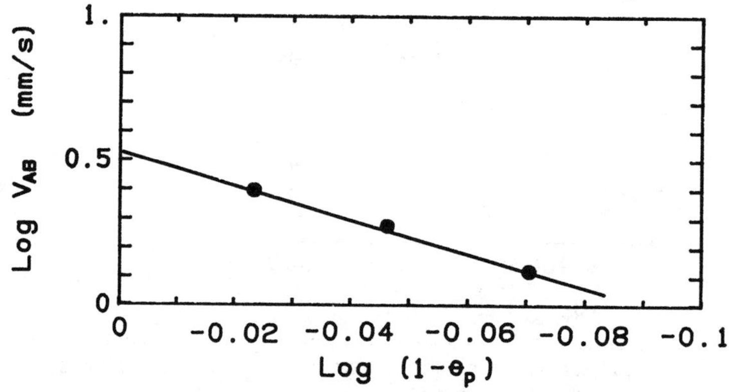

Fig. 11: The velocity of the shock between regions A and B plotted as a function of the aluminum volume fraction, according to Equation 3.

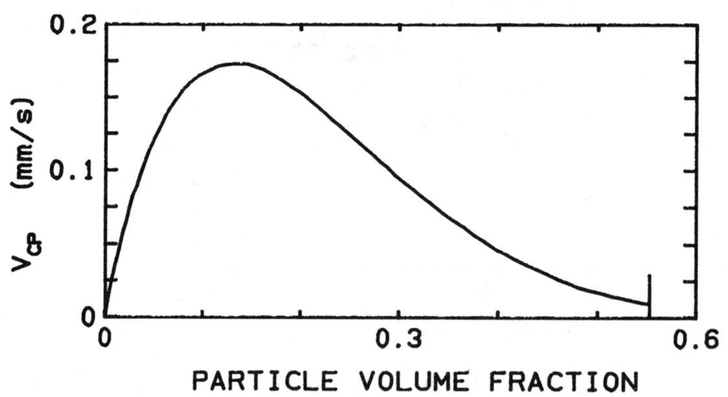

Fig. 12: The characteristic velocity, V_{cp}, for 88 µm silicon carbide in A356 alloy as a function of volume fraction.

The influence of processing parameters on the structure of particle reinforced Al base MMC

P.C.R. Nunes and L.V. Ramanathan
Instituto de Pesquisas Energeticas e Nucleares, C.P. 11049 Cidade Universitaria, São Paulo, Brazil

ABSTRACT

Liquid metal processing to obtain discontinuously reinforced metal matrix composites is considered to be a promising technique. However many aspects related to this technique require attention. In this paper the influence of various processing parameters such as Mg content of the alloy, alumina particle size and quantity, melt temperature, stirring rate, cooling rate and others on the structure of the composite has been presented. The addition of $\geqslant$ 0.7%Mg to the Al-7%Si alloy has been found to be essential to obtain composites. Melt temperature above the liquidus and pretreatment of the alumina particles with borax have resulted in more homogeneous distribution of the particles in the matrix. A high cooling rate of the composite prevents segregation due to solidification fronts. In general, for each particle size/matrix alloy combination there are optimum values for each of the processing parameters.

KEYWORDS

Metal matrix composites; reinforcement; liquid metal processing; alumina particles; processing parameters; microstructure.

INTRODUCTION

Work on metal matrix composites (MMCs) was initially stimulated by the high performance requirements of the aerospace industry, which placed performance ahead of price. After many years of developmental work, it became clear that significant commercial application of MMCs was possible only if it could compete with other materials in terms of cost. Thereafter search increased for cheaper reinforcement and near net shape processing technologies. MMCs reinforced with chopped fibre, whiskers or particulates are referred to as discontinously reinforced metal matrix composites or DMMCS. In general, the properties of DMMCS are more isotropic than those of continously reinforced MMCS. This enables DMMCS to be processed using conventional metal working techniques like rolling, extrusion etc. The potential for commercial utilization of Al base DMMCS is especially high.

The liquid metal technique to produce Al base alloy DMMCS is considered to be simple and economical. The properties of these MMCs are controlled by many variables such as reinforcement distribution, wetting of reinforcement by matrix alloy, reactivity at reinforcement/matrix interface etc. Wetting of ceramic reinforcements by liquid metal is necessary for proper bonding. Various procedures have been recommended to improve the wetting of ceramic particles by liquid metals and include increase in liquid metal temperature(1), pretreatment of reinforcement (2-4), coating of the reinforcement (5-9) and addition of alloying elements like Mg,Li and others to the Al alloy melt (10-13). Preparation of MMCSat temperatures low enough for the melt to be in a semisolid state has been discussed by some workers (3,4), while others have reported obtaining MMCS successfully by stirring reinforcements in melts at temperatures well above their liquidus (14). The reactivity between the reinforcement and matrix which increases with time of contact and temperature has to be maintained at a low level in SiC containing Al base alloy MMC to avoid the formation of Al_4C_3. On the other hand higher temperatures increase fluidity and facilitate particle wetting. The particle/matrix interfacial reaction is beneficial to the extent that adhesion between the phases increases and this is a requirement for efficient load transfer(14). Specific processing details such as stirring rate(15), the use of baffles, single or twin blade stirrer, distance of blade from crucible walls and others to obtain particle reinforced Al base MMCs have also been reported(14-17).

In order to extend the information available about MMC preparation and to throw more light on specific aspects of MMC processing, in this paper the influence of various processing parameters such as Mg content, pretreatment of reinforcement, melt temperature, crucible material, type and rate of agitation, and others on the microstructure of alumina reinforced Al base alloy MMCs has been presented.

METHODS AND MATERIALS

Alloy preparation was carried out in a coated steel crucible inside a resistance furnace. A master alloy containing Al-7%Si was used and different amounts of Mg added to this alloy. After homogenization the alloy melt was degassed using gaseous nitrogen, cleaned and transferred to the compositing crucible. The apparatus for composite preparation is shown schematically in figure 1 and consisted of a coated crucible(graphite or alumina) 120mm high and 80mm in diameter, located in a temperature controlled resistance furnace. A variable speed motor attached to a coated steel stirrer with two sets of propellar type blades was used to stir the melt. The blade lengths were designed to move very close to the walls of the crucible. A steel tube positioned as shown in the figurepermitted the melt surface to be in contact with an inert atmosphere. A vibratory feeder device was used to add the reinforcement to the melt at a predetermined rate.

The overall procedure used to produce the composite consists of preparing the melt as described above while the compositing crucible is maintained at $\sim 720^{\circ}C$. The melt is transferred to the compositing crucible at the desired temperature while the stream of nitrogen is turned on and the stirrer is also switched on.

The reinforcement in particle form was weighed, dried in an oven at $250^{\circ}C$, mixed intimately with predried sodium tetraborate and stored at $250^{\circ}C$ prior to addition to the melt. The particles were added either manually or with the vibratory feeder to the stirred melt in the region of the vortex at a predetermined rate. The melt was maintained stirred during addition of the particles and for about 60 seconds afterwards. The composite slurry was subsequently poured into cold copper molds or allowed to solidify in the crucible outside the furnace.

A number of processing parameters have been varied and their influence on the microstructure studied. The parameters include alloy composition, crucible material, melt temperature, quantity of alumina, rate of addition of alumina, mode of addition, alumina size, agitation of the melt, borax addition, type of stirrer blade and cooling rate of the composite. Table I summarizes the conditions under which the various melts or MMCs were prepared.

The cooled composite was sectioned and specimens for macroscopic and microscopic examination prepared using conventional metallographic techniques. Microhardness measurements were carried out on the matrix phase of the composite.

TABLE I. Summary of process parameters during preparation of melts and/or composites.

Process parameters		1	2	3	4	5	6	7	8	9	10	11	12	13	14	15	16	17	18
		MELT NUMBERS																	
Mg Content (%)		0	0	0.3	0.3	0.5	0.7	0.7	0.7	0.3	1.0	1.0	1.0	1.0	1.0	1.0	1.0	1.0	1.0
Crucible Material G-Graphite A-Alumina		G	G	G	G	G	G	G	G	G	G	G	A	A	A	A	A	A	A
ALUMINA	Grain size (μm)	100	100	100	100	100	100	100	100	100	100	100	100	100	100	100	20	100	100
ALUMINA	Volume Percent	5	5	5	5	5	5	5	10	5	5	5	5	5	5	20	5	5	5
ALUMINA	Borax Pre-treatment Y-YES N-NO	N	N	N	N	N	N	N	N	N	N	N	N	Y	Y	Y	Y	Y	Y
ALUMINA FEEDING	Method M-MANUAL F-FEEDER	M	M	M	M	M	M	M	M	F	F	F	F	F	F	F	F	F	F
ALUMINA FEEDING	Rate V-VARIABLE (g/min)	V	V	V	V	V	V	V	V	6	6	6	6	6	6	6	6	12	6
Stirring rate (rpm)		200	750	750	750	750	750	750	750	750	750	750	750	750	750	750	750	750	750
Melt Temperature (°C)	INITIAL	720	720	720	760	720	760	760	760	760	760	760	750	750	720	750	750	750	750
Melt Temperature (°C)	FINAL	700	700	700	750	720	660	750	750	700	700	620	750	750	630	740	740	700	750
Cooling rate F-FAST S-SLOW		S	S	S	F	S	S	F	F	F	F	S	F	F	S	S	F	S	F

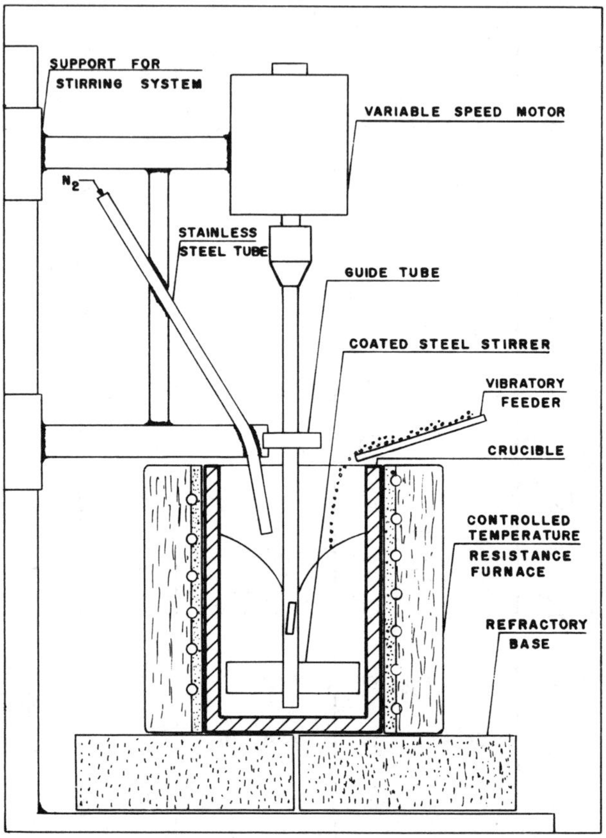

Figure 1. Schematic illustration of equipment used to prepare composites.

RESULTS AND DISCUSSION

A large number of melts were prepared and composites attempted or obtained thereof. The influence of specific parameters are discussed separately.

Magnesium Content

Upon adding 5volume percent alumina particles to a melt with 0.3%Mg, it was observed that most of the alumina first seemed to enter the melt. Soon after, almost all the alumina was rejected by the melt, even at temperatures significantly higher than the liquidus. An increase in the Mg content of the alloy to 0.5% resulted in a small increase in the quantity of particles retained in the melt. These particles were sporadically distributed as shown in figure 2. Further increase in the Mg content to 0.7% or 1.0% resulted in the melt retaining almost all the particles added, eventhough as agglomerates. Figure 3 reveals typical agglomerates of alumina particles and figure 4, the microstructure of Al-7%Si-0.7%Mg containing both individual and agglomerated particles. An examination of the microstructure of the composite produced from melt no. 10, wherein the melt temperature was higher and particle addition was controlled with a vibratory feeder revealed no significant difference from that in figure4.

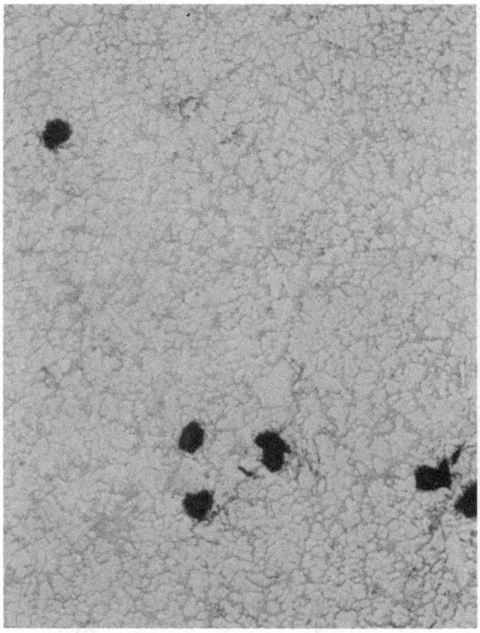

Figure 2. Micrograph of alloy Al-7%Si-0.5%Mg containing some alumina particles. (50X)

Figure 3. Micrograph revealing agglomerates of alumina particles. (20X)

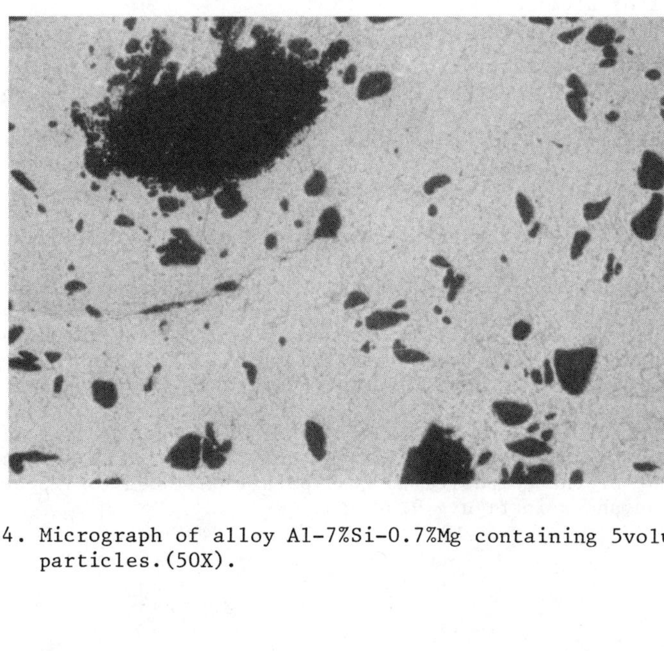

Figure 4. Micrograph of alloy Al-7%Si-0.7%Mg containing 5volume% alumina particles. (50X).

Crucible Material

Coated graphite and alumina crucibles were used in this investigation. The micrographs of composites made in the graphite crucible revealed graphite inclusions and these are due to removal of graphite either accidentally by the stirrer blade or as a result of the abrasive action of the alumina particles on the crucible walls. The use of both graphite and mullite crucibles in other investigations have been reported(14,15,18). Prevention of graphite inclusions depends not only on the quality of graphite but also on the composite preparation procedure. The microstructures of composites made in the alumina crucibles on the other hand were clean. However it has been observed that prolonged use of alumina crucibles results in roughened internal wall surfaces. The roughening of the surfaces is due to the abrasive action of the fast moving alumina particles. Continued use of these crucibles with rough surfaces for preparing composites results in formation of lumps or agglomerates at the roughened surfaces.

Stirring of the melt

Many of the early attempts to prepare composites by stirring melts at ~200 rpm were unsuccessful . The stirring speed was insufficient and the vortex into which the particles were added was very shallow. At higher stirring speeds, ~ 750rpm, the vortex formed (for the dimensions of the crucibles used in this investigation) was sufficiently deep and the extent of agitation adequate to permit the formation of a composite slurry.

Particle size

Alumina particles with an average grain size of 100μm were used in almost all the heats to prepare composites. Melt no. 16 was used to attempt the preparation of composites with 20μm alumina. All the other variables such as Mg content, melt temperature, stirring speed etc. that had been optimized for alumina of 100μm were maintained constant. The 20μm alumina appeared to enter the melt but was subsequently rejected and found to remain on the melt surface. This may have been due to a significant increase in surface area of the solids for the same volume percent. Work is in progress to optimize the conditions to retain the 20μm alumina in the alloy matrix.

Volume percent of alumina

Increase in volume percent of alumina from 5 to 10 when all other parameters were maintained constant resulted in formation of a greater number of agglomerates as shown in micrograph of figure 5. Upon addition of 15 volume% alumina the melt became very viscous although it was maintained at 750°C. Periodic rupture of the surface film and its incorporation into the melt was also observed.

Borax addition

Starting with melt no.13 and therafter, dried alumina particles were mixed with 10wt%dry borax (weight percent of alumina content) before being added to the melt. The composites obtained revealed a more homogeneous particle distribution and the extent of agglomeration was considerably low as shown in figure 6. The influence of borax addition without alumina to the melt was also studied. The micrographs in figure 7 reveal no significant change in the overall microstructure upon addition of borax, although inclusions can be observed in the interdendritic regions. The role of borax in deagglomerating alumina particles is as yet not clear although its effect on modification of silicon(shown in figure 8) and on grain refinement is well known. Even upon increasing the alumina content of the composite to 20%, borax continued to improve particle distribution as shown in figure 9.

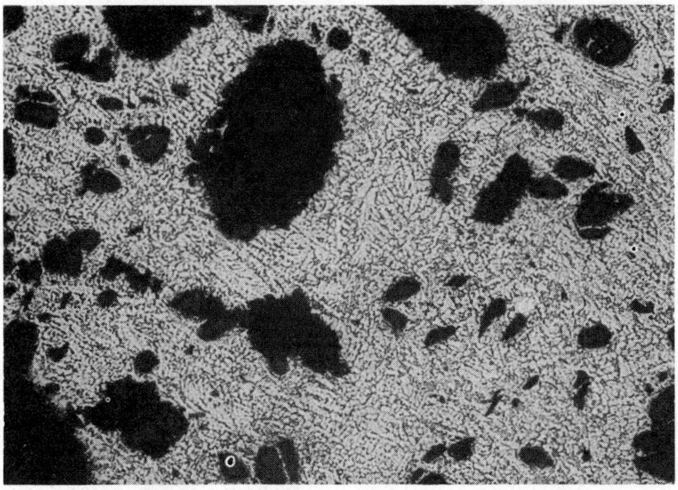

Figure 5. Micrograph of composite containing 10 volume% alumina particles. Note the increase in number of agglomerates.(50X).

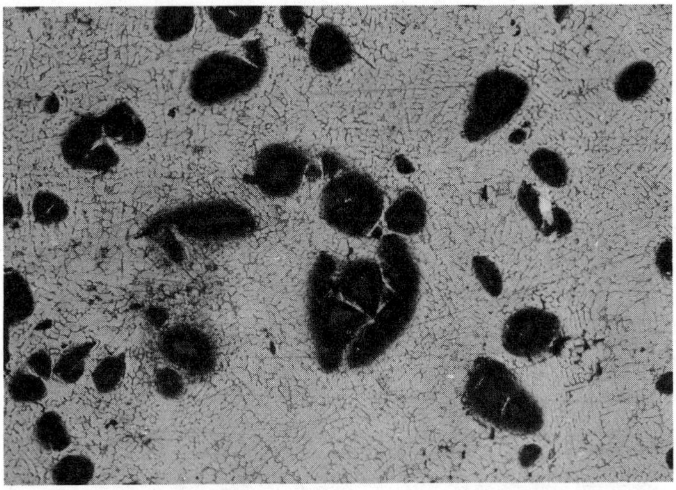

Figure 6. Micrograph of composite with 5volume% alumina pretreated with borax.(50X).

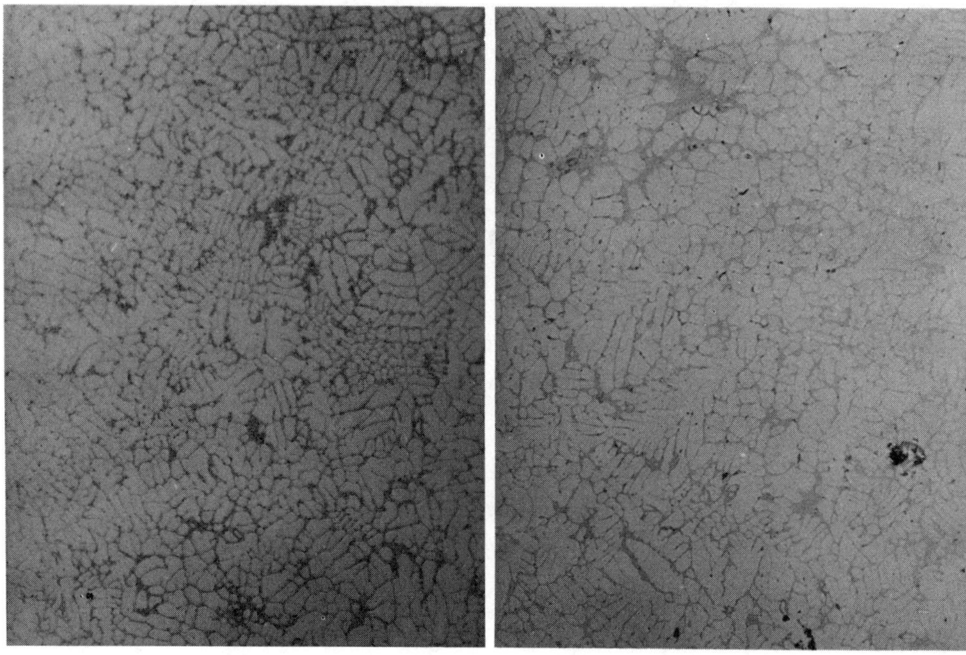

(a) (b)

Figure 7. Micrographs of Al-7%Si-1%Mg alloy (a) without borax and (b) with borax. (50X).

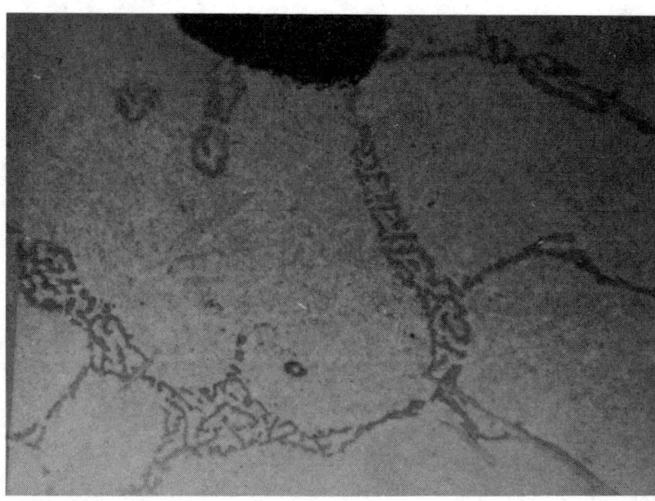

Figure 8. Micrograph revealing modified silicon in the Al-7%Si-1%Mg alloy containing borax. (1200X).

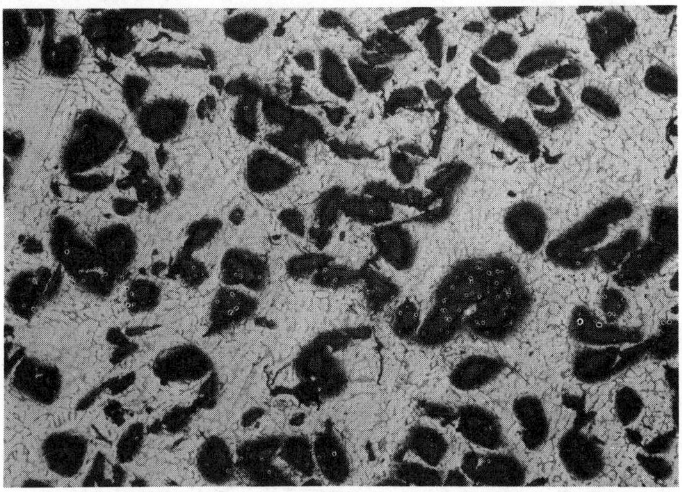

Figure 9. Micrograph of composite with 20 volume percent alumina pretreated with borax. (50X)

Melt temperature

In the early stages of this investigation the melt was transfered at ~720°C to the compositing crucible and gradually cooled, while being agitated. The reinforcement was added when the melt reached a semi solid state. The alumina did not enter the melt, it was rejected and floated on the surface or was found entrapped in pockets as large agglomerates. Thereafter, the melts were maintained at ~750°C, well above the liquidus while the particles were added. The particles were found to enter the melt and remain uniformly distributed. In melts no. 11 and 14 wherein variables such as Mg content, borax treatment, agitation and others were maintained at optimized values, reduction in melt temperature resulted in a viscous composite melt and consequent nonuniform distribution of alumina in the cast composite ingot. Increase in the rate of particle addition resulted in rapid cooling of the melt and consequent non-uniform particle distribution.

Cooling rate of melt

In general, two different cooling rates have been studied: fast, when the melt was poured into a cold copper mold and slow, when the melt was allowed to solidify in the crucible outside the furnace. Macroscopic examination of a longitudanal section through the cast ingot revealed uniform particle distribution in most of the regions, except for a very narrow region close to the outside surface in contact with the ingot mold. A slowly cooled specimen revealed a very large concentration of particles in the upper part of the specimen, and is thought to be due probably to segregation brought on by a slowly moving solidification front.

Other variables

The shape of the stirrer blade is considered to be an important factor. On several occasions, composites attempted with improperly designed blades resulted in nonuniform distribution of alumina particles. In melt no.18, although 5 volume percent alumina was added to the melt under controlled conditions, it was observed that after stirring and pouring, the first poured ingot was almost free of any particles, the second poured ingot had some particles and the melt left behind in the crucible was too viscous to pour, not because the temperature was low but because it contained over 20 volume% alumina, and fairly uniformly distributed. The hydrodynamic conditions existent within the melt during composite preparation are quite important. Hence changes in these conditions lead to nonuniform or layered distribution of particles.

Microhardness measurements

The microhardness of the composite matrix containing different volume percent of alumina are shown in figure 10. A significant increase in the hardness of the alloy matrix can be seen with addition of the particles. Increase in the particle content of the composite did not alter significantly the hardness.

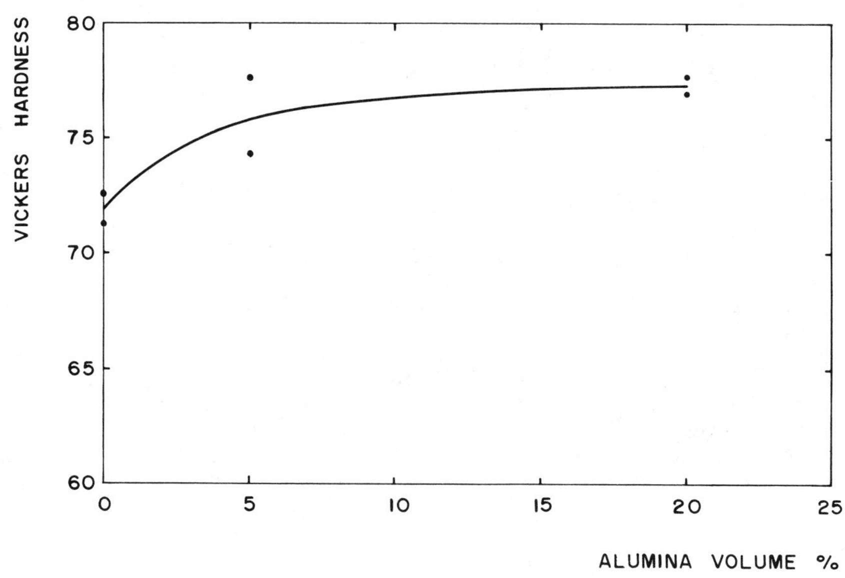

Figure 10. Variation of microhardness of the matrix phase with increasing volume percent of alumina.

GENERAL COMMENTS

A metal matrix composite of alumina particles in an AlSiMg alloy matrix can be fabricated by stirring the particles in a liquid metal bath. Formation & homogeneous distribution of the particles depend on many process variables. Temperatures above the liquidus permit uniform distribution of particles because of higher fluidity. The formation of $MgAl_2O_4$ spinels at the particle/matrix interface ensures strong bonding that permits load transfer from matrix to particle(15). Composite fabrication by addition of particles to viscous melts has not been successful and often leadsto nonuniform particle distribution. The nonuniform distribution of particles is to be found in the form of particle clustering, that is formation of segregated regions of concentrated particles surrounded by regions of matrix devoid of particles. The tendency increases with reduction in the particle size and also in the absence of pretreatment with borax. In the case of fine powders, the nonhomogeneous distribution can also be attributed to inadequate deagglomeration of the reinforcement before mixing with the matrix material, or the inability to disperse the particles in molten metal due to sedimentation, brought on by density differences and surface tension of the melt, or to confinement of the particles at boundaries of the matrix/particles.

The participation of borax in improving the particle distribution is not completely understood. Sodium tetraborate plays a twin role in any normal Al-Si alloy. The sodium modifies the Si and B acts as a grain refiner. Homogeneous distribution of particles previously treated with borax in the matrix has been observed in this investigation and elsewhere with SiC particles (19).

The increase in hardness of the matrix observed upon addition of the reinforcement is not so much due to the effect of the reinforcement, which is inert, but to a change in the volume fraction of the phases precipitated and probably the modifying and grain refinement effects of borax.

A number of parameters play an important role in the fabrication of particle reinforced Al base MMCs. The exact condition of each parameter is specific to each particle type/ particle size / matrix alloy combination.

CONCLUSIONS

1. The presence of $\geqslant$ 0.7% Mg in the Al-7%Si alloy has been found to be essential to produce alumina containing composites.
2. Stirring speeds of $\sim$750 rpm and pretreatment of the alumina particles are necessary to obtain a homogeneously distributed particulate composite.
3. Stirring alumina particles at higher temperatures permits better distribution and increased reactivity at the particle/matrix interface, which forms strong bonds.
4. Other variables such as crucible material, stirrer design, feeding rate etc. need to be controlled to obtain clean, homogeneously distributed particulate MMCs.
5. Increase in volume percent of particles depends on acceptable increase in melt viscosity, which in turn depends on available casting facilities.
6. Cooling rate of the cast composite should be sufficiently high to prevent segregation of particles by the solidifying front. This aspect becomes more important with decreasing particle size.
7. Selection of optimum values for specific parameters are necessary for each particle size-matrix alloy combination to obtain MMCs.

REFERENCES

1. C.G.Levi,G.J.Abbaschian and R.Mehrabian, Metall.Trans. A.,9A,697,(1978).
2. A.Banerjee,P.K.Rohatgi and W.Reif, Metall.,38(7),656,(1984).
3. P.K.Rohatgi,R.Asthana and S.Das, Int.Met.Rev.,31(3),115,(1986).
4. B.P.Krishnan,M.K.Surappa and P.K.Rohatgi,J.Mat. Sci.,16,1209,(1981).
5. A.G.Kulkarni et.al.,J.Mat.Sci.,14,592,(1979).
6. B.C.Pai and P.K.Rohatgi, Mat.Sc.and Eng., 21,161,(1975).
7. J.P.Rocher,F.Girot,J.M.Quinisset and R.Naslain," Developments in the Science and Technology of Composite Materials",Proc. 1st European Conf. on Comp. Mat., ECCM-1, Bordeaux, France,Sep.1985, A.R.Bunshell,P.Lamicq and A.Messiah, Ed. 634,(1985).

8. N.I.Abdul Latef,A.R.Ismael Khedar and G.K.Goel, J.Mat.Sci. Lett, 4,385,(1985).
9. J.P.Rocher,J.M.Quenisset and R.Naslain,J.Mat.Sci.Lett.,4,1841,(1985).
10. F.Delanney,L. Frozen and A.Deryttere,J.Mat.Sci.,22,1,(1987).
11. A.R.Champion,W.H.Krueger,H.S.Hartman and A.K.Dhingra,Proc. of the Int.Conf. on Comp. Mat. ICCM-2,AIME,883,(1978).
12. W.H.Hunt,Interfaces in Metal Matrix Composites. A.K.Dhingra and S.G.Fishman, Eds., Proc. Conf. New Orleans,TMS-AIME.,3.(1986).
13. Y.Kimura et.al., J.Mat Sci.,19,3107,(1984).
14. B.F.Quigley,G.J.Abbaschian,R.Wunderlin and R.Mehrabian, Met.Trans. 13A,93, (1882).
15. F.M.Hosking,F.Folger,R.Wunderlin and R.Mehrabian,J.Mat.Sci.,17,477,(1982).
16. F.A.Girot,L.Albingre,J.M.Quenisset and R.Naslain,J.of Metals,Nov.,18,(1987).
17. C.Milliere and M.Suery, Mat.Sci. and Tech.,4,41,(1988).
18. P.K.Ghosh and S.Ray, J.Mat.Sci.,22,4077,(1987).
19. T.B.Cameron,W.W.SwansonJ.M.Tartaglia and T.B.Cox, U.S.Patent no.4713111, (1987).

Properties and applications of particulate reinforced aluminum MMC

P.L. Morris and C. Baker

Abstract

Particulate reinforced aluminium based metal matrix composites are emerging as a class of engineering materials which exhibit many of the properties of aluminium alloys but with significant increases in modulus,wear resistance and in some applications strength and with reductions in coefficient of thermal expansion. Several low cost manufacturing routes have been developed to yield billet or ingot which can be fabricated by extrusion forging or rolling or can be remelted and shape cast. One of these (the Duralcan process)is now being commercialised. The production and properties of a number of wrought and foundry MMC reinforced with ceramic particulate are discussed. Finally some initial applications of these materials are discussed.

Keywords

Metal matrix composite, particulate, fabrication, shapecasting, extrusion, forging, rolling, machining, strength, modulus, thermal expansion, thermal conductivity.

Introduction

Development of aluminium alloys has led to materials which can challenge many other engineering materials in terms of specific strength. The most recent development of Al-Li alloys has also provided improved specific stiffness. Further development of conventional aluminum alloys is however limited and significant further property improvements are unlikely. For this reason metal matrix composites which can offer significantly increased modulus, wear resistance, strength and decreased thermal expansion are meriting serious attention in materials selection. While MMC reinforced with high aspect ratio ceramic such as fibers offer the most significant property improvements, they have the disadvantage of high cost and are unable to be processed by conventional casting or metal working processes.

The use of particulate ceramic (eg. Silicon Carbide or Aluminium oxide) as a reinforcement offers good property improvements over aluminium alloys with the ability to fabricate using conventional methods with very little modification. While the property improvements are not as significant as those obtained with high aspect ratio ceramics, the lower cost and versatility open up many new applications for aluminum based materials. The improvements may be summarised as:

Increased modulus-

> Depending on the volume fraction of reinforcement the modulus of aluminum alloys may be increased by up to 50%

Increased strength

> The increase in strength is dependent on the alloy and heat treatment. Typically the increase is greater in low strength alloys and is particularly attractive in foundry applications. In 6000 series wrought alloys in the T6 temper an increase of about 20% is realised.

Improved Wear resistance-

> The improvement in wear resistance is very dependent on the type of application and test method. Generally wear is evaluated in application oriented testing but for example in a standard block on ring test the volume loss was over two orders of magnitude less than in a comparable un-reinforced alloy.

Lower thermal expansion-

> The coefficient of thermal expansion is decreased by up to 20% which facilitates uses in mixed metal applications which are subject to temperature changes.

The improvements in properties combined with the low density and good corrosion resistance of aluminium are leading to applications particularly in the transportation sector.

Production Methods

A number of methods are available for the production of particulate reinforced metal matrix composites. They include powder metallurgical techniques,infiltration methods or methods which involve direct incorporation into molten metal. Of these the latter offers a combination of large scale and low cost production of MMC with between 10 and 20 V.% of particulate.Two such methods are under development by Alcan. These are:

Molten metal mixing. In this patented process ceramic particles are mixed into molten aluminium alloys in conditions which promote dispersion and wetting of the particulate. The composite is then cast either as foundry ingot in the case of casting materials or as D.C. billet for further fabrication. Currently large scale quantities are available from Duralcan U.S.A. in San-Diego and a full scale plant in Quebec with a capacity of 25 million pounds per year of composite, has already cast its first composite.

The process is capable of incorporating a wide range of ceramics into most aluminium alloys. Materials currently offered are Al-Si foundry alloys with 6-10% Si reinforced with SiC and 6061 or 2014 billet for wrought applications reinforced with alumina.

A second process which is at an earlier stage of development is based on the Osprey process. In this method which is termed Co-Spray molten aluminium alloy is atomised and SiC is injected into the stream of droplets. The resulting composite is collected on a substrate to produce a billet for extrusion. The conditions of atomisation and particulate feeding have to be carefully controlled to ensure a uniform distribution of particulate with low porosity. The process allows production of alloys which cannot be conventionally produced both with and without reinforcement. Also because of the short contact time between the ceramic and melt it permits the use of combinations which conventionally would react. MMC based on 2000 and 7000 series and Al-Li are currently under development.

Fabrication

Shape Casting

Foundry ingot of DURALCAN MMC can be remelted and cast using all of the conventional casting methods, namely investment,sand,permanent mold,low and high pressure die casting and squeeze casting. There are however some differences in remelting practices and handling. Overheating the melt can give rise to a reaction between SiC and the aluminium matrix leading to aluminium carbide formation and loss of fluidity. The temperature at which this will occur in practice increases with the Si content of the alloy., With 6% Si the maximum temperature is about 750 °C. Since the ceramic is more dense than the molten alloy the particulate settles in the melt and accordingly gentle stirring is necessary to maintain suspension.

An important factor in foundry economics is the ability to recycle scrap gates risers and castings. Procedures are in development to clean and degass melts containing scrap with good results.Although the viscosity of the melt is higher than that of aluminium the material pours and fills molds almost as readily as aluminum. The increased viscosity does however lead to an increased tendency to entrap atmospheric gases as bubbles. Accordingly care must be taken to avoid turbulent mold filling and some re-design of the feeding system may be necessary. A surprising result is that high pressure die-castings display excellent integrity which may be due to shear thinning effects.

Extrusion

A large quantity of ingot has been extruded to a variety of shapes including complex shapes such as tubes and hollow sections. Extrusion is performed using conventional presses for aluminium employing shear face dies. The peak pressure is only slightly increased over a comparable aluminium alloy and extrusion speeds are only slightly decreased. Two additional considerations arise in the case of extrusion of MMC. The first is a phenomenon of cracking at low speeds not experienced with aluminium which appears to arise due to a transition between sticking in the die and slipping. The phenomenon can be minimised by appropriate extrusion conditions. The second phenomenon relates to die wear, which as would be expected is much more severe than in aluminium. Fortunately there is a wide choice of novel die materials which offer increased wear resistance.

Forging and Rolling

In contrast with the large amounts of casting and extrusion relatively little forging or rolling has been performed. Some connecting rod blanks were forged from extruded rod in 2014-15v% SiC, 2618-15v%SiC and 2219-15%Al_2O_3.
An aircraft linkage test section was forged in 6061-15v%Al_2O_3 (figure 1) without any cracking. A more severe test was performed by Cameron Iron Works in the U.K. who produced the test forging shown in figure 2 in four stages from billet of similar material.

Preliminary triaxial forging
Unsupported upsetting to 300mm diameter
Blocking into a 300mm diameter ring and
Final closed die forging.

Two billets were forged to the final shape without fracturing (even though one showed a crack on the face after the triaxial forging). A third was forged successfully without the initial triaxial stage.

Laboratory hot and cold rolling trials have indicated that the composite can be rolled satisfactorily to sheet and plate but with some greater tendency to edge cracking. Commercial scale trials using ingots up to 600 Kg in weight are planned.

Measurements of mechanical properties at elevated temperatures over a wide range of strain rates indicated comparable behaviour to the matrix materials.

Machining

As with extrusion die material,hardened tool steel is not suitable for machining particulate reinforced MMC. Either carbide or polycrystalline diamond tools can be used depending on the type of operation, number of parts and the MMC. A detailed study was conducted by Cleveland Twist Drill in the USA under contract to Alcan and results are being used to prepare sets of guidelines for conventional machining. A further study is in progress by Metcut Research Associates to yield additional handbook data.

Structure and Mechanical Properties

Castings

The microstructure of typical castings are shown in figures 3 and 4. The particulate is SiC with a low aspect ratio and a diameter of about 12 µm. While at low magnifications the particulate appears to be uniformly distributed, closer examination reveals that the ceramic is concentrated in interdendritic regions. This is believed to be due to particle pushing by the solidification interface combined with gravity settling of the particulate.

The resultant clustering of the particulate appears to have little effect on physical properties but is believed to influence failure initiation in tensile tests.

Figure 5 shows the properties of foundry composites based on A356, obtained from permanent mold cast test bars heat treated to the T61 (peak aged) temper. The results show increases in yield strength, tensile strength and modulus over the range 0-20 V% SiC. Tensile elongations however fell from 5-6% for A356-T61 to less than 1% in the MMC. Similar results were obtained for sand cast bars. Recently a composite, F3D.XXS, has been developed for use in die casting applications where no formal heat treatment is given. Die Casting results in uniform dispersion of SiC as shown in Fig 4.

Tensile values are shown in figure 6, they indicate excellent increases in yield strength over conventional die castings with modulus increases of up to 60%.

High temperature properties are shown in figure 7. In F3A.10S excellent high temperature properties are found after short exposures but after 100 Hrs. exposure at high temperatures any advantage is lost. A number of alloy variants have been examined which would provide enhanced stability to a composite and materials which exhibit yield stress values of 120MPa. at 250° C after 100 hrs.exposure are in development.

Wrought composites

Composites for wrought applications fall into two main types, those reinforced with Alumina (DURALCAN) and those reinforced with SiC (Cospray). The matrix alloys include 6061,2014,2618,7075 and 8090. The following results are a representative sample of the properties of such materials.

Figure 8 shows the tensile properties of DURALCAN W6A.XXA T6. As with the foundry alloys the yield strength is increased markedly with a less significant increases in UTS. This appears to be due to a very high initial work hardening rate which falls to conventional levels at relatively low strains. The tensile elongation falls however to about 8%. If one is prepared to make compromises in strength some improvement in elongation can be realised.

The reduction in tensile elongation is believed to result from the complexity of stress state in clustered areas of particles and the consequent difficulty of maintaining stability. This is specific to thc typc of test and for example plane strain fracture toughness does not show such a marked deterioration as shown in figure 9.

The tensile properties of 8090 (Al-Li) reinforced with 12% SiC are given in figure 10 . While the strength improvement is relatively modest the modulus ,even with this relatively modest volume fraction is further boosted to over 100GPa. This coupled with the low density of 2.62 gm/cc offer a specific modulus some 50% higher than aluminum or steels and 20% higher than 8090. The use of higher lithium contents with higher volume fractions of SiC offers further enhancement of specific modulus.

Other Properties

While many applications benefit from the higher stiffness and strength a majority of the initial applications of the material exploit the excellent wear resistance of the material or its low coefficient of thermal expansion. Wear resistance can be measured in a number of ways each seeking to simulate conditions encountered in different applications. Figure 11 shows conforming block on ring wear data for F3A.15S compared with other foundry alloys in use in high wear applications. Thrust washer tests (ASTM - D3702) showed at least two orders of magnitude less wear rate in comparison with unreinforced matrix alloy.

Coefficients of thermal expansion are reduced as shown in figure 12. This can be particularly important where it is necessary to lessen differences in CTE in mixed metal applications. An interesting finding is that while electrical conductivity is reduced thermal conductivity is actually improved in MMC with SiC (resulting from the high thermal conductivity of SiC). At room temperature the thermal conductivity of F3A.15S is 0.415 cal/cm.sec.K compared with 0.360 for the unreinforced matrix (A356).

Applications

A number of applications in several market sectors are in development. Many of these are shape castings, using the unique ability of these types of MMC to be remelted and cast. Several applications for extruded composites are also in development.

Some examples of initial commercial applications are given below.

1) Fig 13 shows an investment cast aircraft camera gimbal 780mm in diameter and weighing some 17Kg. It was produced from F3B.20S (A357 with 20%SiC) by Cercast, Inc. in Montreal for Ball Aerospace Inc. It is one part of a six part assembly made from the composite and requires a combination of the higher stiffness and lower CTE offered by the material

2) Figure 14 is a brake rotor for a racing-car application. It was cast in green sand from a foundry alloy composite by Texas Metal Casting C.,Inc. for Race Car Products Inc. The composite rotor weighs 3Kg.less than the cast iron rotor it replaces significantly reducing the unsprung weight and offers excellent thermal performance and wear resistance. This application is also in investigation for road vehicles by a major automotive manufacturer.

3) Figure 15 illustrates die cast components including a gearbox housing made from F3D.10S by high pressure die casting by Metallic A/S (Skive, Denmark). The integrity of this part is indicated by the absence of porosity in the machined surfaces.

4) An example of an application for wrought alloys in the sporting goods market is depicted in figure 16. The bicycle frame is constructed of W6A.10A and offers excellent stiffness:weight ratio. The use of welded construction avoids the need for lugs which are used to join tubes in competing materials. As a result even though the specific stiffness of the MMC is somewhat lower the overall stiffness:weight ratio of the frame is higher. A bicycle manufacturer is soon to announce a mountain bicycle with a DURALCAN frame.

Other products, not illustrated, include drive-shafts for light truck applications (specific stiffness),wire for thermal spray applications where corrosion and wear resistance are important and cylinder liners.

Conclusions

The introduction of low cost composites projected to be of the order of $2 when in commercial production which can be fabricated with conventional equipment (usually with some modification to tooling) offers a new range of engineering materials for the metals industry. A number of products have been introduced and more are in development.

Acknowledgements

The authors thank Alcan International for authorisation to publish this paper and Duralcan USA and its customers who provided much of the data and several photographs of products which are reproduced with their permission.

They also wish to thank their collegues in the MMC groups in the Banbury and Kingston research centres of Alcan International for their contributions to many aspects of the work.

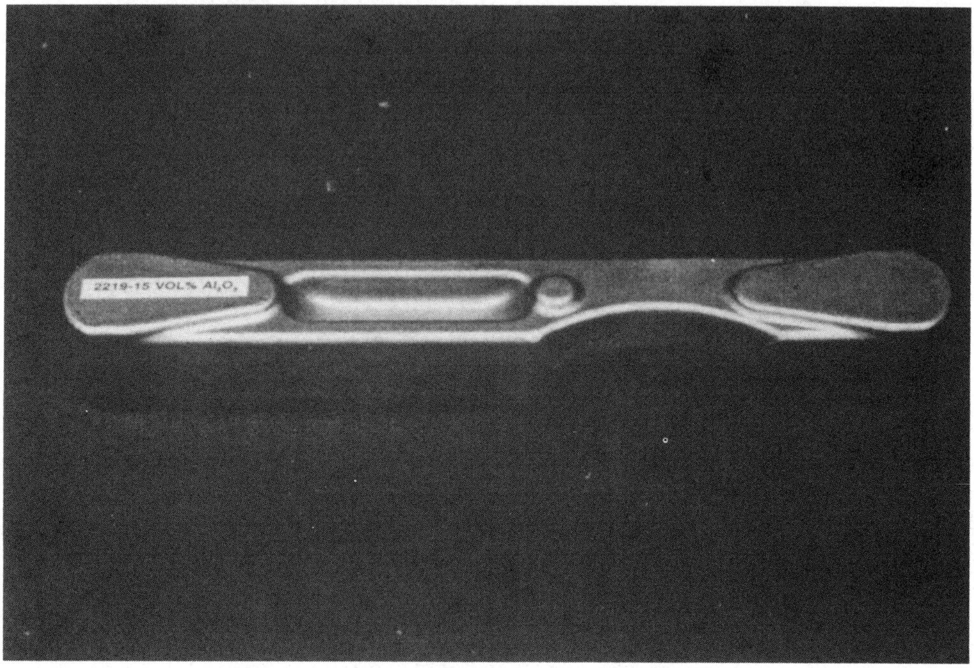

Fig.1

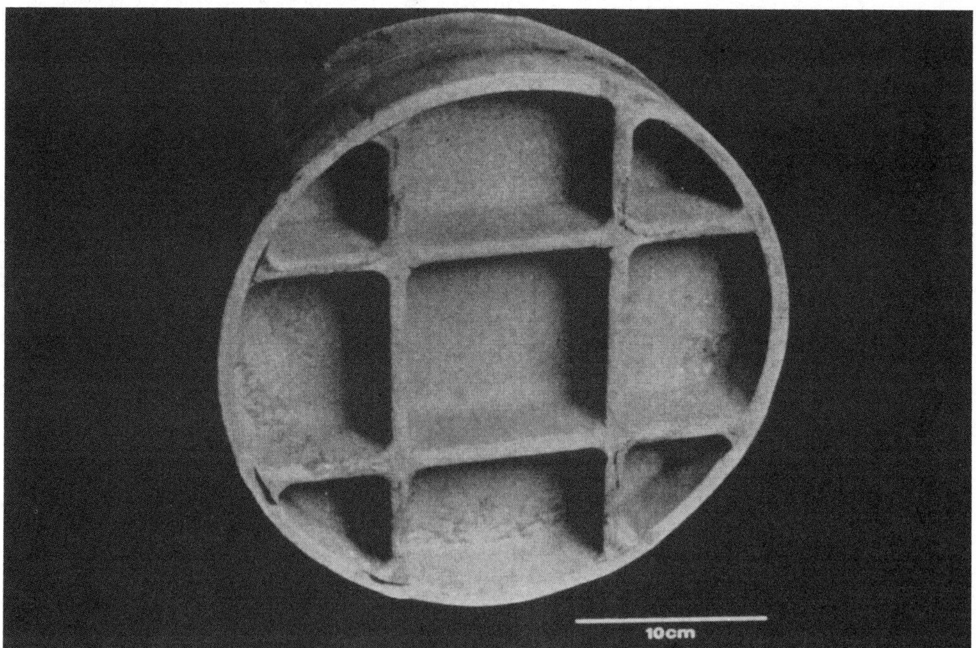

Fig. 2

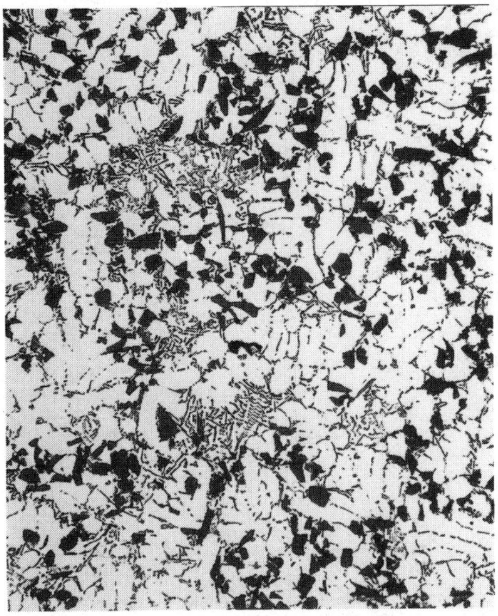

Fig. 3.

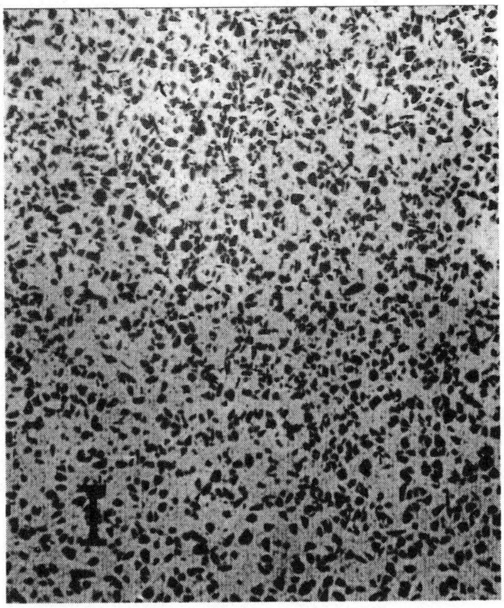

Fig. 4.

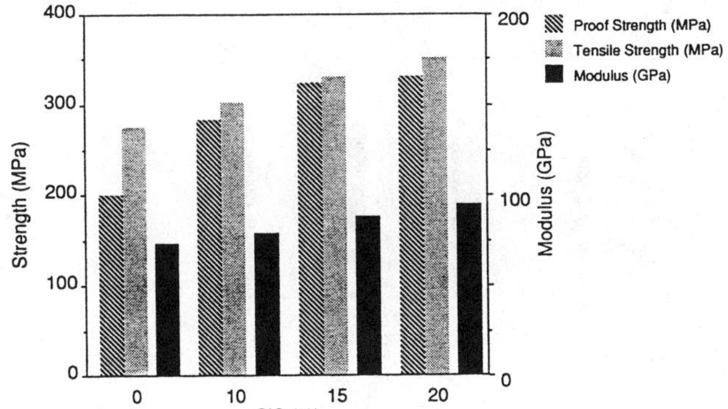

Fig. 5

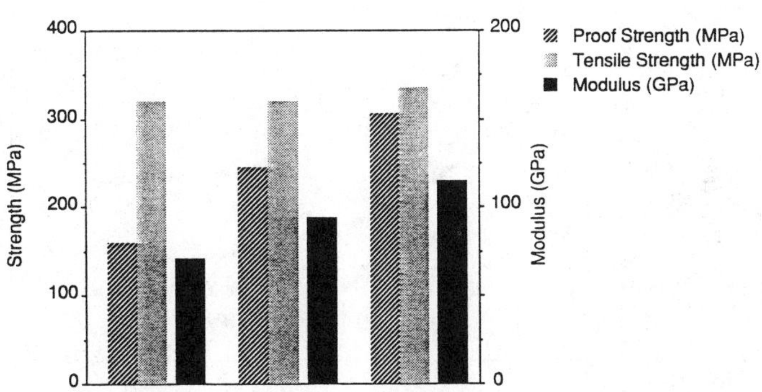

Fig. 6.

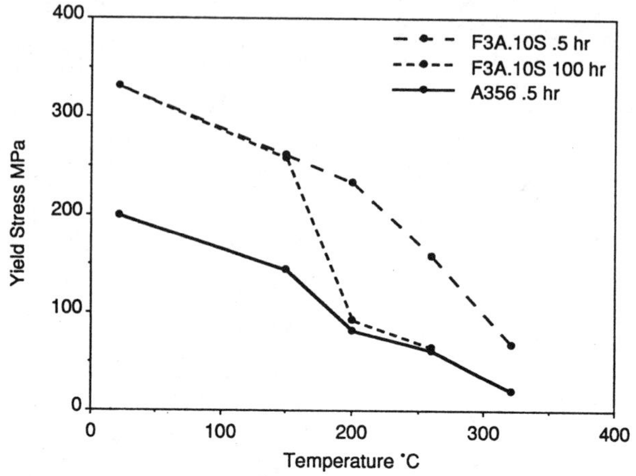

Fig. 7.

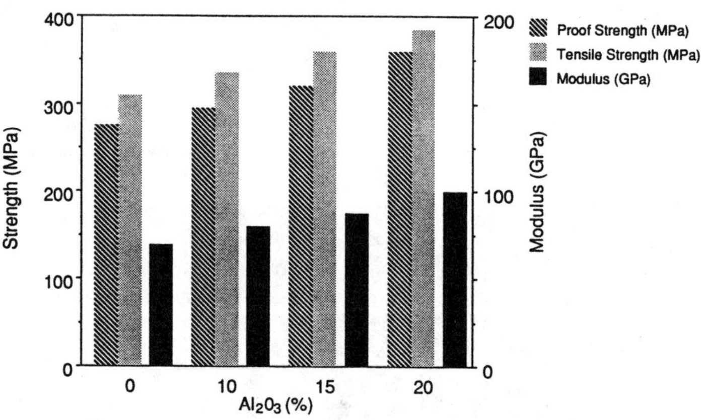

Fig. 8.

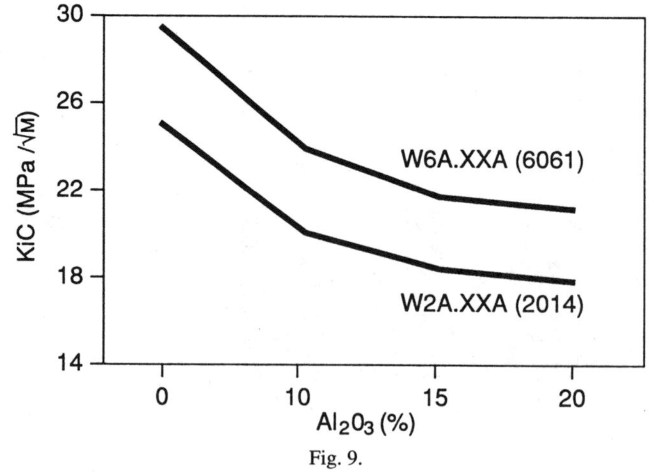

Fig. 9.

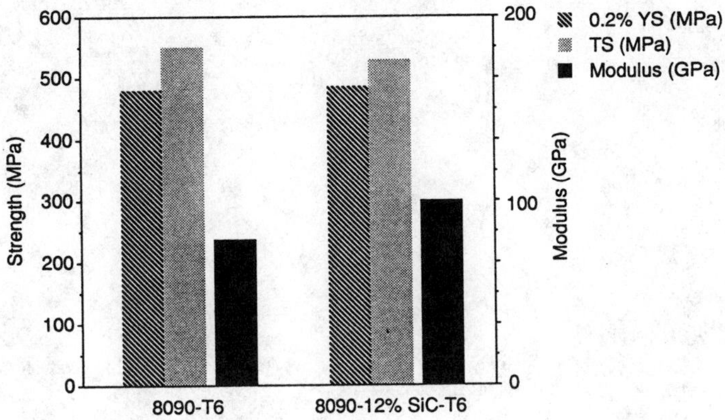

Fig. 10.

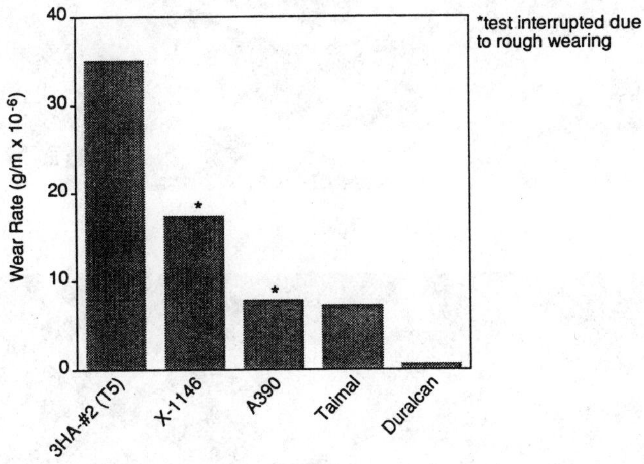

Fig.11.

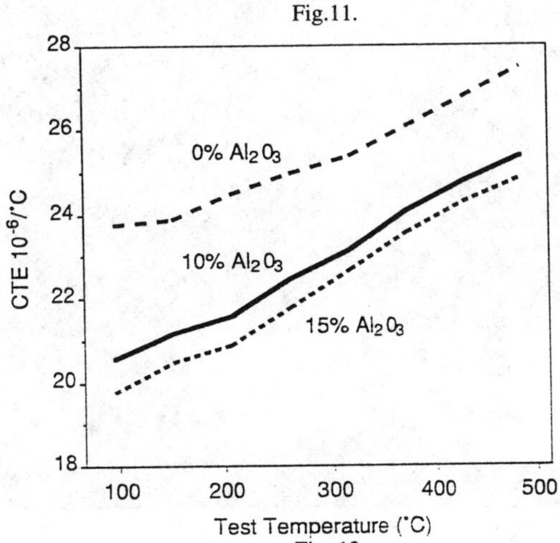

Fig. 12

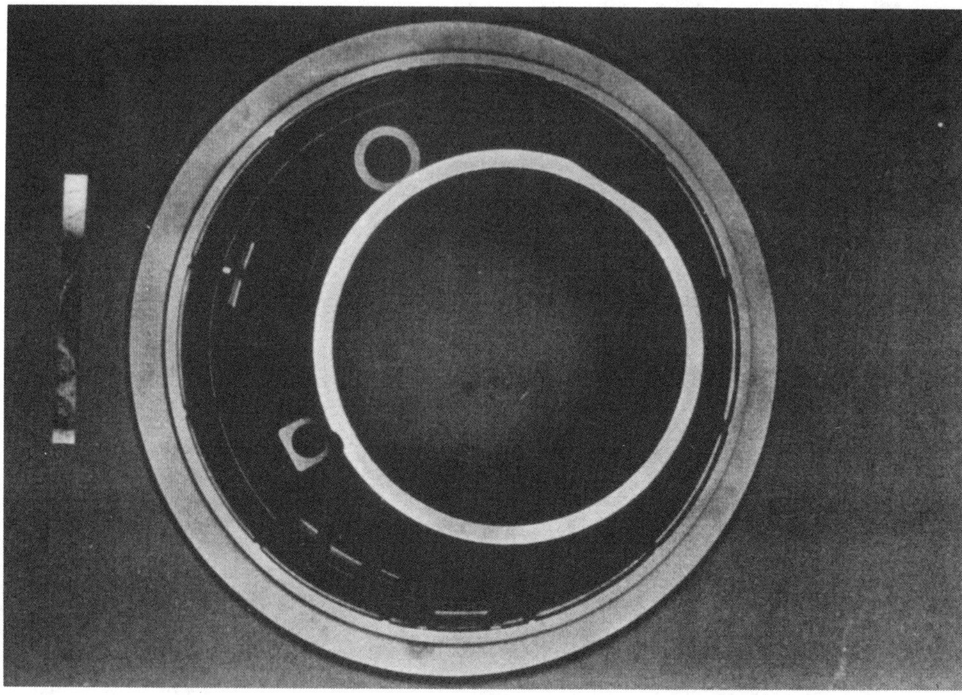

Fig.13

Fig. 14.

Fig.15

Fig. 16.

Melting, alloying and treatment of light metals

Inclusion separation by electrochemical dissolution on 3004 alloy: a reference method

Ph. Gimenez, H. d'Hondt and G. Pignault
Pechiney Recherches, CRV S.A., B.P. 27, 38340 Voreppe, France

Abstract

Based on an adaptation of steel industry practice, a new method allowing selective dissolution of the metallic matrix, leaving embedded oxides untouched, has been designed and applied for the highly inclusion sensitive body stock AA 3004 alloy. Nature, sizes and distribution of inclusions have been determined in two extreme cases of good and bad metal, and compared with another in-plant quality test. Results are in agreement with published values, and set limits on the capability of inclusion measurement systems.

Keywords

Aluminum alloy, AA 3004, body stock, quality, inclusion, dissolution, oxides, inclusion measurement.

Introduction

The continuing development of the aluminum can market in the US, and now, in Europe, has been boosted by a continuous improvement of can making technology, such as reducing sheet thicknesses and increasing drawing rates. Top performance facilities now run down to thicknesses of 305 µm and attain 300 strokes/minute. At the same time, quality requirements have become more and more stringent, due to both the highly sensitive food industry and competition between aluminum producers. It is now usual for can makers to require a rejection rate of less than 1 can per million, due to inclusion caused split flanges. This is less than 1 reject per 16 t slab. A 100 µm inclusion may be enough to cause a split-flange.

The large amount of metal handled during the process between the castshop and the final light tester at the can maker's, and the potentially dramatic effect of a bad cast house quality have generated considerable efforts among aluminum producers to develop ways of testing the inclusion content of the metal as early as possible, the earliest being when it is still liquid just before solidification.

Numerous methods were developed [1-7]; most of them were tried at the Pechiney Research Centre, CRV S.A. Although some of these methods give valuable information, it is clear that there is a nedd for an absolute reference, that would quantify the amount of inclusions directly on a piece of solidified metal. This method could then be used as a calibration for other instruments, and assess their sensitivity.

One of the simplest ideas, although not that simple to implement, is to use the different electrochemical behaviours of the aluminum alloy matrix and the oxides that are contained in it. As will be explicit further, the very low inclusion content of high quality aluminum slabs requires that large samples be tested. An electrodissolution method derived from research done for the French steel industry that leads to reasonable dissolution times has thus been chosen.

In subsequent sections, a brief discussion of the aspects of sampling and the electrochemical setting of the process is first given, followed by results on the reproducibility of the method and the inclusion type and size distribution. A comparison of the method with other inclusion tests is also made.

1. Experimental Method and Procedure

1.1 Sampling

Inclusion testing can proceed in 2 ways:

i) cut-surface examination and counting of sliced inclusions (such as in metallographic techniques);

ii) volume testing (such as in analytical methods).

Using Statistics, one can determine the size or extent of a sample required to obtain a certain level of precision regarding the real number of objects being counted. This is essentially a variation of the law that a measurement of N discrete items is a Gaussian distribution whose standard deviation is $\sqrt{N}$.

For a monodispersion of spheres with a diameter d, if p is the surface or volume fraction of the dispersed second phase (inclusions), the required surface (S) or volume (V) to test for a given relative precision ε reads [8-9]:

$$S = \frac{4}{5} \frac{\pi d^2}{p \varepsilon^2}$$

or

$$V = \frac{2}{3} \frac{\pi d^3}{p \varepsilon^2}$$

As an example, Table 1 gives the side of the square of a cut or the mass of metal one must test, as a function of surface or volume fractions and desired relative precision, for 30 μm diameter inclusions.

TABLE 1

Precision \ Volume fraction ppm	0.1	1	10
10%	1.5m/151.2 g	48 cm/15 g	15 cm/1.5 g
30%	50 cm/ 17 g	16 cm/1.7 g	5 cm/0.17 g
60%	24 cm/4.2 g	8 cm/0.42 g	2.5 cm/0.042 g

Figures 1 and 2 illustrate the same relations. For a given precision of 20%, the required surface or volume of a sample can be read directly as a function of inclusion fraction, depending on the average inclusion diameter.

For high quality aluminum alloys, the level of oxides at the extreme end of the process (settling, AlPur treatment, filter) is much lower than 1 ppm. Dissolution of samples of the order of 100 g weight is thus required to yield some precision in the measurement.

Another calculation of the required sample size was done by Simensen [10], based on a different hypothesis of the inclusion distribution. The formulae derived give surfaces or weights of the same order of magnitude as the simple ones presented here.

1.2 Dissolution method

1.2.1 Basics

Several dissolution methods have been developed for similar applications, using solvents such as brominated methanol [11] or a mixture of hydrochloric and nitric acids [12]. The first one leads to a dissolution time of over two weeks, and involves a large amount of very dangerous products and has been disregarded. The second one actually dissolves oxides such as MgO, which we considered to be a major drawback for isolating inclusions in magnesium containing alloys.

An electrochemical dissolution cell, developed by the IRSID for steel inclusion isolation, was bought [13], as this system would give the shortest dissolution time. For steel, the potential had to be kept constant during the experiment to avoid dissolving searched inclusions; these considerations do not apply to aluminum oxides that are totally inert and we decided to use the system with a fixed current density, rather than a fixed potential.

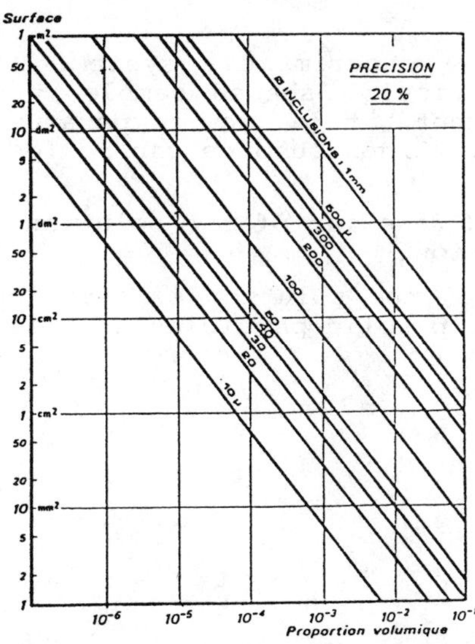

Figure 1: Surface of metal required to be examined as a function of average inclusion diameter and volume fraction [13]

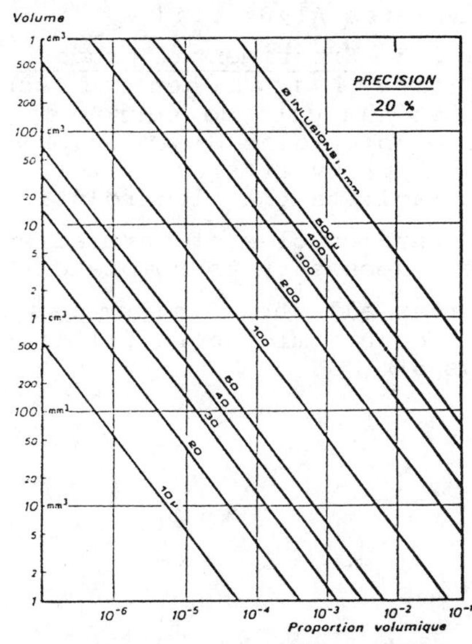

Figure 2: Volume of metal required to be examined as a function of average inclusion diameter and volume fraction [13]

1.2.2 Electrolyte setting

As can be seen in Figure 3, aluminum ions have a high solubility in both acidic or basic media. Thus either of them could be used.

We next examined the action of these two media on the expected inclusions. Numerous tests were done on synthetic oxides and poly-oxides such as spinels (Mg_2 Al04). These tests showed that spinel dissolution in an acid such as HNO3 is considerable: a 10-15% weight loss was obtained in about 3 h, whereas it was less than 0.1% in a sodium hydroxide solution. The latter was then chosen as the main component of the electrolytic solution. Other species like depassivating ions and a complexing agent were added to improve the dissolution and to prevent any oxide precipitation.

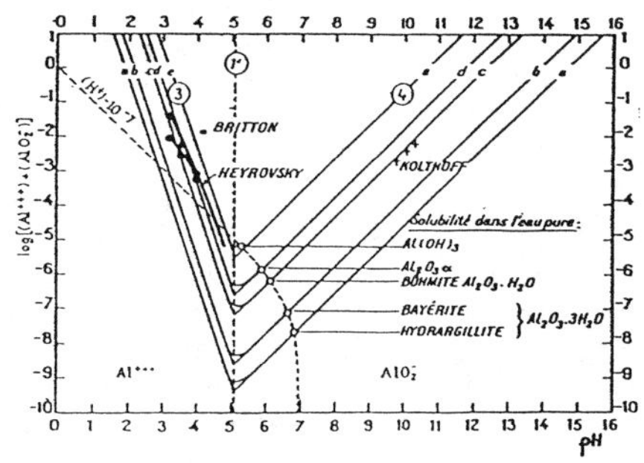

Figure 3: The influence of pH on the solubility of Al_2O_3 and hydrates Al_2O_3 [14]

1.2.3 Experimental procedure

Figure 4 is a general scheme of the experimental system. A platinum cathode surrounds the cylindrical - shaped sample, and the plexiglas dissolution cell is kept at a low temperature (~ 5°C) by external water circulation. Argon bubbles can be fed to agitate the electrolytic liquid.

Several cycles of partial emptying and renewing of the solution are necessary to complete the dissolution.

After all the aluminum has disappeared, the solution is filtered through a micropore filter that can stop solid particles as small as 5μm.

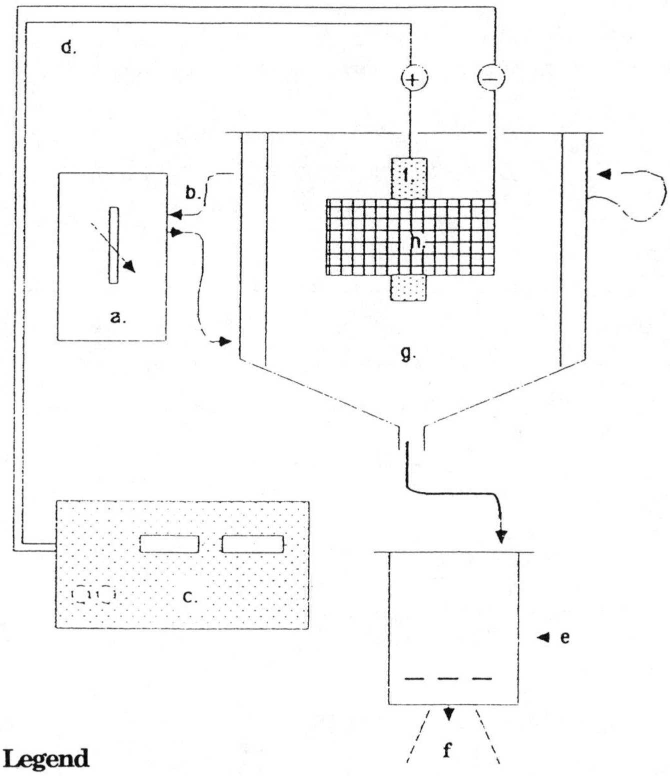

Legend

a Cell cooling system
b cooling fluid circulation
c dull current generator
d electrical circuit
e filter
f evacuation
g dissolution cell
h cathode
i aluminum alloy sample

Figure 4: General scheme of the setting

Mg and Mn muds, formed during the dissolution have to be eliminated by an appropriate chemical treatment.

The filter is then soaked and calcinated at high temperature.

Precide written procedures were established; following these procedures leads to the isolation of distinct individual oxide particles that can be examined and analyzed under a Scanning Electron Microscope.

2. Results

2.1 Reproducibility

For all these trials we used metal from two sources:

i) A: as a reference, giving an acceptable metal related split-flanges rate;

ii) B: as a "bad" metal, giving a substantially higher rate.

Table 2 shows how reproducible the counting of isolated inclusions was:

TABLE 2

	Number of particles of Ø >20μ isolated in 100 g of metal			
A metal 4 determinations	57	33	40	50
B metal 3 determinations	558	612	600	--
Blank (no sample) 2 determinations	0	2	--	--

These numbers can be considered as measurements of parent distributions:

A metal: 45±10 inclusions/100 g of metal
B metal: 590±25 inclusions/100 g of metal

Average diameters were also very similar from one run to another:

TABLE 3

	Average diameters of the inclusions of Ø >20μ (μm)	
	1st trial	2nd trial
A metal	35±5	35±6
B metal	63±10	51±7

(confidence level 95%)

2.2 Kind of inclusions observed

As expected, the inclusions isolated were mostly oxides. Different stoichiometries of the 3 basic constituents:

i) Al
ii) Al-Mg
iii) Al-Mg-Si

were detected, through EDAX analysis.

Pictures (Figures 5, 6, 7) representing all of these types of particles show that some contamination of bayerite remained; it could be distinguished from inclusions (and inclusion counting) by the observed morphological characteristics: the oxides are small massive particles, whereas the contaminant is a sort of fiber system.

Very small (~1 μm) Ti rich particles were found to spot the surface of the oxides; they are most likely TiB_2 grain refiner particles.

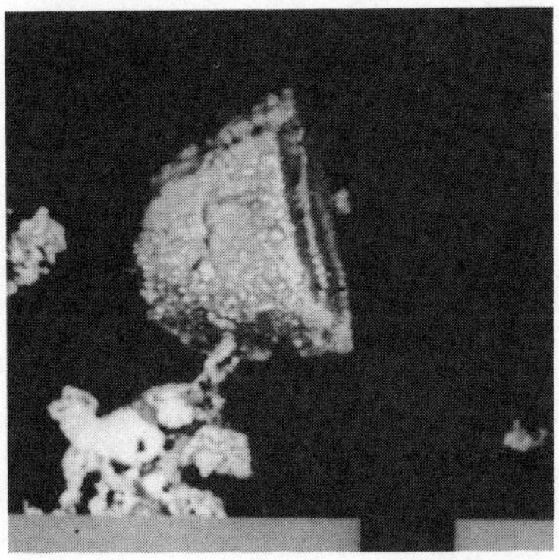

Figure 5: Al-Mg oxide particle, spotted by Ti rich particles (x 250)

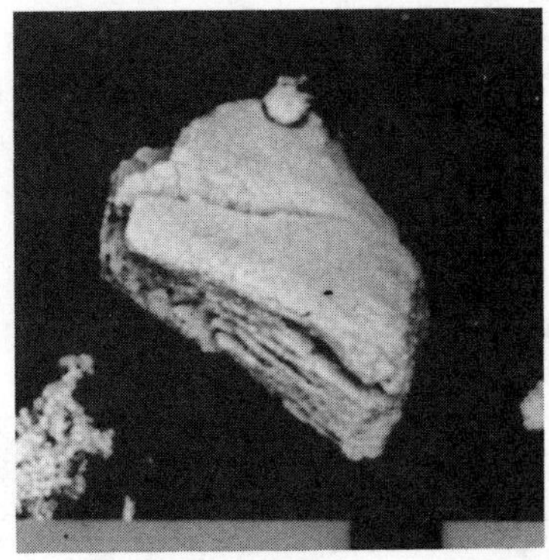

Figure 6: Complex Al-Mg-Si oxides (x 250)

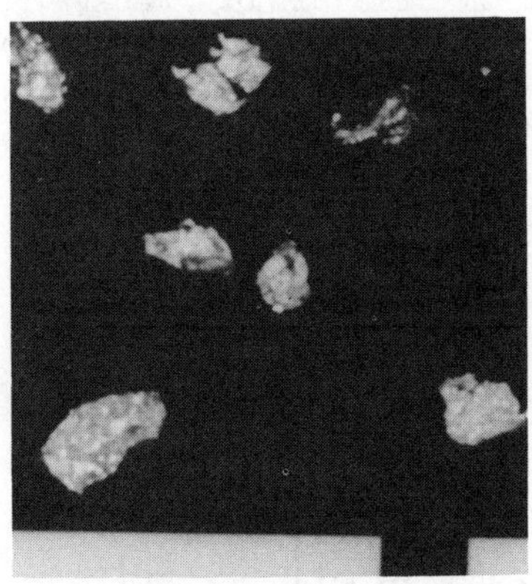

Figure 7: General view of Al and Al-Si oxides (x 100)

2.3 Size distributions of inclusions

Figures 8 and 9 show typical size histograms for both A and B metal. It is clear that B metal contains not only more, but also larger inclusions. Distribution functions are presented in Figure 10; these functions are known to converge more quickly than single histograms to the parent function. It is to be noted that, on a logplot, we observe 2 parallel lines. This shows that the distribution functions are exponential in both cases, with the same exponent.

Fits to these curves yield:

good metal:

$$N\ (\emptyset > \emptyset o) = 105\ e^{-0.058\ \emptyset o}$$

(number of inclusion per 100 g of metal with diameter > Øo)

bad metal:

$$N\ (\emptyset > \emptyset o) = 1320\ e^{0.043\ \emptyset o}$$

precision: 105±50, 1320±160
0.058±0.025, 0.043±0.003

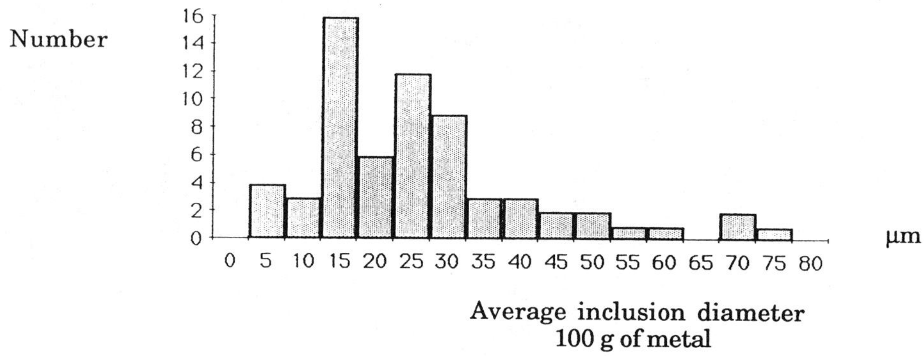

Figure 8: Size histogram of inclusion in A metal

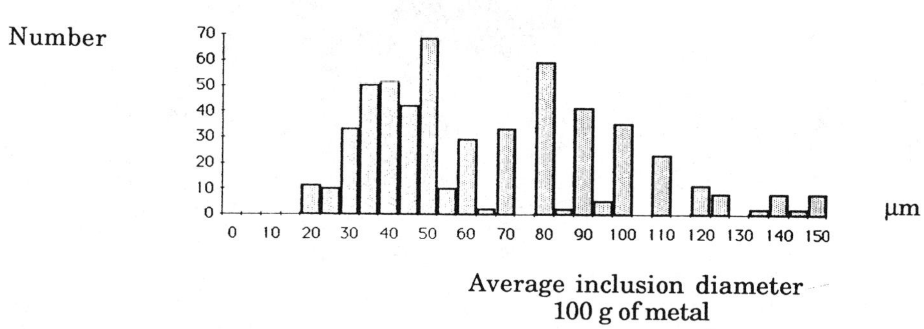

Figure 9: Size histogram of inclusion in B metal

Number of inclusions per 100 g of size > x µm

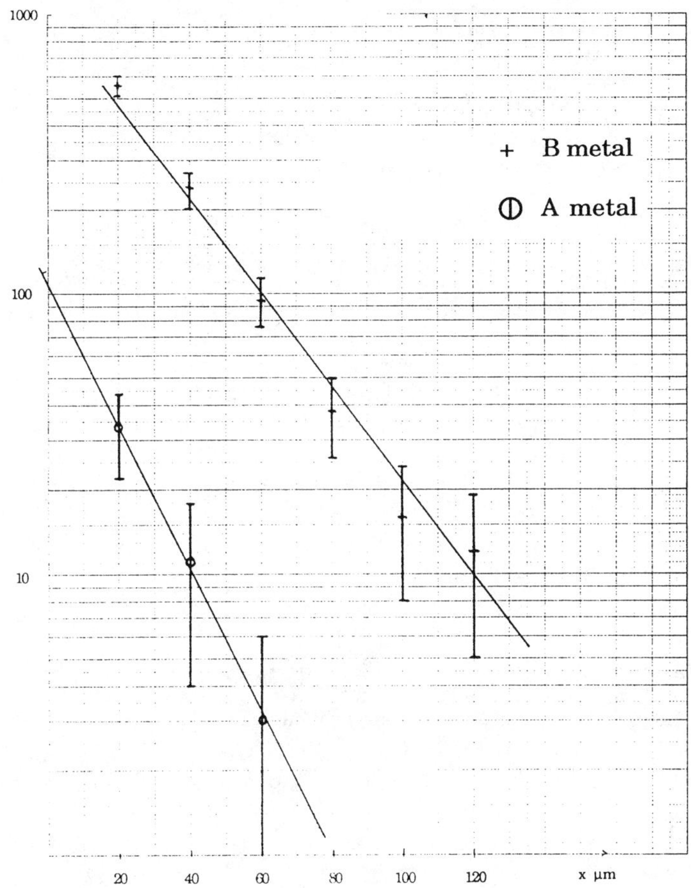

Figure 10

3. Comparison with other tests

3.1 Direct observation of harmful inclusions

As was mentioned in 2.1, the B metal came from a coil that gave excessive split-flange rates. Direct observation of the fracture surface on split-flanges revealed the presence of the same kind of inclusions that were isolated by electrodissolution.

3.2 Oxygen content determination

CRV operates a 14.6 MeV neutron generator [15], which offers great possibilities for O content determination, through the ^{16}N reaction. Measurements on the two original metals yielded:

A metal: O<0.5 ppm
B metal: O=2.2±0.4 ppm

From the distribution curves showed in 2.3, it is possible to calculate an oxygen level, assuming a specific stoichiometry for the Si, Mg and Al oxides.

For all these oxides, the ratios of oxygen mass to the total mass are very similar:

Al_2O_3 :	0.47
SiO_2 :	0.53
MgO :	0.4
$MgAl_2O_4$:	0.45

Taking a 0.5 ratio as typical, oxygen concentrations deduced from inclusion spectra are thus:

A metal: ~0.1 ppm
B metal: ~2.8 ppm

As this oxygen content depends mainly on the number of large inclusions (~60 μm), the B/A ratio (28) is much higher than the same ratio for total inclusion content (~10).

3.3 In-plant-test

Results of electrochemical dissolution can be compared with the results of a routine production test, consisting of counting surface defects on the sheet after a special chemical treatment (razor-streak type).

On A metal, no defect was seen, and statistical considerations show that this result give us a [0,5] confidence range.

For B metal, on which the test was done after, a number of 54±5 defects could be counted.

From this very limited set of data itself, the in-plant and electrodissolution tests are seen to give correlating answers.

3.4 Values obtained from literature

[1] and [16] quote values for the number of inclusions having sizes larger than 20 μm, measured with Alcan's LIMCA.

[1] : 50-100 inclusions >20 μm/100 g after Deep Bed Filter.
[16]: 4000 inclusions >20 μm/100 g at the exit of a casting furnace.

If one adds an AlPur (90% removal efficiency) and a filter (50%), that number is reduced to around:

$$4000.\ (1-90\%).(1-50\%)\ \sim 200$$

Both numbers fall within the range of what we measured by electrodissolution.

Conclusion

The numerical and morphological considerations in section 3 support the idea that the electrodissolution of 3004 samples gives us access to the true oxide content of the metal. It is clear that routine production inclusion levels are lower than 0.1 ppm and that, above 20μm, the size distribution of inclusions is an exponential shaped curve.

Big inclusions (100 μm) are very infrequent, and all inclusion testing systems have to take into account small harmless particles in order to yield a non-zero answer, even if several kilograms of metal are involved. The confidence that they can deliver a quality related answer lies in the belief that, if there are more small inclusions, then there have to be more large ones, too.

As most treatment devices (filter, degassing box) or procedures (settling) work in a proportional mode, this may sound reasonable at first. However, two objections must be raised:

i) most treating devices are very sensitive to the size of the inclusions to be eliminated. Efficiencies can be very different for large and small particles [17];

ii) it is not known if inclusion creation (metal oxidation, alloying, interaction with refractories) also generates such populations.

These heuristic reflections should lead us to be very careful when using inclusion sampler responses. I do not believe that a device capable of indicating the quality of a slab while still in the liquid state exists anywhere, and will not in the coming years.

Progress will proceed by using technologies available today to better understand inclusion behaviour and by carefully following customers' feed-back regarding quality. This is a long term work program that will require good cooperation between research, production and sales sectors, but then, has there been a field in which working together did not yield greater interest and benefit?

References

1. D. Doutre, B. Gariépy, J.P. Martin and G. Dubé, "Aluminium Cleanliness Monitoring: Methods and Applications in Process Development and Quality Control", (Paper presented at the 114th AIME Annual Meeting, New York, February 1985) 1179.

2. S.A. Lévy, "Applications of the Union Carbide particulate tester", Light Metals 1981, p. 722-733.

3. D.A. Bates, L.C. Hutter, "An evaluation of aluminum filtering systems using a vacuum-filtration sampling device", Light Metals 1981, p. 707-721.

4. J.F. Grandfield, D.W. Irwin, S. Brumale, C. J. Simensen, "Mathematical and physical modelling of melt treatment processes", Light Metals 1990, p. 737-745.

5. C.J. Simensen, "Sedimentation analysis of inclusions in aluminum and magnesium", Metallurgical Transaction B, Vol. 120 B, p. 733-748. (1981).

6. Buxmann, Furrer, "Zum problem der "Grauen Zeilen" auf anodish oxidierten aluminium - produkten", Metall 34, (3), p. 222-228, Mar 1980 (8101).

7. Mansfield, "Molten metal quality measured with Reynolds 4M system", Light Metals 1984, p. 1305-1327.

8. J. Serra, "Introduction à la morphologie mathématique", Les cahiers du Centre de Morphologie Mathématique de Fontainebleau, 1969.

9. T. Hersant, D. Jeulin, "L'échantillonnage dans les analyses quantitatives d'images - Exemples d'application aux mesures de teneurs de phases dans les agglomères et des inclusions dans les aciers IRSID, p. 241 - Octobre 1975.

10. C.J. Simensen, "Sampling and analysis of impurities in aluminium", International Seminar on Refining and Alloying of Liquid Aluminium and ferro-alloys, August 26-28, 1985,

Trondheim, Norway.

11. G. Kraft and A. Kahles, Erzmetall 23 (1976) 365.
J. Fisher, W. Kästner, H. Aschehoug, GieBerei 58 (1971) 680.

12. C.J. Simensen, G. Strand, "Analysis of inclusions in Aluminium by dissolution of the sample in Hydrochloric/Nitric Acid, Fresenius Z, Anal. Chem. 308, 11-16 (1981).

13. D. Henriet, P. De Gelis, "European Community Report", 7210/GA322 Eur 8885 FR - Steel Technical Research.

14. M. Pourbaix, Atlas d'équilibre électrochimique à 25°C.

15. G. Beurton, R. Pillon, "Experimental improvements of the sensitivity of neutron activation analysis for oxygen in metal", Journal of Radioanalytical Chemistry, Vol 40 (1977), p. 189-201.

G. Beurton, "Determination of nitrogen, oxygen, fluorine, phosphorus, sulfur and chlorine in aluminum alloys by means of neutron activation", Ibid, Vol. 77, no 1 (1983), p. 123-139.

16. J.P. Martin, G. Dubé, D. Frayce, R. Guthrie, "Settling phenomena in casting furnaces: a fundamental and experimental investigation", Light Metals 1988.

17. J.P. Desmoulins, H. d'Hondt, J.M. Hicter, P. Netter, "Which filter for what plant? The Pechiney research approach", Light Metals 1989.

Grain refinement with a boron-free master alloy

P.J. Read
Anglo Blackwells Limited, Widnes, Cheshire, England

Abstract

Current commercial grain refining practices and some of the problems associated with them are briefly reviewed.

There is a requirement in some critical applications for an effective, non-boron containing master alloy which will reduce downstream fabrication problems such as razor streaks, pin-holes and hard spots, which are often associated with TiB_2 particles from the AlTiB grain refiners now widely used. Work carried out at Anglo Blackwells Limited, and Liverpool University, has shown that effective binary AlTi6 rod master alloys can be consistently produced. Results of Alcoa grain refining tests on a variety of commercial aluminium alloys demonstrate the effectiveness of this test procedure and the AlTi6 rod master alloy for some of these critical end uses.

Key Words

Grain Refining: AlTi6 Rod: Boron free: Aluminium alloys: Alcoa test

Introduction

Grain refining of aluminium alloys using AlTiB or AlTi master alloys is now a well established practice worldwide. Additions of low levels of these master alloys, typically 0.5 kg to 2 kg per tonne of aluminium, in either waffle plate or 9.5 mm rod form, eliminates columnar and twinned-columnar grains and gives a fine equiaxed structure with the following advantages in D.C. casting:

- faster casting rates before hot cracking is encountered
- improved tensile and mechanical properties
- improved surface finish

In shape castings there are other benefits notably better distribution of intermetallics and improved feeding to reduce shrinkage porosity.

Master alloys containing titanium and boron (with various Ti:B ratios such as 5/1, 3/1, 5/.2 and 10/.4) are generally acknowledged to be more effective grain refiners than binary AlTi alloys (containing typically 5, 6 or 10% Ti). However, the presence in AlTiB alloys of insoluble TiB_2 intermetallics, which are retained in the final cast structure, are undesirable in certain critical applications such as foil, canstock, architectural, lithographic, copy drums and aerospace. In addition, as molten metal filtering becomes both more common and demanding, AlTiB alloys can cause premature filter plugging.

There have been major improvements in recent years in processing and quality control procedures by the major master alloy producers, which have greatly reduced the incidence of large TiB_2 defects and it is generally claimed that the majority of TiB_2 defects are less than 2 microns, with a maximum size of 5 microns. A detailed study by Schneider (1), however, on a large number of commercially produced AlTi5B1 and AlTi3B1 alloys, has shown that TiB_2 particles with sizes in the range 8 microns to 20 microns can sometimes be found.

Thus, there is an increasing demand in the industry for an effective, non-boron containing master alloy. Anglo Blackwells have developed a more effective boron-free AlTi6 alloy in rod form which can be used to grain refine certain critical commercial alloys.

Mechanisms of Grain Refining

The mechanism of grain refinement in aluminium alloys with AlTiB or AlTi master alloys is still a subject for dispute. There are several theories including the peritectic theory (2), the carbide/boride theory (3) and, more recently, a 'hypernucleation' theory (4,5). The latter gives a more satisfactory explanation of many of the observed aspects of AlTiB grain refining. It is not part of this paper to review these theories in detail, but readers are referred to a review of the subject recently published by McCartney (6).

What is clear is that grain refining of commercial aluminium is achieved by the presence of very effective heterogeneous nuclei which can nucleate alpha aluminium grains at small undercoolings below the liquidus temperature. Growth conditions are then important to prevent rapid growth of a small number of grains.

In AlTiB alloys the heterogeneous nuclei probably consist of TiB_2 or a more complex $(AlTi)B_2$ phase (7,8), but in AlTi alloys the nature of the heterogeneous nuclei which must exist is much less clear. Early theories (8), suggested that the morphology of the $TiAl_3$ phase controlled the grain refining effectiveness of the master alloys - 'blocky' $TiAl_3$ being more effective than plate-like or flaky-type $TiAl_3$ particles. In a series of publications (9,10,11) Hadlet, McCartney and Thistlethwaite showed that this was clearly not the case and effective grain refining could be achieved with all kinds of $TiAl_3$ morphologies. The processing route was the critical factor in developing potent heterogeneous nuclei in AlTi binary alloys. Work is continuing to fully identify the effective nuclei in these AlTi alloys, but they are believed to be sub-micron particles of either titanium carbide or titanium oxide.

It should also be noted that titanium in solution in an aluminium melt is one of the most effective elements in providing constitutional effects to limit growth of nucleated grains (12).

Commercial use of AlTi Alloys

Commercial AlTi master alloys are available in a number of forms. The most common ones are AlTi6 or AlTi10 in waffle ingots and AlTi6 in 9.5 mm rod. The use of these alloys depends partly on the end product being treated, but there are also significant preferences in different geographic areas. Typically in Europe, AlTi10 waffle, often based on secondary aluminium is used to give a low cost titanium addition to foundry alloys. A higher quality AlTi6 or AlTi10 waffle is used to grain refine foilstock or to increase titanium base levels before adding AlTiB rod in hard alloys (7xxx series) by certain producers. AlTi6 rod is not used widely in Europe and Japanese producers tend to follow similar practices.

In the U.S.A., however, much larger quantities of AlTi6 rod are used. Again, applications vary from producer to producer, but include use on hard alloys (7xxx and 2xxx), architectural products (5xxx series), canstock (3xxx series) and foilstock (1xxx). Addition rates are normally higher than when using AlTiB alloys, but this is considered to be less important compared to the advantages in reducing TiB_2 related problems in downstream fabrication such as razor streaks, pin-hole defects and hard spots.

Future trends in grain refining practice are likely to be either:

i) To improve the quality of AlTiB alloys so that large individual TiB_2 particles and the occurrence of TiB_2 defects associated with oxides/flux inclusions will be at such low levels that downstream quality will not be adversely affected (so that, for example, ceramic tube filters can be used with the grain refiner added after the filter), or

ii) To enhance the grain refining effectiveness of binary AlTi alloys to such an extent that they can be used effectively at similar addition levels to AlTiB alloys.

Quality Control Procedures

The main quality criteria demanded (13) from any grain refiner should be grain refining effectiveness and metal cleanliness (i.e. freedom from large oxides and insoluble hard particles such as borides, carbides, etc.). Chemical analysis, metallurgical structure, rod diameter and mechanical properties should all be controlled but are generally of less importance in determining the performance and true value of the master alloy. Measurement of all these parameters is an important part of the master alloy producers' quality control procedures.

There are a number of laboratory grain refining test procedures, including those developed by Alcan, Alcoa, L.M.G., KBI, Reynolds, VAW and, for AlSi foundry alloys, the thermal analysis technique. The test which is most commonly used in the wrought aluminium industry for quality control purposes is the Alcan test (14), and this test is used by Anglo Blackwells on every heat of AlTi6 rod.

Statistical Process Control (S.P.C.) is also used during the production of this and other alloys to control critical stages of the process and to improve the overall product quality in terms of grain refining, metal cleanliness, chemical analysis and rod diameter. Improvements in the vital areas of grain refining and micro oxide levels which have been achieved over the past four years are shown in Figures 1 and 2. It is believed that these levels of grain size are far superior to those of other commercially available, non-boron containing AlTi6 alloys, and are the direct result of special process techniques developed by Anglo Blackwells.

Whilst the Alcan test is a very useful laboratory scale Q.C. test, the thermal gradients, cooling rates and fluid flow characteristics are not typical of large scale D.C. cast billets or slabs. The Alcoa test (15) is believed to be more representative of these conditions and can be used to predict grain structures, columnar to equiaxed transition points and especially conditions which suppress feather crystals (twinned columnar growth). The two test moulds are shown schematically in Figures 3 and 4. A detailed analysis of the solidification conditions in the Alcoa test and its use on a range of commercial aluminium alloys has recently been carried out (16).

Using the Alcoa test under the specific conditions applicable to a particular producer (i.e. alloy composition, trace element levels, casting temperature, residence time, grain refiner type and addition level) allows reasonably accurate predictions and recommendations to be made to achieve the desired results. The predictions obviously need verifying with plant trials. Typical examples of this approach are given in the following section.

Test Results on Commercial Alloys

Grain refining is generally more difficult to achieve in pure-based alloys and in large cast sections (i.e. those with slower cooling rates). Also, the presence of certain elements, notably Zr and Cr, are known to 'poison' AlTiB grain refiners (14) probably by some form of chemical reaction with the TiB_2 intermetallics. Ingot cracking is more pronounced with faster casting speeds and in certain high strength alloys. The high thermal stresses set up during solidification and cooling can cause centre line cracking or hot tearing in the shell zone.

Ahmady, McCartney et al (16) have shown that the results from Alcoa tests on a range of commercial alloys agree well with theoretical predictions for the influence of phase diagram parameters on suppression of columnar growth.

AA7050

AA7050 is normally considered difficult to grain refine due to the poisoning effect of zirconium and has a pronounced tendency towards hot tearing. It is probable that very fine grain sizes promote hot tearing due to easier crack propagation paths (17) and ideally a relatively coarse but uniform grain structure, free from feather crystals is preferred.

Grain refining results showing the effect of Anglo Blackwells AlTi6 rod on AA7050 are shown in Figure 5. The photomicrographs indicate that uniform grain sizes of between 150 microns and 200 microns, free from feather crystals can be achieved with an addition rate of 0.01% Ti. The effect of contact time (time after rod addition to solidification) is critical in these alloys and Figure 6 compares AlTi6 rod with AlTi5B0.2 rod at holding times of 1 and 15 minutes. This shows that the grain refining effect is reduced with increasing time, after addition, but to a lesser extent for the AlTi6 compared to the AlTiB alloy.

AA5657

This alloy is based on higher purity (99.85%) aluminium and is used in various automotive applications in a chemically brightened and polished condition. Razor streaks are a major problem and it is common practice to grain refine the alloy with a boron-free master alloy such as AlTi6 rod. The low impurity content, especially iron, in the base metal makes the alloy relatively difficult to refine adequately and thereby eliminate feather crystals and prevent ingot cracking without very high titanium addition levels. High titanium levels are, however, undesirable since they tend to reduce brightness and reflectivity. Alcoa tests carried out on AA5657 at addition rates of 0.01% Ti using Anglo Blackwells AlTi6 rod and that from an alternative source are shown in Figure 7. This photomicrograph clearly shows the superior effectiveness of the Anglo Blackwells AlTi6 product. These laboratory scale findings have also been confirmed in a number of plant trials.

AA6061

The AlMgSi 6xxx series alloys are generally easy to refine, and at low addition levels. For most applications too, TiB_2 related defects are not critical and typically AlTi5B1 rod is added at 0.5 kg/t.

AA6061 is a medium strength structural alloy used for ladders, masts and tubing. The grain refining treatment is therefore more critical in this alloy due to the surface finish requirements and the presence of Cr. Alcoa test results on AA6061 are shown in Figure 8, comparing AlTi6 rod with AlTi5B1 and AlTi3B1. The results were surprising in that they show AlTi6 to be superior to both AlTi5B1 and AlTi3B1. This is believed to be due to the poisoning effect of chromium on the TiB_2 in the AlTiB alloys.

AA3004

AA3004 is used in very large quantities for can body stock. It is relatively easy to grain refine due to the high Mg and Mn contents. Because of the critical nature of downstream fabrication processes, the alloy must be as clean as possible to prevent splitting at the forming stages. For this reason, extensive melt filtration is normally carried out using deep bed filter systems, ceramic foam filters or, more recently, high efficiency ceramic tube filters. AlTiB grain refiners can promote premature filter plugging which has a high replacement cost in deep bed filters and can prevent the use of ceramic tube filters. AlTi6, free from coarse borides could reduce filter plugging problems.

Alcoa tests carried out with AlTi6 rod on AA3004 are shown in Figure 9. These results indicate that satisfactory grain sizes can be achieved at cooling rates comparable to large slab castings at titanium addition rates of 0.005% to 0.01%.

Conclusions

1. AlTi6 rod is an effective grain refiner at low addition levels in certain alloy systems.
2. The use of AlTi6 rod can eliminate boride problems associated with the use of AlTiB alloys.
3. The Alcoa grain refining test is a useful laboratory technique in predicting results in full scale D.C. castings.

References

1. W. Schneider. 'Aluminium', 64 (1988), 157 - 160.
2. F.A. Crossley and L.F. Mondolfo. Trans A.I.M.E., 191 (1951) 1143 - 1148.
3. A. Cibula. J. Inst. Met., 76 (1949/50), 321 - 360.
4. G.P. Jones. Proc. Int. Seminar on "Refining and Alloying of Liquid Aluminium and Ferro Alloys", Trondheim, Norway (1985).
5. G.P. Jones. 'Solidification Processing 1987', London (1988, 496 - 499).
6. D.G. McCartney. Int. Materials Rev., 34 (1989), 247 - 260.
7. L. Arnberg, L. Backerud and H. Klang. Met. Technol., 9 (1982), 7 - 3.
8. M.M. Guzowski, G.K. Sigworth and D.A. Sentner. Metall. Trans., 18A (1987), 603 - 619.
9. D. Hadlet, D.G. McCartney and S.R. Thistlethwaite. "Aluminium Technology 1986", London 1986, 125 - 132.
10. D. Hadlet, D.G. McCartney and S.R. Thistlethwaite. "Solidification Processing 1987", London 1988, 141 - 144.
11. D.G. McCartney. Metall. Trans., 19A (1988), 385 - 387.
12. I. Maxwell and A. Hellawell. Acta Metall., 23 (1975), 229 - 237.
13. D.A. Granger. Light Metal Age, (June 1987), 17 - 26.
14. G.P. Jones and J. Pearson. Metall. Trans., 7B (1976), 223 - 234.
15. D.A. Granger. Proc. Int. Seminar on "Refining and Alloying of Liquid Aluminium", Trondheim (1985), 231 - 244.
16. S.M. Ahmady, D.G. McCartney and S.R. Thistlethwaite, Light Metals 1990, 837 - 843.
17. D. Warrington and D.G. McCartney. Cast Metals, (January 1990), in press.

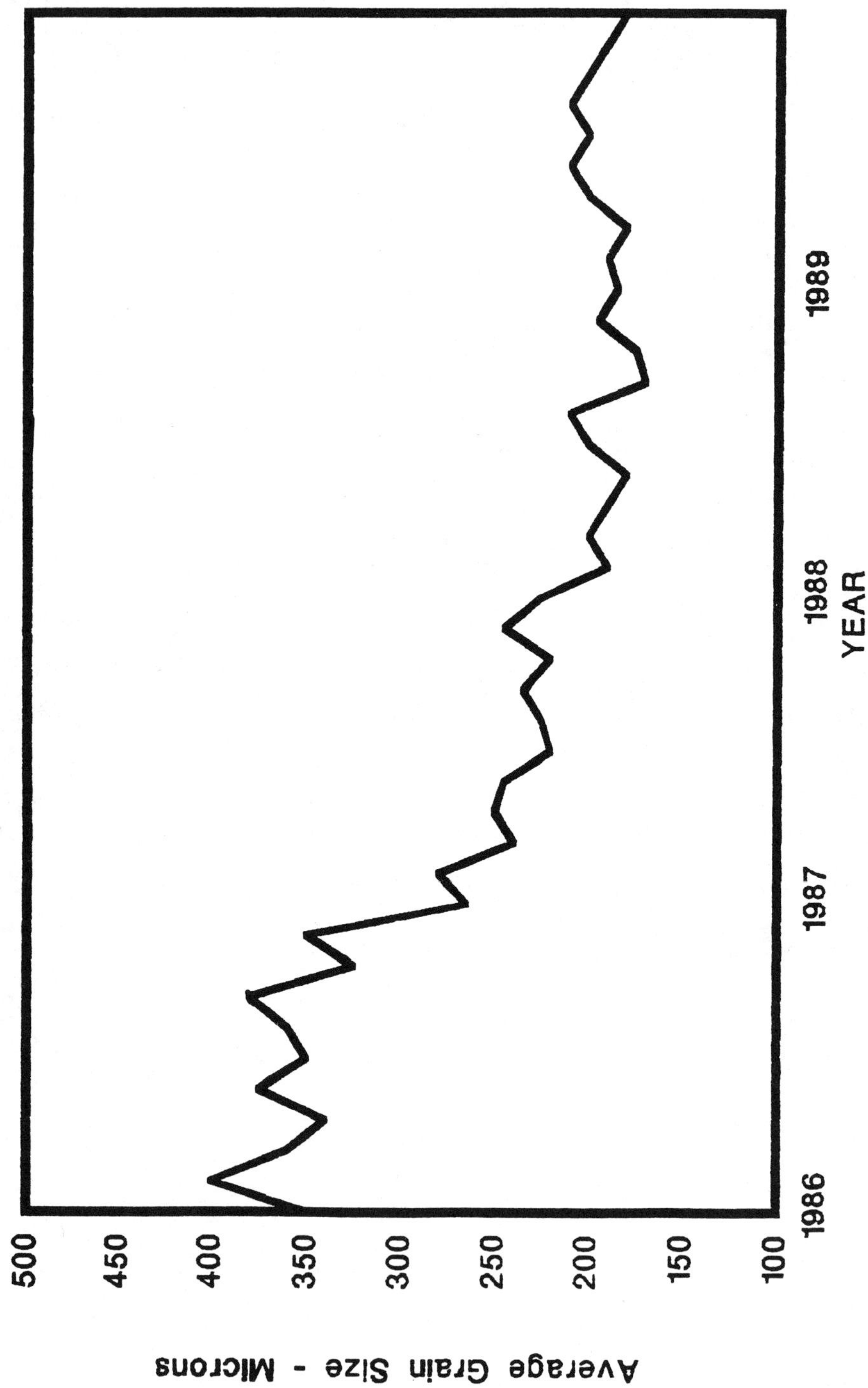

Fig.1 Improvements in Grain Size of AlTi6 Rod
Alcan Test - 0.01%Ti - 1 Minute Contact - 99.7%Al

Fig.2 Improvements in Cleanliness of AlTi6 Rod Micro Oxides seen at X50 in 1 Sq.Cm Area

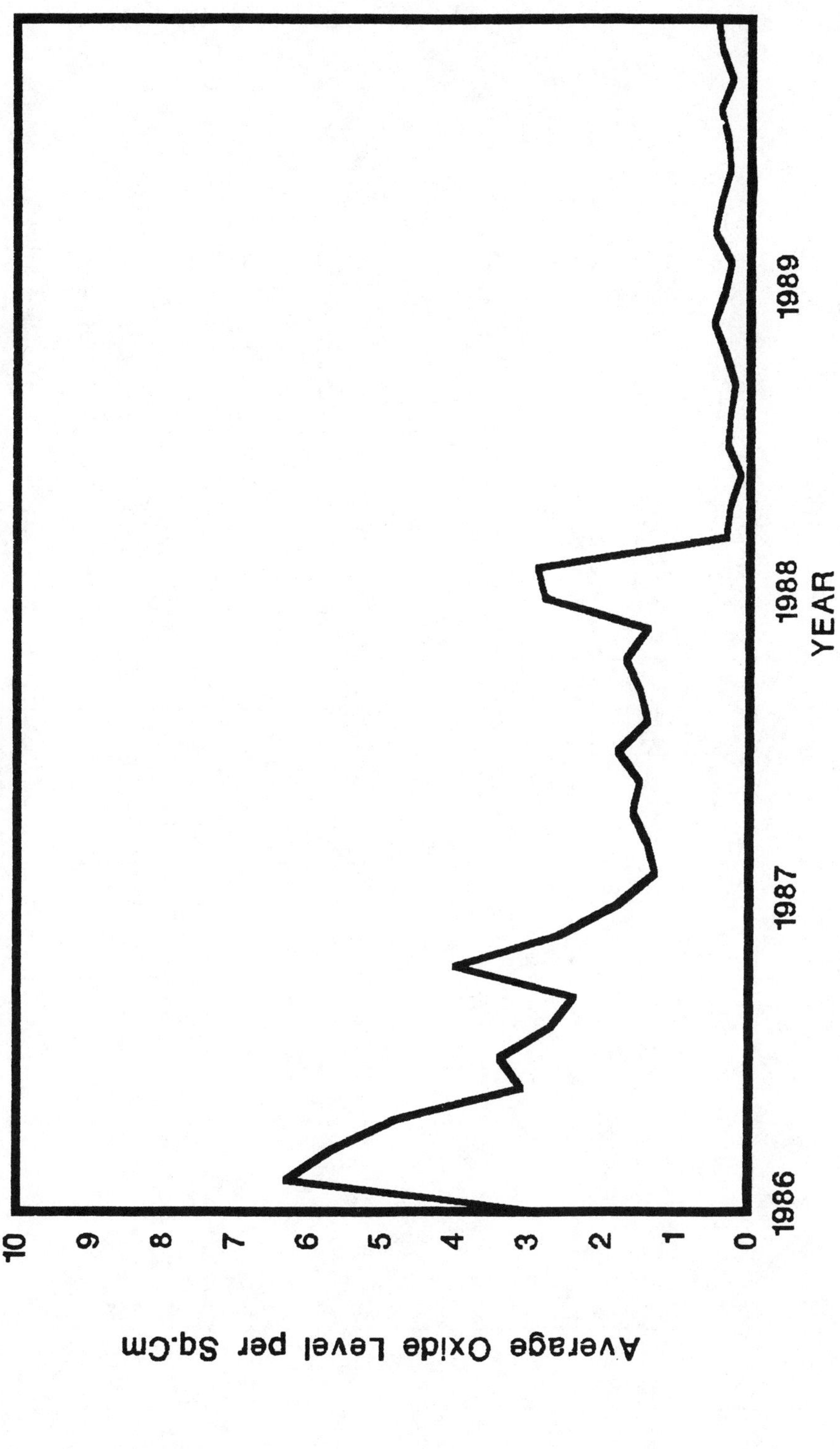

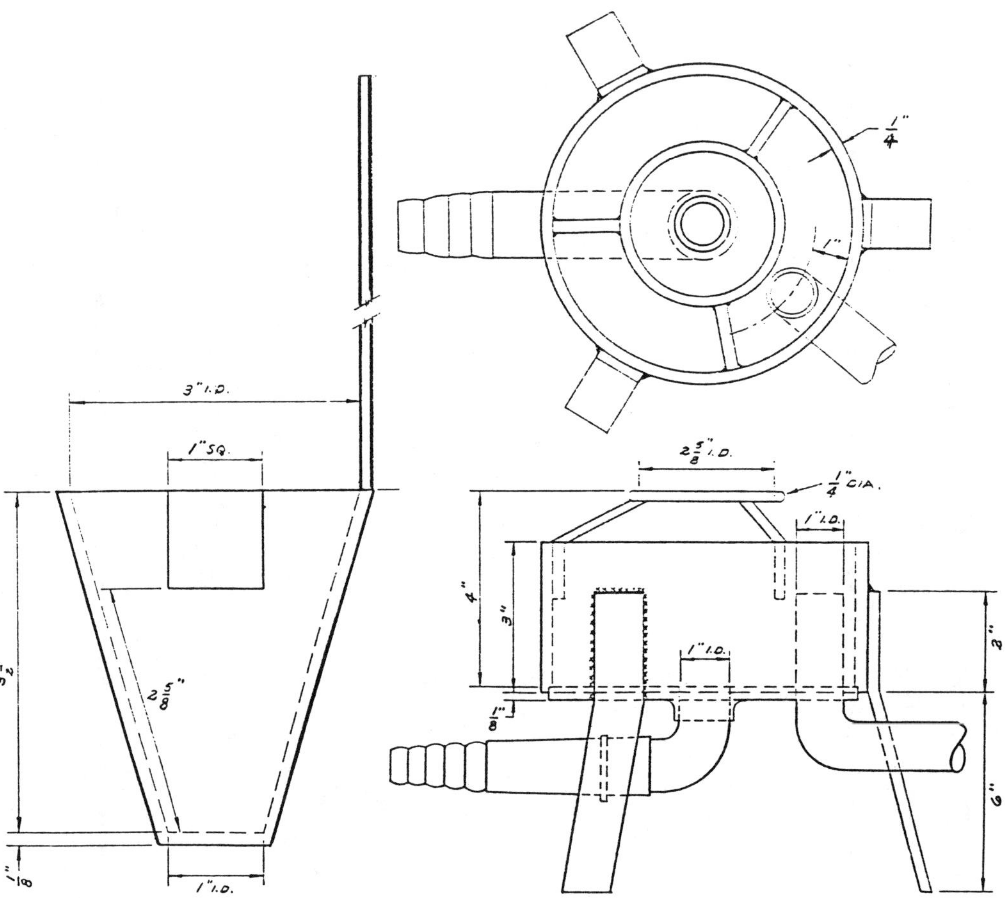

FIG.3 Schematic Illustration of the Alcan Grain Refining Test

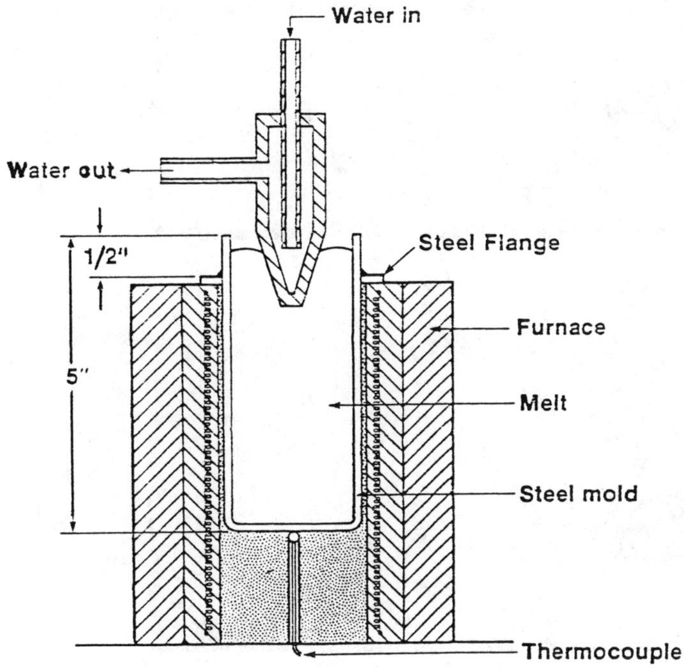

FIG. 4 Schematic Illustration of the Alcoa Grain Refining Test

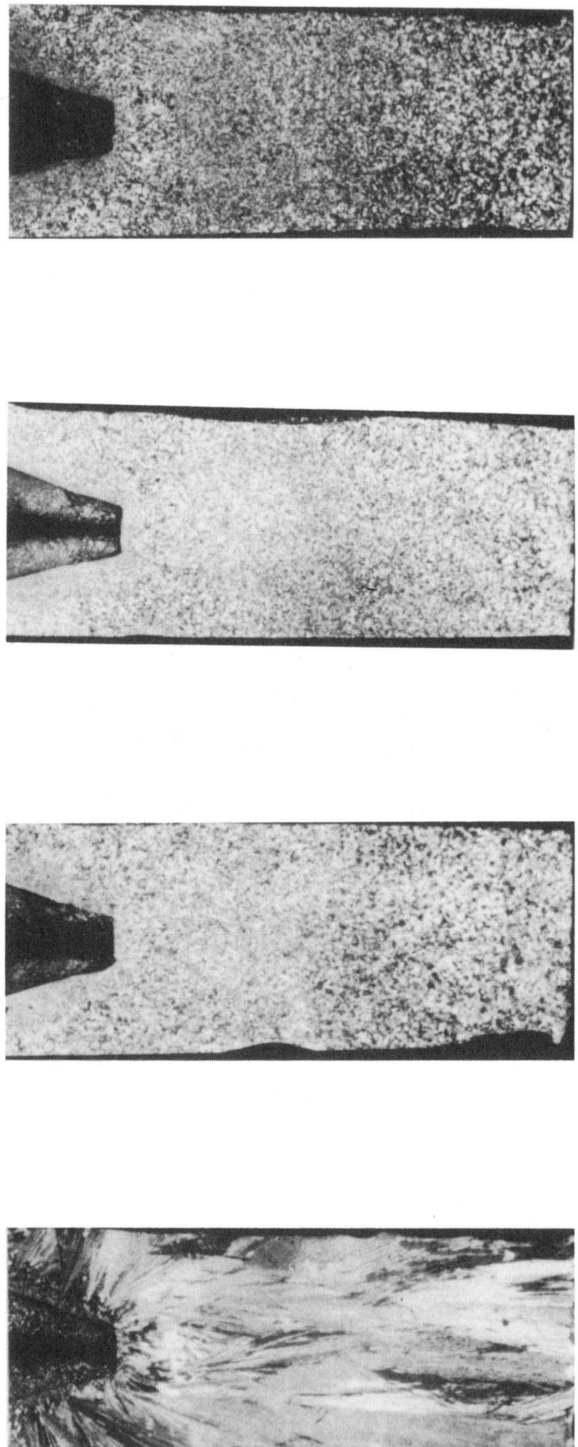

No Grain Refiner   0.005% Ti   0.01% Ti   0.02% Ti

Fig. 5 Etched longitudinal macro sections of AA 7050 alloy grain refined with AlTi6, showing the effect of increasing Ti additions

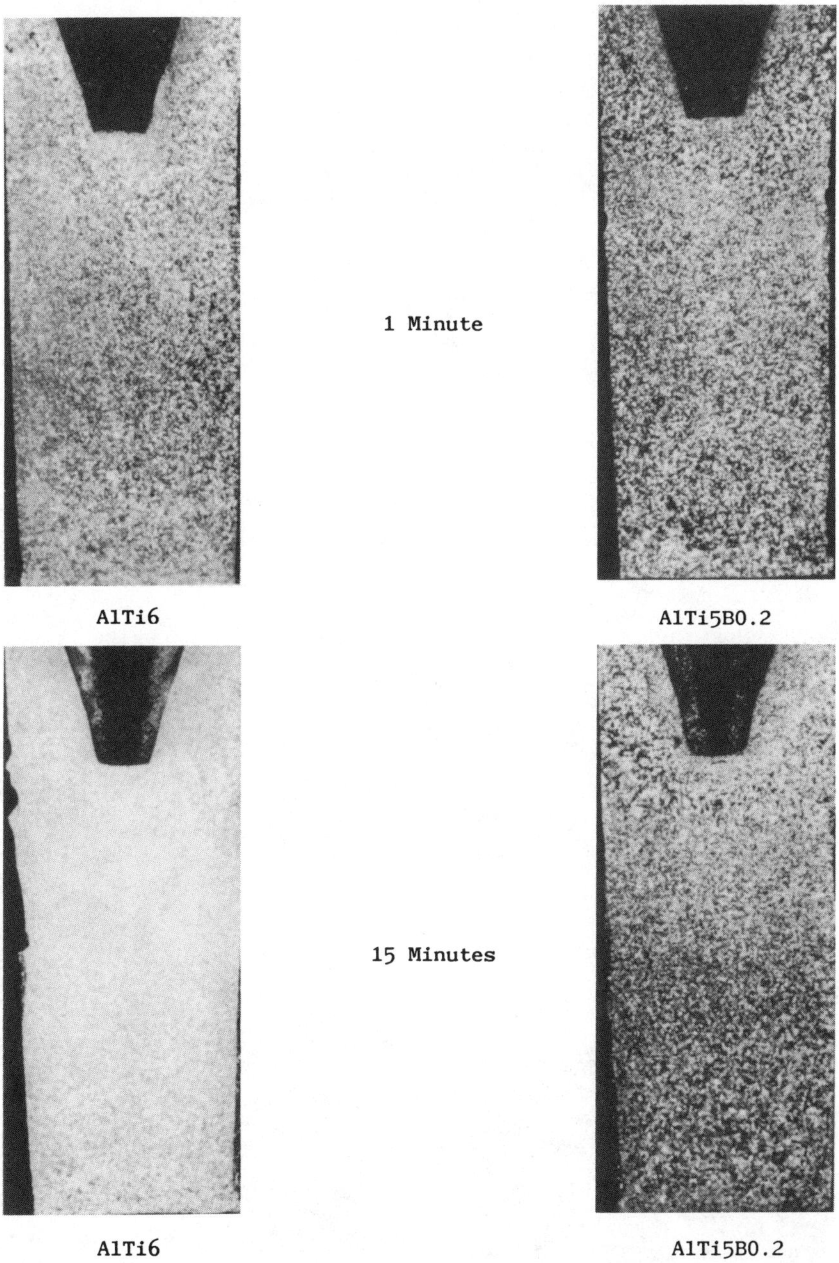

Fig:6 Comparison of grain refining efficiency at 1 and 15 minutes for AlTi6 and AlTi5B0.2. Alloy 7050, Ti addition level 0.01% Ti in all cases

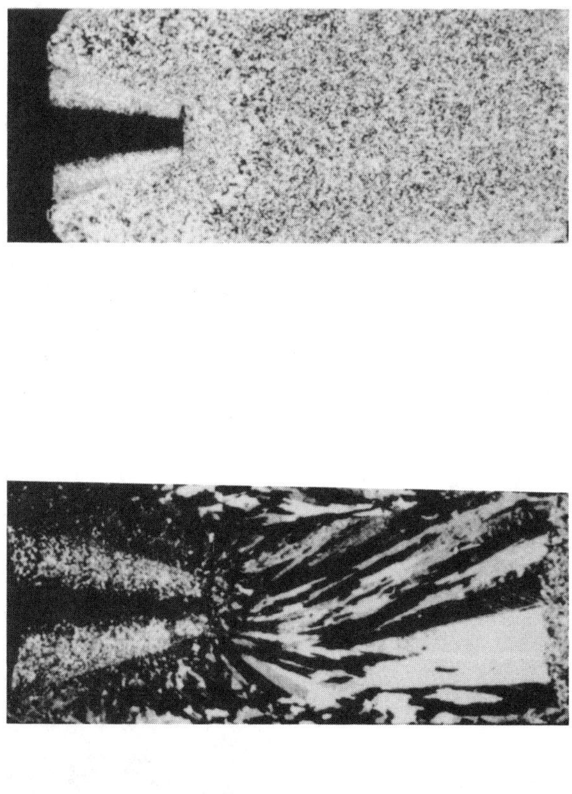

No Grain Refiner

Supplier "x" AlTi6

Anglo Blackwells AlTi6

Fig. 7 Etched longitudinal section of AA 5657 comparing Anglo Blackwells AlTi6 rod with an alternative supplier at the same addition level of 0.01% Ti

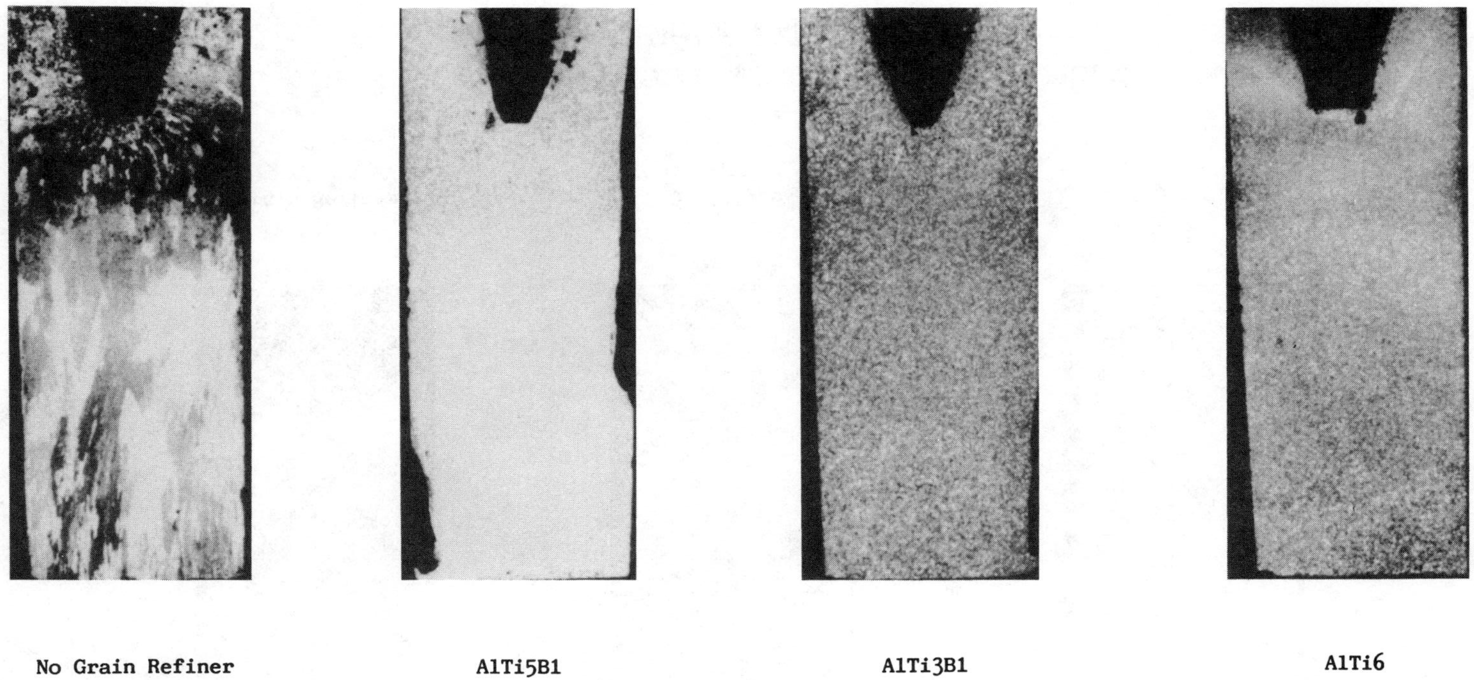

Fig. 8 Etched longitudinal macro sections of AA 6061 comparing the grain refining efficiency of AlTi5B1, AlTi3B1 and AlTi6 all at addition rates of 1 kg/MT and at 2 minutes contact time

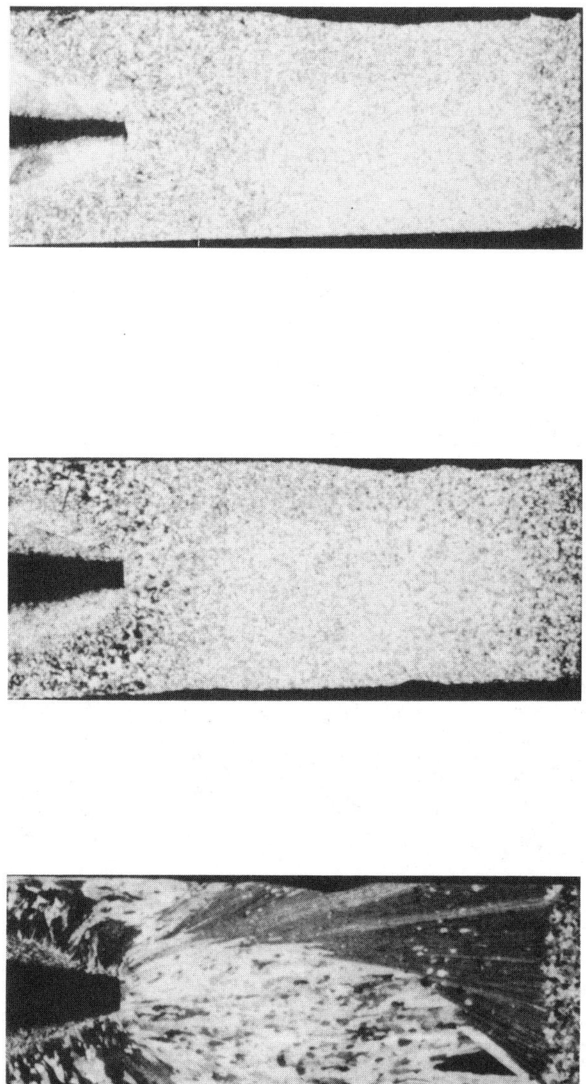

No Grain Refiner

0.005% Ti

0.01% Ti

Fig. 9 Etched longitudinal macro sections of AA 3004 grain refined with AlTi6 at two addition levels

Effect of alloy composition of microstructural refinement formation of quasi-equilibrium phases and related properties in melt spun alloys

F.H. Samuel, A.M. Samuel
Department of Applied Sciences, Université du Québec à Chicoutimi, Chicoutimi, Quebec, Canada, G7H 2B1
G. Champier
Lab. de Physique du Solide, École des Mines, Parc Saurupt, 54042, Nancy, France

Abstract

The present work was performed on two series of aluminum alloys, namely Al - 2.8% Li - (0.1-0.8)% Co and Al - (15-27)% Mn, produced by melt spinning. For the Al - Li - Co series, the as melt spun microstructure consisted of very fine grains (about 1-2 µm vs. 150 µm for classical solidification). Increasing the Co content and, therefore, the volume fraction of the Al_9Co_2 dispersoids produced, improved the yield strength and ductility and promoted transgranular fracture instead of the intergranular fracture usually observed in Al-Li binary alloys. The as melt spun ribbons of Al-15% Mn exhibited the icosahedral structure ($m\bar{3}\bar{5}$ symmetry) in an Al-matrix. Increasing the Mn content upto 27% led to the formation of the decagonal phase (10/m or 10/mmm symmetry) in addition to the icosahedral one. Annealing in the temperature range 200-600°C revealed that while the icosahedral phase became unstable at 400°C and crystallized into Al_6Mn, the decagonal phase persisted even after 500°C/100h, and at 600°C was seen to coexist with Al_4Mn. These observations were supported by electrical resistivity and DTA measurements.

Introduction

Strengthening in Al-Li alloys is associated with the precipitation of a high volume fraction of the metastable phase $\delta'(Al_3Li)$ owing to its resistance to dislocation motion (1). This effect is reduced when the precipitates are sheared by dislocations. This, in turn, leads to a successive decrease in resistance of the δ' particles for further dislocations, and consequently results in planar slip and a tendency towards strain localization and distortion of the Al_3Li. The slip distribution in Al-Li alloys depends on the magnitude of local work-softening that occurs during deformation.

In rapidly cooled alloys of Al with 10 to 14 at% Mn, Fe or Co, a new phase (now commonly known as icosahedral) is observed. The remarkable sharpness of the diffraction spots given by the icosahedral phase is comparable to that usually encountred in crystals (2). Field and Frazer (3) have explained the experimentally obtained five-fold diffraction patterns in Al-15% Mn on the basis of twinning model. Shechtman and Blech (4), however, have refuted this proposal. According to them, the microstructures that result in the observation of five-fold symmetry in the TEM diffraction patterns arise due to a structure consisting of multiple polyhedra having limited orientations.

The work reported here was directed at:

1- determining the role of rapid solidification and precipitation of fine Al_9Co_2 phase particles on promoting homogenized slip and improving the alloy ductility of melt-spun ribbons without prior consolidation; and

2- studying the relationship between the microstructures formed in rapidly solidified Al-Mn ribbons and precipitation processes following annealing in the temperature range 300 to 600°C for times varied between 15 min. and 100 hr.

Experimental Procedures

The present work was performed on different aluminum alloys produced by melt-spinning (MS) and prepared from high purity elements under an inert atmosphere of helium (for alloy composition see Table I). The alloys were spun into ribbons by conventional melt spinning using a Cu-2% Cr wheel (27 cm in diameter) as the rotating chill substrate. The melt spinning was done in a chamber under He atmosphere. The ribbon width was typically 1500-2000μm with a thickness lying in the range 30-40μm. Ribbons of Al-Li-Co alloys were aged at 473 K for various times up to 100 h without prior solution heat-treatment, whereas those of Al-Mn alloys were annealed in the temperature range 300 to 600°C for times up to 100 h.

TABLE I
Alloy Chemistry

Code	Fabrication Technique	Designation	Form	Alloy Composition (wt%)
10ALCR	Melt spinning	MS	Ribbon	Al-2.85 Li-0.1 Co
40ALCR	Melt spinning	MS	Ribbon	Al-2.85 Li-0.4 Co
80ALCR	Melt spinning	MS	Ribbon	Al-2.85 Li-0.8 Co
10ALC1	Ingot metallurgy	IM	10 mm round ingot	Al-3 Li-0.1 Co
15ALMn	Melt spinning	MS	Ribbon	Al-15Mn
27ALMn	Melt spinning	MS	Ribbon	Al-27Mn

Detailed metallographic examinations were made on the alloys in different conditions, employing standard light and transmission electron microscopy (TEM). The foils were investigated in a Jeol-JEM 200CX electron microscope operating at 200 kV and various specimen inclinations and rotations were adopted.

Modulus measurements were carried out on the Al-Li-Co ribbons, after careful polishing, using a bending test technique developed in the laboratory. Values reported are the average of at least five readings. The yield strength (0.2% offset), ultimate tensile strength and the elongation were measured from longitudinal tensile test ribbons (20 mm gauge length and 1.5 mm width) using a strain rate of 6 x 10^{-4}/s. All tests were conducted at room temperature and the average of at least five readings are reported here.

Fracture surfaces were studied by a scanning electron microscope operating at 25 kV. X-ray diffraction patterns were obtained from both wheel and gas surfaces using a Co Kα source for each working condition.

For the Al-Mn alloys electrical resistivity measurements were made on the ribbons using an apparatus assembled at the Lab. de Physique du Solide, and equipped with computer control. Differential thermal analysis was carried out using the same unit. For the latter, a 100 gm specimen in powder form was used.

Results and Discussion

Al-Li-Co

Figure 1 represents the microstructures from through-thickness cross-sections of two alloys that are hypo-eutectic (with respect to Al-Co binary diagram). The solidification started by columnar formation that extended throughout the entire thickness. These columns are often found making an angle of 75°C to the ribbon direction.

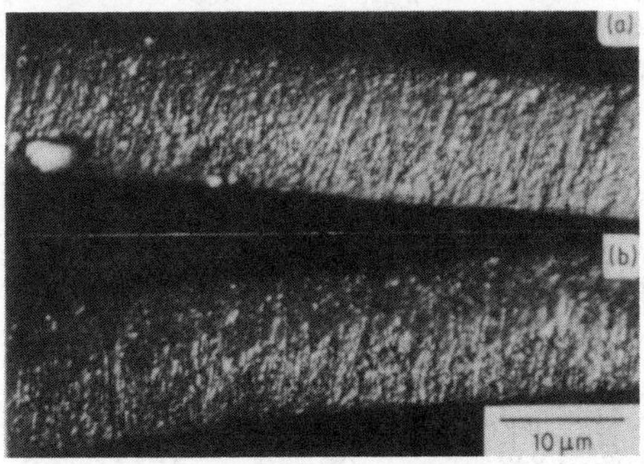

Figure 1 Light optical micrographs of Al-Li-Co ribbons in the as melt-spun condition: (a) 10ALCR, (b) 80ALCR.

TEM has yielded valuable information on the structural refinement resulting from the high freezing rate associated with the melt spinning process. This is evident in the very fine grains ~2μm, as displayed in Fig. 2. Scattered Al_9Co_2 particles are seen inside the grains. For the highest cobalt content, 0.8 wt%, which is very close to the eutectic composition of 1 wt%, precipitation of Al_9Co_2 was found to occur at grain boundaries as well as in the matrix.

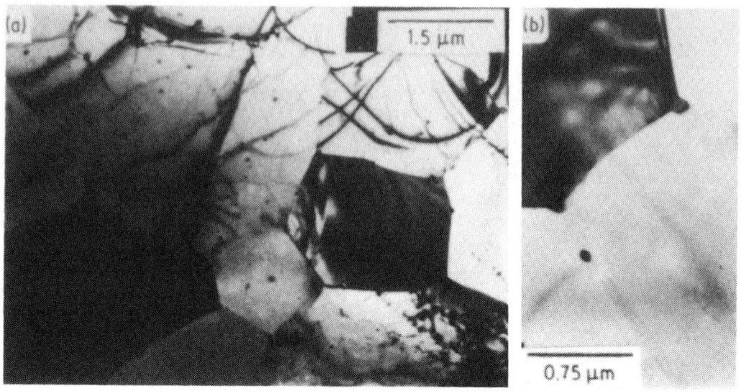

Figure 2 Bright-field micrographs of 10ALCR ribbon in the as melt-spun condition showing (a) the grain size in a section normal to a set of columnar grains, (b) location of Al_9Co_2 particles at triple meeting points.

Figure 3a (10ALCR) indicates the breakdown of the solid solution during melt-quenching, resulting in homogeneous dispersion of fine $Al_3Li(\delta')$ having a size approximately 5 nm. According to Giant (5), the cooling rate used in the present study is estimated at about 10^5 K/s. Thus, the presence of δ'phase particles is possibly due to difficulty in suppressing their homogeneous nucleation even within a very short time (~10min) after quenching and ageing at room temperature. This is explicable in terms of the small lattice misfit (-0.18%) between the δ' and the fcc matrix as well as an excess vacancy concentration that is generated on quenching (6).

Representative electron micrographs for δ' precipitation on subsequent ageing of 10ALCR ribbon at 473K are shown in Figs. 3b and 3c. Well defined δ' precipitates are observed for ageing times from 1 h up to 100 h. A widening of the precipitate - free zones (PFZs) accompanying the precipitation is present. On the basis of the TEM micrographs in Fig. 3, it is found that a spherical shape is the common morphology observed for δ'. After ageing at 473 K for 100 h collinear rows of δ' are observed, sometimes elongated along the line direction. The average coarsening rate at this temperature is about 2.9×10^{-25} m^3/h. With the help of the micrographs in Fig. 3, the best fit to the variation in radius (R) of δ' phase particles with time $(t^{1/3})$ is illustrated in Fig. 4.

The results of the Young's modulus (from bending tests) determination for 10ALCR and 10ALCI alloys are shown in Table II. For Al-3 wt% Li, the modulus increases from 66 GPa to 78.5 GPa due to lithium in solid solution, and from 78.5 to 83 GPa due to δ' (Al_3Li) precipitation on ageing at 473 K. To investigate the influence of δ' volume fraction on the value of E, the rule of mixture equation

$$E_{aged} = E_m V_m + E_{\delta'} V_{\delta'}$$

was utilized assuming E_m at 1.6 wt% Li ≅ 78.5 GPa and $E_{\delta'}$=96GPa (7). The results are listed in Table II. The difference between the calculated and measured values is about 3%.

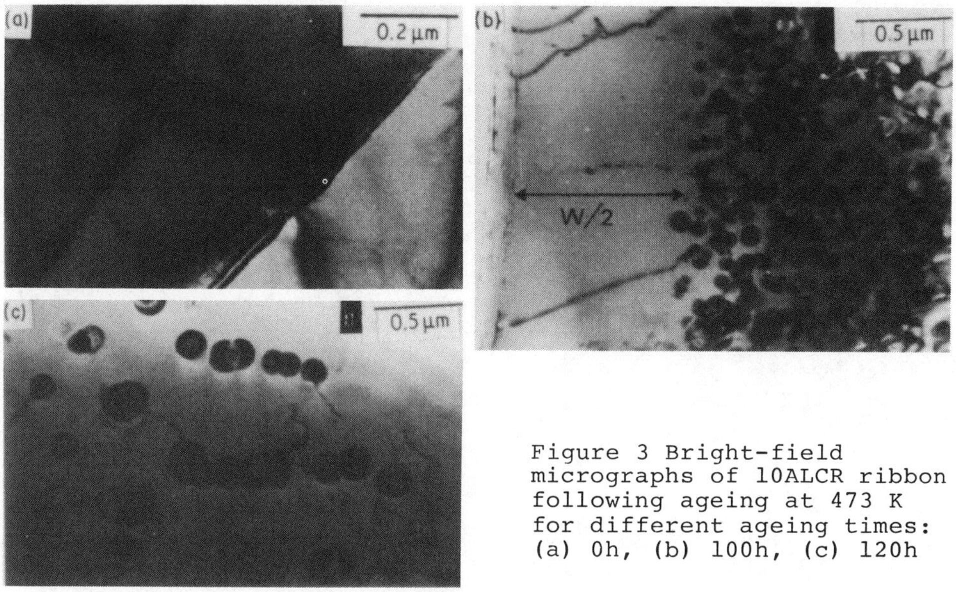

Figure 3 Bright-field micrographs of 10ALCR ribbon following ageing at 473 K for different ageing times: (a) 0h, (b) 100h, (c) 120h

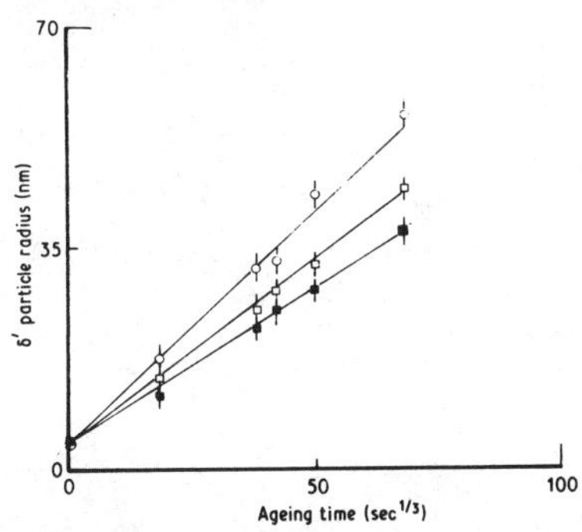

Figure 4 The average δ precipitate radius as a function of the cube root of ageing time at 473 K: (o) ingot metallurgy alloy (10ALCO), (□) melt-spun ribbon (10ALCR), (■) solution heat treated melt-spun ribbon (80ALCR).

TABLE II
Modulus and volume fraction of δ' in 10ALCR and 10ALCI alloys

Condition	δ' (vol%)	E_{calc} (GPa)	E_{meas} (GPa)	Specific Modulus E_q (10^6m)
10ALCR alloy				
MS	14	81.65	79.70	3.237
MS+1h at 473K	22	82.37	81.90	3.329
MS+16h at 473K	28	83.41	82.50	3.321
10ALCI alloy				
IM	12	80.65	80.32	3.346
IM+1h at 473K	18	81.65	80.45	3.258
IM+24h at 473K	24	82.70	81.20	3.288

The yield stress (0.2%YS) and ultimate tensile stress (UTS) for a variety of Al-Li-Co ribbons in either as melt-spun condition or after ageing are listed in Table III. The YS and UTS of as-melt-spun ribbons in the under-, peak-and overaged conditions are increasing with the increase in the Al_9Co_2 phase particles volume fraction which persisted during ageing at 473 K. The relationship between total elongation and cobalt content for MS ribbons is also shown in Table III. The overall improvement in ductility arises from the dispersal of the slip by the dispersoid Al_9Co_2 particles, as well as the coarsening of δ'.

TABLE III
Room temperature tensile properties

Alloy	Treatment	YS (MPa)	UTS (MPa)	El(%)*	YS/ρ (m)	UTS/ρ (m)
10ALCR	MS	70	197	15	28.35	79.80
	MS+16h at 473K	290	385	3.5	117.45	155.93
	MS+100h at 473K	180	308	7	72.90	124.74
40ALCR	MS	94	221	18	30.70	89.51
	MS+16h at 473K	324	400	5	131.22	162.00
	MS+100h at 473K	189	336	9	76.55	136.10
80ALCR	MS	119	246	20	48.19	99.64
	MS+16h at 473K	349	440	6	141.30	178.20
	MS+100h at 473K	205	348	10	83.00	141.00

*El(%) is the total elongation in per cent.

Figure 5 emphasizes the role of rapid solidification in increasing the extent of solid solubility. An important fraction of cobalt was in solid solution after melt-quenching and it decomposed uniformly throughout the grains after ageing.

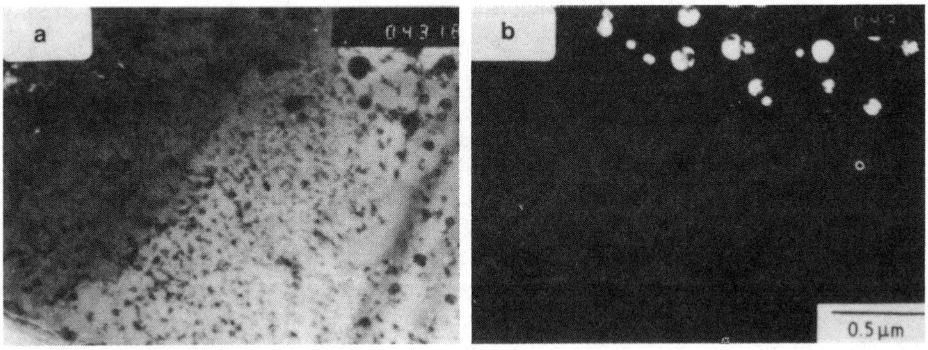

Figure 5 TEM micrograph of melt-spun ribbons following ageing at 473 K: (a) 80ALCR, 100 h, bright field; (b) 80ALCR, 100 h, dark field imaged on a δ' reflection.

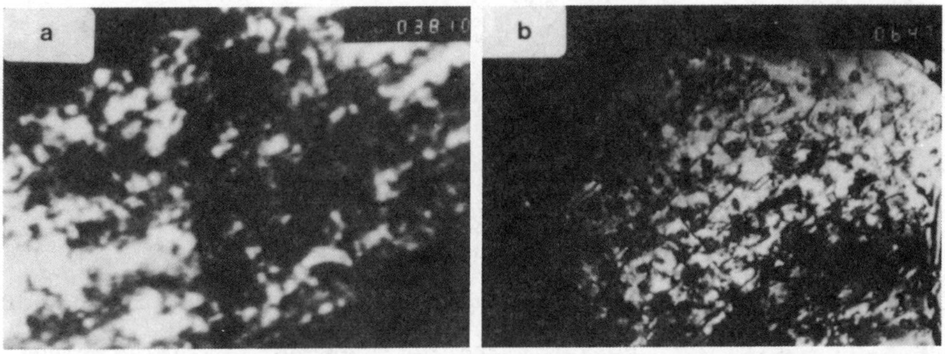

Figure 6 TEM micrographs of ribbons aged at 473 K for 16 h and pulled to failure: (a) 10ALCR (b) 80ALCR.

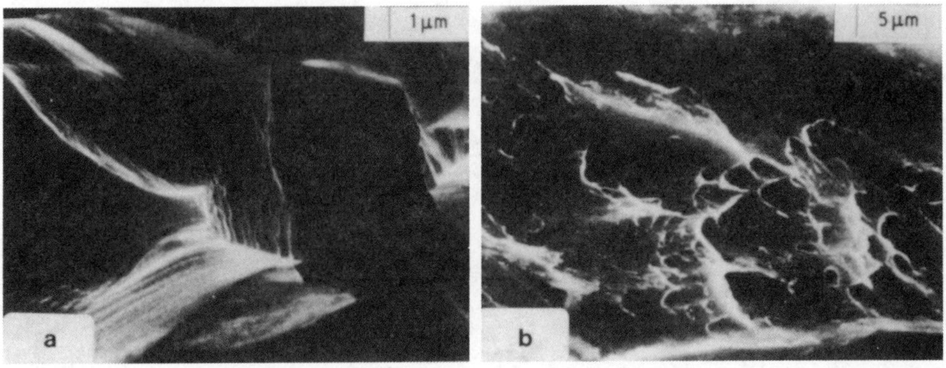

Figure 7 SEM micrographs of tensile fractures of Al-Li-Co ribbons aged at 473 K for 16 h: (a) 10ALCR, (b) 80ALCR.

The substructure of Al-3 wt% Li aged at 473 K for 16 h (peak-aged condition) reveals intense planar slip with the dislocations arrayed in parallel bands (Fig. 6a). A dark field image using a δ' diffracted beam showed that the precipitates within the bands are severely sheared as compared to those in the matrix. These bands, in addition, are causing mutual offsets in the grain boundaries (Fig. 7a). The effectiveness of Al_9Co_2 particles as incoherent dispersoids in promoting homogeneous deformation without planar slip is shown in Fig. 6b. The ability of the dispersoid particles to disperse slip, coupled with an almost complete absence of coarse, heterogeneous grain-boundary precipitates, is conducive to improved ductility in Al-Li alloys containing cobalt as shown in Table III.

Failure of Al-3 wt% Li in the peak-aged condition appears to be completely intergranular with significant grain (columnar) boundary separation below the fracture surface (Fig. 7a). As can be seen, several sharp ledges, are visible on the grain boundaries when the slip bands intersect the grain boundaries. Addition of cobalt appears to change the mode of failure into transgranular through dimple formation. The dimples suggests a ductile type of fracture behaviour (Fig. 7b).

Al-Mn alloys

Figure 8 shows a representative plot (curve A) of the electrical resistivity (at constant current and gauge length 1 cm) of as melt spun 15 ALMN alloy ribbon as a function of temperature. The ribbon was heated from room temperature up to 600°C. The resistivity is seen to be nearly constant from room temperature to about 320°C, after which the crystallization process presumably starts to take place, as evidenced by the decrease in resistivity. This decrease continues in a gradual manner until about 500°C, after which there is an increase up to 600°C. Cooling results in a continuous, linear decrease in the resistivity, approaching a room temperature value of ~40Ωcm. The DTA curve (curve B) in Fig. 8 shows also a region of interest corresponding to the temperature range ~320 to 450°C, the area where the recrystallization process is taking place. The small peak in the crystallization isotherm (26.5 mV) allows us to presume that the icosahedral phase is not too far away from the stable phase Al_6Mn.

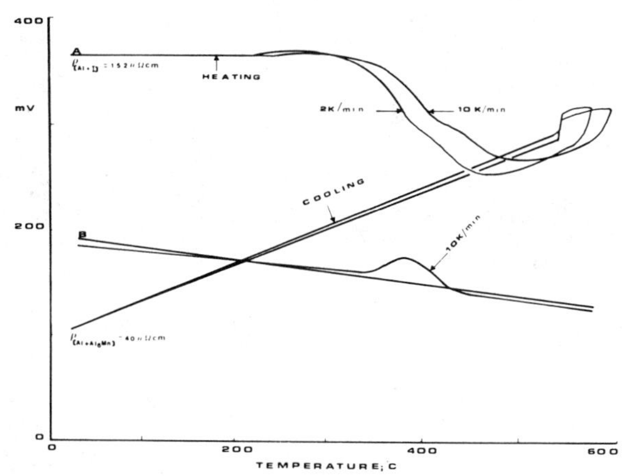

Figure 8 Variation of the electrical resistivity (A) and DTA (B) on heating from room temperature up to 600°C, as measured by the variation in mV (at constant current, using Cr-Ni/Cr-Al thermocouple); I:Icosahedral phase.

Figure 9a exhibits the typical 1-1.5μm sized, spherulitic/ dendritic shaped particles of the icosahedral phase in the as-spun Al-15% Mn ribbons. Each particle is surrounded by a band of aluminum at its outer edges. These observations were supported by corresponding electron diffraction patterns. Figure 9b is the SAD corresponding to Fig. 9a, and exhibits the typical 5-fold symmetry pattern obtained for the icosahedral phase.

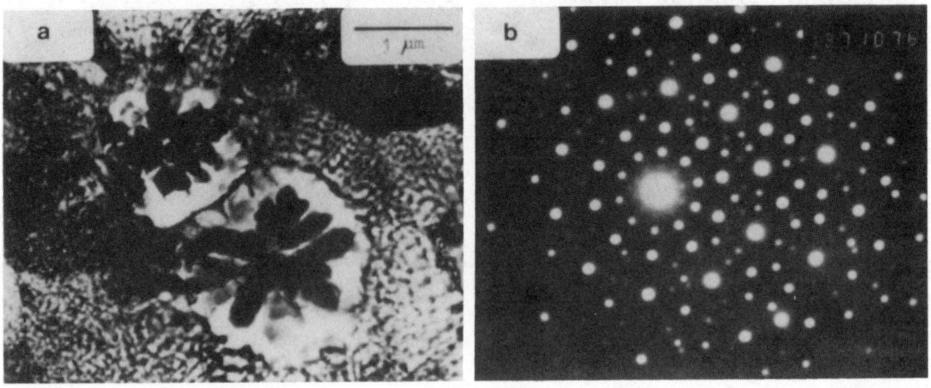

Figure 9 TEM micrographs representing the microstructure of: (a) spherulites characterized by elongated branches that stem from their central cores, (b) SAD exhibiting the five-fold symmetry zone axis (15ALMN).

Figures 10a and 10b, are micrographs, taken from 27ALMN as-spun ribbons, representing wheel side and gas side, respectively. Figure 10a exhibits many small grains (about 0.2μm diameter) where Moiré fringes are frequently observed at boundaries between neighbouring grains, indicating some overlapping of these grains.

In Fig. 10b, corresponding to an area near the gas side, coarse particles of Al_6Mn are observed, which occur due to direct precipitation from the melt, followed by solidification of the remaining melt liquid into the icosahedral phase. Small islands of Al_4Mn are also seen in this figure.

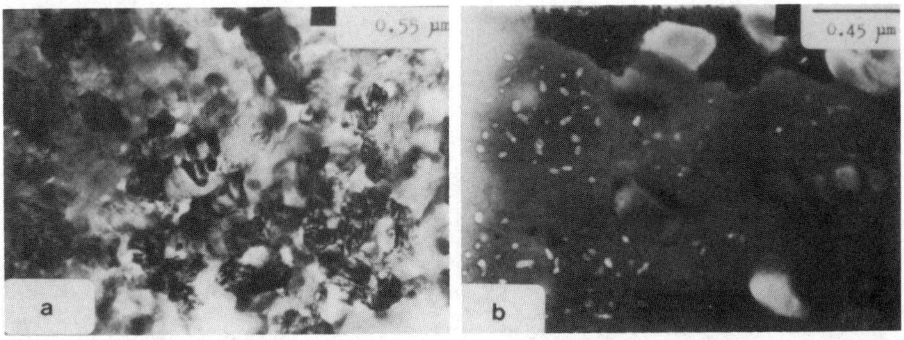

Figure 10: Transmission electron micrographs for the as melt-spun ribbons: (a) wheel side, (b) gas side (27ALMN).

The electron diffraction patterns obtained from the two areas, however, show interesting differences. Representative SADs are given in Figs. 11a and 11b. Comparing the two patterns shows that, while Fig. 11b exhibits a true five-fold pattern, typical of the icosahedral phase ($m\bar{3}\bar{5}$ symmetry), Fig. 11a exhibits an almost five-fold symmetry.

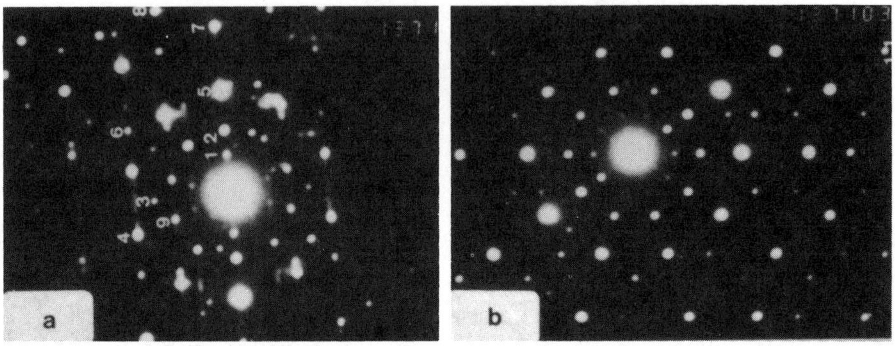

Figure 11 SADs corresponding to Figs. 10a and 10b showing (a) almost five-fold symmetry, (b) true five-fold symmetry.

Considering only the strongest spots in the two photographs, one can easily discern that while the angles between the radial rows in Fig. 11b are exactly 36°, they are approximately 35° and 37° in Fig. 11a and the spots, in addition, are not exactly collinear. The d-values measured for various spots in this pattern are given in Table IV.

TABLE IV
Measured d-values for spots shown in Fig. 10a

Spot no.	d(nm)	Spot no.	d(nm)
1	0.528	5	0.2067
2	0.342	6	0.188
3	0.243	7	0.1308
4	0.2096	8	0.110

These values have been identified as corresponding to the so-called decagonal phase as reported by Shechtman et al (2) and by Bendersky (8). According to Bendersky, this phase has a non-crystallographic point group 10/m (or 10/mmm) together with long-range orientational order and one-dimensional translational symmetry.

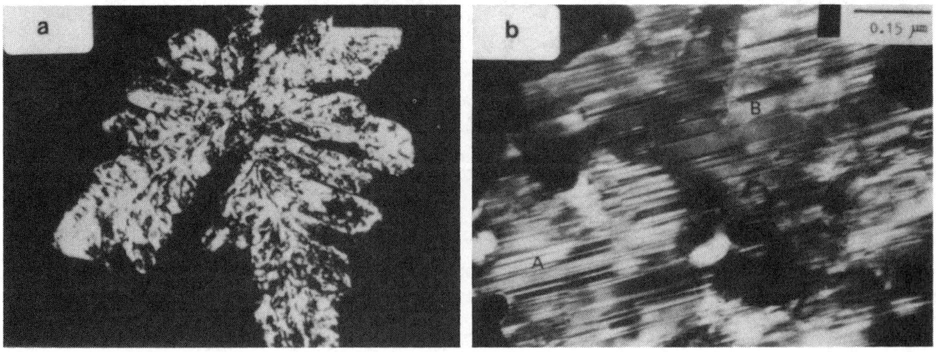

Figure 12 Dark field micrographs corresponding to (a) icosahedral and (b) decagonal phases.

Figure 12 displays dark field micrographs corresponding to the two quasi phases. Figure 12a, a typical representation of the icosahedral structure, exhibits no visible details (no twins or microcrystallinity observed). On the other hand, the decagonal phase, Fig. 12b, reveals the presence of a high density of planar defects. We believe these correspond to microtwins in the form of sheets normal or inclined to the plane of the micrograph.

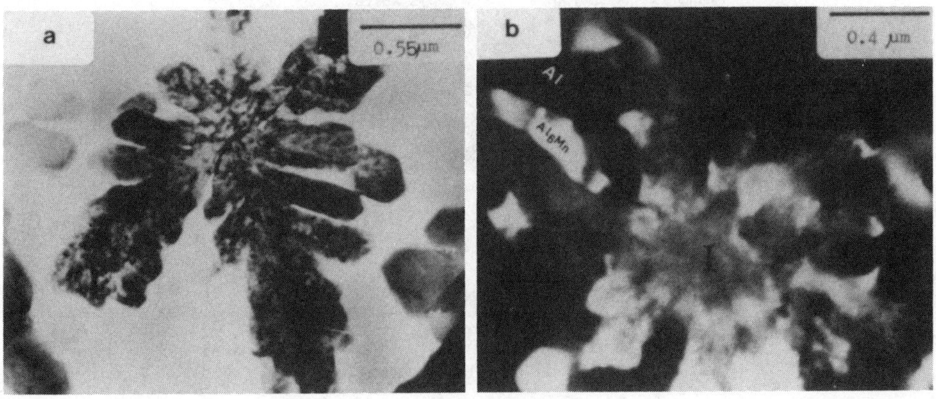

Figure 13 TEM micrograph of 15ALMN (MS) (a) as melt spun ribbon and (b) after annealing at 400°C/30min.

Effect of annealing

The 15ALMN and 27ALMN ribbons were annealed under various conditions of time and temperature (300 to 600°C; 15 min to 100 h). The changes (starting from the as melt-spun condition) were monitored using X-ray diffraction.

For the 15ALMN alloy ribbons, the following observations were made:

Starting from an initial icosahedral + Al structure in the as-spun condition (Fig. 13a), heat treatment at 300°C produced little change in the icosahedral phase. However, at 400°C/30min, narrow bands of Al_6Mn started to form at the outer edges of the icosahedral spherulites (Fig. 13b). After 400°C/1 h, Al_6Mn was the main structure observed. Annealing at 500°C and 600°C increased the abundance of the Al_6Mn phase but increasing the annealing time between 400 and 600°C only coarsened the Al_6Mn particles.

For the 27ALMN ribbons, annealed for 1 to 100 h at temperatures ranging from 400 to 600°C, the following observations were made:

After 1 h annealing, there was a complete disappearance of Al-peaks, indicating the transformation to Al_6Mn as a result of the rejection of excess of manganese atoms. The amount of icosahedral phase was greatly reduced corresponding to the increase of Al_6Mn phase. However, the phase was still visible after 500°C/1 h. In comparison, the decagonal phase persisted, even after 500°C/1 h. The Al_4Mn phase was observed, along with Al_6Mn and decagonal phases. Figure 14 is a diffraction pattern of Al_4Mn produced from a specimen annealed 1 h at 600°C.

Reflections due to the decagonal phase in five-fold symmetry are clearly visible in the same diffraction pattern.

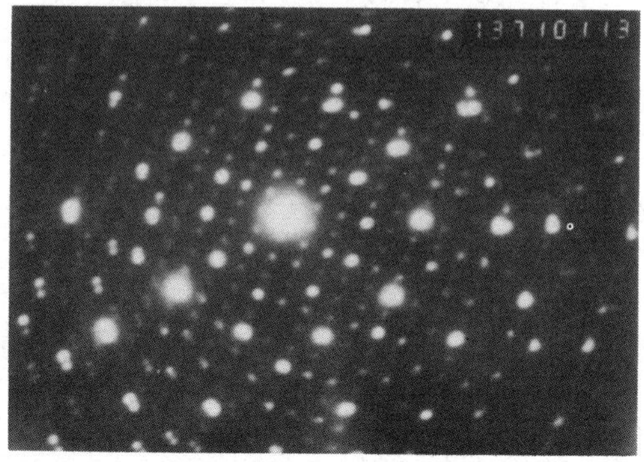

Figure 14 SAD obtained from 27ALMN ribbon annealed 1 h at 600°C showing Al_4Mn phase along with the decagonal phase oriented near five-fold symmetry.

After 100 h annealing, however, there was complete disappearance of the icosahedral phase. The decagonal phase persisted upto 500°/100 h.

Conclusions

From the results obtained in the present investigations, we draw the following conclusions:

I Al-Li-Co alloys

1- Increasing the cobalt content increases the volume fraction of Al_9Co_2 phase particles. Their precipitation occurs in the matrix as well as at grain boundaries.

2- Suppression of the δ' (Al_3Li) phase particles nucleation reaction on melt-quenching is very difficult. The average coarsening rate is about 2.9 x 10^{-25} m³/h. Coarsening of δ' particles follows a linear relation with $(time)^{1/3}$.

3- In the peak-aged condition, the values of the Young's modulus are 80 to 83 GPa for MS alloy.

4- Increasing the Co content up to 0.8% and, in turn, the volume fraction of Al_9Co_2 dispersoids, is sufficient to disperse the slip bands and to promote transgranular fracture.

II Al-Mn alloys

1- Increasing the Mn content from 15 to 27% results in the formation of another quasi structure, the decagonal phase, in addition to the icosahedral one. This phase exhibits an almost five-fold symmetry similar to the true five-fold symmetry of the icosahedral structure.

2- It can form from the melt or grow on the surfaces of the icosahedral phase and is seen to exhibit different morphologies. Within the grains of the decagonal phase, several domains are observed, with a high density of planar faults. In comparison, the icosahedral phase occurs directly from the melt and exhibits no morphological difference, no faults or twins being observed within its grains.

3- On annealing (400-600°C/1-100h), the icosahedral phase starts to crystallize into Al_6Mn at 400°C/1 h, while the decagonal phase persists even after 500°C/100 h. The existence of the decagonal phase along with Al_4Mn at 600°C suggests that this phase arises from a commensurate to incommensurate transformation due to atomic displacement within the Al_4Mn phase.

4- Essentially, therefore, the presence of the icosahedral and decagonal phases are related to the equilibrium structures present at a particular concentration of Mn in the Al-Mn alloy. Thus, at 15% Mn only the icosahedral phase is observed (Al_6Mn being the only equilibrium structure), while at 27% Mn, both quasi-structures exist (along with Al_6Mn + Al_4Mn in equilibrium).

References

1. T.H. Sanders Jr. and E.A. Starke Jr., Acta Metall. 30, 927, 1982.
2. D. Shechtman, I.A. Blech, D. Gratias and J.W. Cahn, Phys. Rev. Lett. 53, 1951, 1984.
3. R.D. Field and H.L. Frazer, Mater. Sci. Eng. 68, L17, 1985.
4. D. Shechtman and I.A. Blech, Metall. Trans. 16A, 1005, 1985.
5. N.J. Grant, J. Metals 35, 20, 1983.
6. H. Tamler and O. Kanert, Acta Metall. 32, 1205, 1984.
7. B. Noble, S.J. Harris and K. Dinsdale, J. Mater. Sci. 17, 461, 1982.
8. L. Bendersky, Phys. Rev. Lett. 55, 1461, 1985.

Performance characteristics of metallurgical grain refiners in hypoeutectic Al-Si alloys

G.W. Boone, R.F. Carver and W.C. Setzer
KB Alloys, Inc., Corporation Technology, 3293 McDonald Road, Robards, Kentucky 42452, U.S.A.

Abstract

This paper presents grain refining performance data using the Aluminum Association, the preheated Golf Tee mold, and the Calibrated Ring Tests. The grain refining response of two commercial hypoeutectic Al-Si alloys, A356 and A319, as well as commercial purity aluminum (P1020A) are reported. Grain refining master alloys evaluated were 5%Ti-1%B, 2.5%Ti-2.5%B and a recently developed aluminum 6% titanium - carbon alloy. The grain refining responses were developed for each of the three alloys for a range of addition levels, incubation times, stirring actions, and residual titanium contents.

Effectiveness of the grain refiner alloys varied depending upon the opportunity and magnitude for each of several mechanisms to be influential. A grain refiner with good intrinsic effectiveness may be more or less effective depending on the interactive time effect for settling, for any change (either positive or negative) of effectiveness with time and for resistance to dissolution. Residual titanium content is most important when a boron free refiner alloy is utilized. For a metallurgical grain refiner with excess boron, residual titanium is of lesser importance.

Key Words

Grain refiner, grain refining mechanism, settling, fade, dissolution rate, residual titanium, titanium, boron, carbon, TiB_2, $TiAl_3$, aluminum master alloy, A356, A319, P1020 aluminum, TIBOR[R], TICAR[R], hypoeutectic aluminum silicon, Al-Ti, Al-Ti-B, Al-Ti-C, metallurgical grain refiner.

Introduction

As part of its ongoing technical effort, KB Alloys has been active in the development of new and improved grain refiner products. These include the first metallurgical grain refiners, several TIBOR (titanium plus boron in ratios from 3:1 to 100:1) alloys, master alloy rod for continuously controlled additions, and recently, an alloy containing 2.5%Ti-2.5%B developed specifically for grain refining hypoeutectic Al-Si foundry alloys (1). Even more recently an alloy has been commercialized which has proven as effective on a pound for pound basis as 5%Ti-1%B in grain refining wrought alloys. This alloy is boron free and uses $TiAl_3$ and titanium carbide as the active heterogeneous nucleants (2, 3) This TICAR alloy is being considered where it is desirable to minimize the amount of titanium required to grain refine but where the presence of boron might adversely impact either ductility, fracture toughness or other dynamic properties. The purpose of this paper is to report additional laboratory work performed using both 5%Ti-1%B and 2.5%Ti-2.5%B alloy as bench marks for the performance of the TICAR alloy in the grain refining of P1020, A356 and A319 foundry alloys.

The Aluminum Association Test (4) represents a condition where directional solidification is occurring very rapidly (5°C/second). The Golf Tee Test, wherein the mold is heated to 320°C, represents a condition where the cooling rate is reduced to 0.8°C/second. The KBA Calibrated Ring Test with a ceramic "chill" has an even lower solidification rate, simulating conditions approaching the center of much larger casting sections (0.2°C/second). Thus, comparison between the tests would show sensitivity of the grain refiner alloys to casting conditions.

The majority of the tests were conducted with A356, to show the effect of residual titanium content. However, tests were also run on P1020 and A319 to provide some basis for selecting effective grain refiners for these alloys (at a high residual titanium level). Data for 5%Ti-1%B and the TICAR alloy in P1020 aluminum with low titanium residuals has already been reported, but will be reviewed (5).

Experimental Procedure

Table 1 gives the chemical composition of the alloys used in this work. Although A319 and A356 are frequently modified with Na or Sr, no modification was used in this study. The TICAR, 2.5%Ti-2.5%B and 5%Ti-1%B alloys were in waffle form. These forms necessitated cutting thin wafers based on the nominal heat chemistry in order to provide grain refining additions of 0.5, 1 and 1.5 kg/1000 kg of heat weight as given in Table 1. The total heat weight used for each test was nominally 5 kilograms. The Aluminum Association Test (TP-1), the Golf Tee Test and the KBA Calibrated Ring Test were used to produce the castings which were examined for all the grain refining results reported. Test procedures are reported elsewhere (4, 6, 7). Specifics of test conditions are given in Table 2.

Grain refining specimens were cast after different incubation periods; samples tested at 5, 10, 20 and 205 minutes were stirred. Unstirred samples were cast at 50, 100 and 200 minutes.

TABLE 1. Chemical Composition of Alloys

ALLOY	Si	Fe	Cu	Mn	Mg	Ti	B	C
P1020A	0.06	0.19	----	----	----	0.005	---	----
A356	6.6	0.16	----	----	.35	0.005	---	----
A319	6.0	0.66	3.2	.30	.09	0.005	---	----
5/1 TIBOR	0.11	0.13	----	----	----	5.1	1.0	----
2.5/2.5 TIBOR	0.10	0.17	----	----	----	2.6	2.5	----
TICAR	0.06	0.15	----	----	----	6.0	---	<0.1

TABLE 2. Experimental Casting Parameters

Test Method	Casting Temperature (°C)	Mold Temperature (°C)
Aluminum Association	720	720
Golf Tee	704	320
KBA Ring	720	22

Following casting, the as-cast specimens were sectioned, mechanically polished and etched or electrochemically polished. Prepared samples were then compared to known standards according to TP-1 or anodized and measured on an optical metallograph. Grain sizes were determined on the anodized surfaces under polarized light at 100X by measuring the average intercept distance (AID) in microns using the linear intercept technique. The AID grain size was then converted to grains per cubic millimeter as described in ASTM E-112.

Test variables such as ladle temperature, water temperature, flow rate, heat weights, etc. were closely controlled. However, there was considerably more spread in the grain refining responses of the Al-Si alloys than had been anticipated. Several factors contributed. The thin wafer sections, though accurately weighed, were subject to local segregation. Also, the Al-Si alloys have a more complex microstructure which makes these alloys more difficult to etch or anodize to reveal the grain structure.

Results and Discussion

Commercial hypoeutectic Al-Si foundry alloys do not grain refine as well with TIBOR alloys as wrought aluminum alloy systems. To overcome this, fairly large residual titanium contents as well as larger grain refiner additions are typically used in the industry. Depending upon casting conditions and/or chemistries, a commercial casting may contain regions with a columnar or a twin columnar grain structure (TCG). These structures can lead to very directional properties, poor casting quality or cosmetic defects. Where it may be desirable to have a relatively fine grain size, it is beneficial to achieve this with the minimum grain refiner addition so as to minimize cost and runaround buildup of residual titanium while insuring a consistent product. A fine grain size may be used to control homogeneity, porosity size and distribution, and increase casting density.

Aluminum Association test results obtained for P1020 aluminum with both the TICAR and 5%Ti-1%B additions have been previously reported (5). In those tests the P1020 material did not have titanium added to build residual titanium above the 0.004 - 0.005% level in the base metal. The TICAR alloy was as effective as 5%Ti-1%B, but faded slightly at 30 minutes, whereas the 5%Ti-1%B alloy did not. This would not be a problem for wrought alloys, since any refiner addition is continuously fed into a launder during the casting operation.

The results reported here were obtained from two separate experiments. In one experiment tests were conducted in P1020 aluminum, A356 and A319 containing nominally 0.2% residual titanium, a level above the peritectic value. It was anticipated that there would be a slight "fade" or loss of grain refining effectiveness with holding time. A grain refiner was added at the rate of

1 kg/1000 kg of alloy. This rate is at a point on the grain refining response curve where the rate of change of grain size with refiner addition rate is low.

In 1985, Sigworth et al showed the effect of holding time on grain size in A319 (8). A319 with 0.15% residual titanium was inoculated with 1.2 kg/1000 kg of 5%Ti-1%B. Even after being held molten for several days, the grain size was essentially the same as was observed after five minutes.

The effect of hold time and stirring action on the various grain refiners and alloys allowed four conclusions to be drawn. For 5%Ti-1%B, holding time and stirring results confirmed Sigworth's observations in A319 and showed no change in grain refining effectiveness with time or stirring in either P1020 or A356 from 5 to 205 minutes. Secondly, all grain refiners became active within 5 minutes of dissolution. The fade noted earlier for the TICAR refiner was observed in all alloys and was generally enhanced with stirring. In all alloys there was a settling effect noted for the 2.5%Ti-2.5%B refiner alloy which was restored to its original effectiveness level, even at 205 minutes, by stirring. Noting the settling effect for 2.5%Ti-2.5%B, when stirring is neither feasible nor contemplated, 2.5%Ti-2.5%B is comparable to 5%Ti-1%B. With stirring or for shorter hold times, 2.5%Ti-2.5%B alloy is superior to 5%Ti-1%B and TICAR alloys.

In any commercial foundry situation there are variations in holding time and stirring action due to setup, handling procedures etc. Consequently, the stirring/hold time data (7 samples) was combined to provide an idea of the consistency and average value which might be achieved. It was felt this method of presentation should give a better idea of the refining to be expected within a given casting under a range of solidification conditions and casting parameters. If in practice adequate grain refining was not achieved, the knowledge of the effect of stirring and hold time mentioned above would be useful in selecting the most cost effective grain refiner and practice.

In previous papers (9, 10) it was noted that grain diameter, although being the conventional method for reporting grain size data, does not discriminate very meaningfully between different grain refiner alloys. Consequently, it was concluded that a more effective means of comparing different alloys (and test procedures) would be to convert grain diameter to grains per cubic millimeter, using ASTM E-112. The number of grains per cubic millimeter is a truer measure of the active "sites" since each grain originates from an effective nucleus. Grains/mm^3 counts depict the magnitude of the change when adding a grain refiner, at different levels, of different types or under different solidification conditions (tests). Using this approach, it is possible to demonstrate more graphically what might appear to be relatively small changes in grain size.

Comparison of Figures 1, 2, and 3 showscommercial purity aluminum is much more easily grain refined than either A356 or A319. Only 1/10 the number of sites are generated as compared to P1020.

Figure 2 shows the results of the Aluminum Association Test for A319. TCG predominated the 5%Ti-1%B and TICAR refined samples. Only one of seven samples refined with 2.5%Ti-2.5%B exhibited TCG. Alloy A319 is the most sensitive alloy of the three alloys tested for TCG.

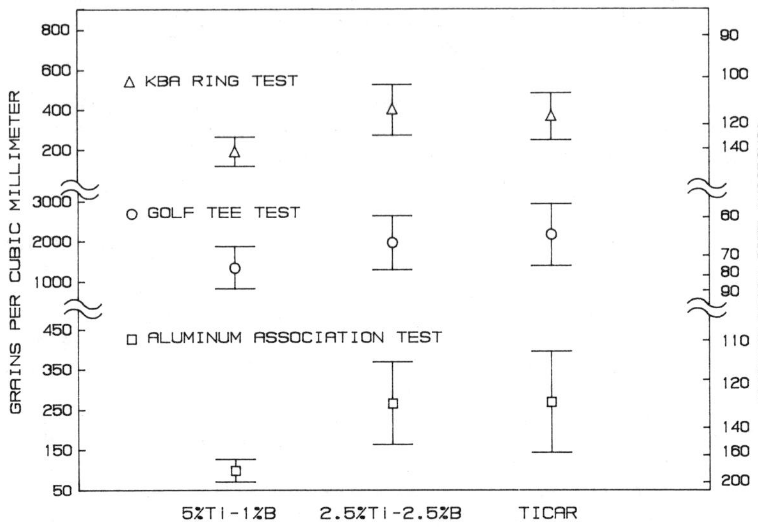

FIG. 1. Grain size results achieved in commercial purity aluminum (P1020) with 0.2% residual titanium, when grain refined with 1 kg/1000 kg utilizing the alloys indicated. Values given are the average of 7 samples, held from 5 to 205 minutes, with or without stirring. See text for details. Note the Golf Tee Test, with the intermediate cooling rate and lower hold temperature produced the finer grain size.

In the case of alloy A356, a review of the Aluminum Association test data showed the following results. TCG was present in four of seven 5%Ti-1%B refined samples. The TICAR refined samples showed one with significant TCG and one other with scattered individual TCG grains. The 2.5%Ti-2.5%B samples exhibited only a few individual grains of TCG in two of seven samples. The data suggests that in castings which are susceptible to TCG, the 2.5%Ti-2.5%B alloy provides more resistance.

In the case of the A356 alloy, Sigworth presented data which showed the effect of residual titanium content on A356 and A319 using the Calibrated Ring test. With no residual titanium, the A356 alloy grain refines to a greater extent than A319 at a residual titanium level above the peritectic, the grain size of A319 was comparable to A356 for the same grain refiner addition. We obtained the same result for 5%Ti-1%B and 2.5%Ti-2.5%B.

Based on this information, the 5%Ti-1%B and the TICAR provide the same grain refining effectiveness. In summary, this data suggests that the 2.5%Ti-2.5%B alloy provides an improved grain refining response as compared to 5%Ti-1%B and the TICAR alloy.

The second study undertaken was to show the effect of residual titanium content and grain refiner addition level on the response of A356. Five residual titanium levels from 0.005 to 0.20% were investigated in increments of 0.05% titanium. Additions of 5%Ti-1%B, TICAR and 2.5%Ti-2.5%B grain refiners were made at the levels of 0.5, 1 and 1.5 kg/1000 kg. The higher addition rates, being on the "flat" part of the response curve, did not show

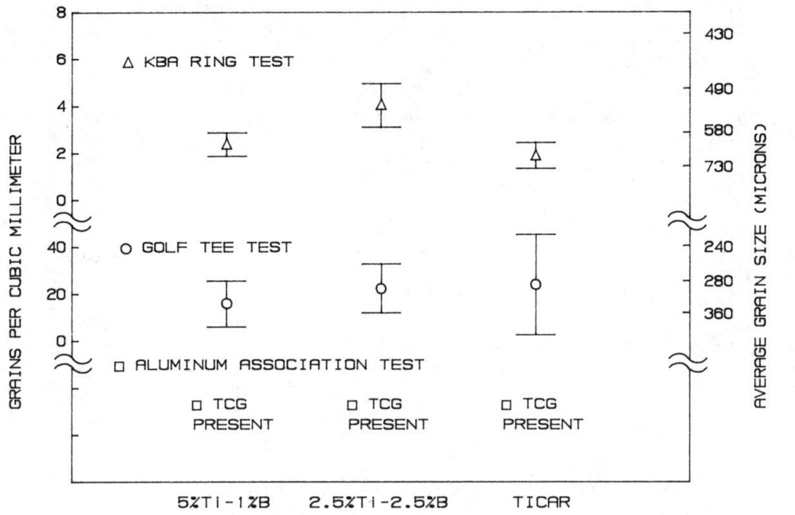

FIG. 2. Grain size results achieved in A319 with 0.2% residual titanium, when grain refined with 1 kg/1000 kg utilizing the alloy indicated. Values given are the average of 7 samples, held for 5 to 205 minutes, with or without stirring. A twin columnar grain structure was observed in the Aluminum Association Test. 2.5%Ti-2.5%B exhibited TCG only in the 200 minute sample. See text for details.

any change in effectiveness. Consequently, these data were averaged together. The results are given in Figure 4.

If one requires only the grain size achieved by a 0.2% residual titanium content, this grain size can be achieved with a 1 kg/1000 kg addition of the TICAR alloy without the need for a 0.2% residual titanium content. Consequently, when there are concerns around building the residual titanium content due to runaround scrap, this approach may be beneficial. This broader data base also suggests that the TICAR alloy is generally not as effective a grain refiner as either 5%Ti-1%B or 2.5%Ti-2.5%B alloy. Basically, the TICAR alloy only becomes truly effective at very high residual titanium levels. This might be expected in that we have observed that stirring causes a reduction in grain refining effectiveness which we have interpreted as being dissolution of the $TiAl_3$ phase. The dissolution should be enhanced when the residual titanium is below the peritectic value. Where effective grain refining is required, the alloy of choice again appears to be the 2.5%Ti-2.5%B alloy.

The best grain refiner at low residual titanium levels was the 2.5%Ti-2.5%B alloy. Based on the number of grain/mm^3, 2.5%Ti-2.5%B was twice as effective as the 5%Ti-1%B alloy and four times as effective as the TICAR alloy. Even with a high residual titanium content, 2.5%Ti-2.5%B was almost twice as effective as the other two alloys. Although the implication is that poor grain refining can be compensated for by doubling the addition rate, this is not true as a site saturation effect appears to be occurring. This concept was introduced by Avrami for describing the process of recrystallization such that further additions cause the grain size to asymptotically approach a

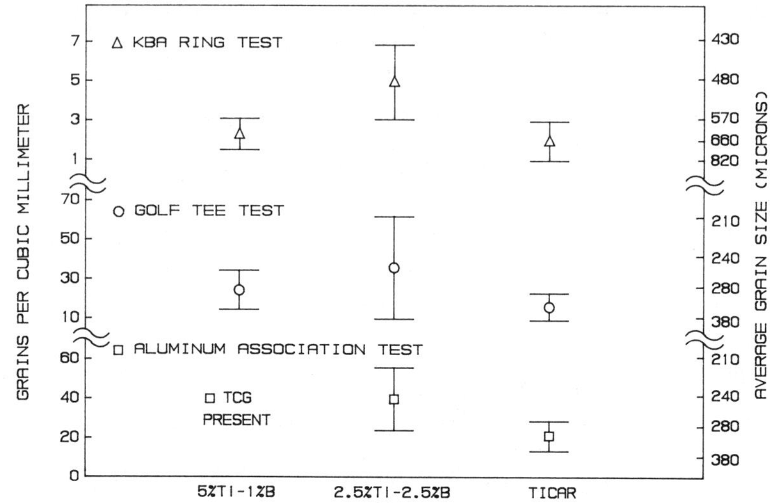

FIG. 3. Grain size results achieved in A356 with 0.2% residual titanium when grain refined with 1 kg/1000 kg, utilizing the alloys indicated. Values given are the averages of 7 samples,held for 5 to 205 minutes, with or without stirring. 5%Ti-1%B exhibited the largest amount TCG in the Aluminum Association Test. See text for details.

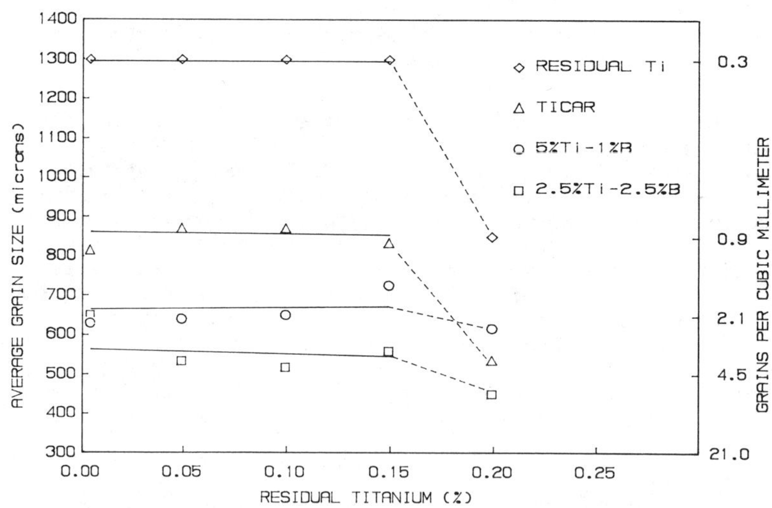

FIG. 4. Grain size results achieved in A356 when grain refined with the indicated alloys as a function of residual titanium content. Note residual titanium alone provides a grain refining effect only when the solubility limit for titanium and aluminum is exceeded (0.15%). All grain refiner alloys appear to benefit from residual titanium at this level.

given "intrinsic" value. This is further supported by the fact that in any system we have examined, we have observed that doubling the addition does not double the number of sites generated.

As reported by Sigworth, the 2.5%Ti-2.5%B provides an alloy with nominally 1.4% excess boron in the form of mixed borides. When this alloy is introduced into the melt, the boron immediately has the opportunity to react with the residual titanium present in the alloy to form TiB_2 or to change the mixed borides to a higher ratio of titanium to boron. This change has been reported as a mechanism for grain refining (11). In the case where 0.2% residual titanium is present, 0.05% is present as $TiAl_3$. Excess boron will have two effects. As above, it will react with the residual titanium in solution to form TiB_2. In addition, it will react with the $TiAl_3$ intermetallic particles by "coating" them with TiB_2 particles, which will render them more stable according to several theories (12). The data shown in Figure 4 is not conclusive but does suggest that there is an additional advantage to being above the peritectic level for titanium. If there is a competing reaction between the silicon, which "tends to coat and poison" or interact with the $TiAl_3$ and TiB_2 phases, the free boron in 2.5%Ti-2.5%B is in competition for titanium which would not exist in the case of the 5%Ti-1%B alloy. Likewise, in the case of the TICAR alloy, one would suspect that the same poisoning mechanism is active and that the surface of $TiAl_3$ particles, which seems to be a necessary component for making the titanium carbon system an effective grain refiner, is altered by the presence of silicon.

It has been noted that commercial grain refiners that exhibit a fine needle type $TiAl_3$ structure require an incubation time for the product to become an effective grain refiner. One explanation is dissolution causes needle particles to break into individual discreet particles so that one particle becomes two particles acting as two nucleation sites. However, at least in the case of the TICAR alloy, this mechanism must not be operative. As previously reported, the grain refining effect for rolled rod product versus as-cast product, where the $TiAl_3$ particle size has been reduced by 75%, does not produce a significant difference in grain refining effectiveness (5). Consequently, in this system, other mechanisms must be operative.

In addition to the incubation time effect, some systems exhibited a fade effect. This effect can come about by several mechanisms. Firstly, one would expect fade if the effective particles were settling. Secondly, one would also expect fade if small effective particles are completely dissolved because of either temperature or concentration effects. Another operative mechanism can be Oswald ripening, in which small particles are dissolving and larger particles are growing, decreasing the total number of available particles. As an adjunct to that effect, the growth of existing particles can lead to a further enhancement of the settling effect. Stirring, of course, will have the effect of reintroducing settled particles into the melt. However, stirring also has the effect of enhancing dissolution if the alloy in question is sensitive to this feature. Stirring also enhances poisoning if surrounding atmospheres are depleted of the offending element in a stagnant liquid. There is also the possibility of agglomeration by which several particles must be considered as an individual particle.

The tests represent different solidification rates but they also represent other features which may influence the results. As noted previously, the directional solidification conditions provided by the Aluminum Association Test, as compared to the Golf Tee Test or Ring Test, indicate sensitivity or susceptibility to TCG or a columnar grain structure in materials marginally grain refined.

Conclusions

This work has confirmed previous study in which it was shown that an alloy containing 2.5% titanium and 2.5% boron provides additional insurance against inadequate grain refining of A356 and A319 alloys. At both high and low residual titanium levels, the data suggests that 5%Ti-1%B is the next best alloy. With high residual titanium, the TICAR alloy is an effective substitute for the 5%Ti-1%B alloy. The 2.5%Ti-2.5%B alloy appears to be substantially superior to the other alloys tested at low solidification rates and where a susceptibility to TCG is indicated by casting conditions. It is also confirmed that A319 alloy is much more difficult to grain refine than A356, except when a high residual titanium level is present.

Acknowledgments

The authors gratefully acknowledge the technical assistance of Messrs. Ty Dunville, Frank Koch, Glenn Mabry, Mike Patton and David Young who assisted in the experimental work and to the management of KB Alloys, Inc. for giving permission to publish this work. Our thanks are also extended to Geoffrey Sigworth for enlightening discussions.

References

1. G. K. Sigworth and M. M. Guzowski: "Grain Refiner for Aluminum Containing Silicon", Patent Pending in the U.S.A. and in other countries.

2. G. K. Sigworth: KB Alloys, Inc., "Third Element Additions to Aluminum Titanium Master Alloys", March 14, 1989, U.S.A. Patent No. 4,812,290.

3. G. K. Sigworth: KB Alloys, Inc., "Third Element Additions to Aluminum Titanium Master Alloys", October 10, 1989, U.S.A Patent No. 4,873,054.

4. Aluminum Association: "Standard Test Procedure for Aluminum Alloy Grain Refiners", TP-1, 1987.

5. R. F. Carver, G. W. Boone and F. P. Koch: "Characteristics of New Generation Grain Refiners", Light Metals 1990, 119th Annual Meeting AIME Proc., February 18-22, 1990, pp. 845-850.

6. J. L. Kirby, R. W. McCarthy and S. A. Levy's: "Grain Size Test Methods Comparisons", Light Metals 1986, 115th Annual Meeting AIME Proc., March 2-6, 1986, pp. 749-757.

7. KB Alloys, Inc.: "Grain Refinement Test", MA-PD9.

8. G. K. Sigworth and M. M. Guzowski: "Grain Refining of Hypoeutectic Al-Si Alloys", AFS Transactions, Vol. 93, 1985, pp. 907-912.

9. W. C. Setzer, G. W. Boone, and B. H. Wilson: "Grain Refining Response Surfaces for Three Commercial Aluminum Alloys", Metallurgical Society of CIM Proc., Vol. 16, 1989, pp. 153-161.

10. W. C. Setzer, G. W. Boone, R. F. Carver and B. H. Wilson: "Grain Refining Response Surfaces in Aluminum Alloys", Light Metals 1989, 118th Annual Meeting AIME Proc., February 27-March 3, 1989, pp. 745-748.

11. M. M. Guzowski, G. K. Sigworth and D. A. Sentner: "The Role of Boron in the Grain Refinement of Aluminum with Titanium", Metall. Trans. A., 1987, Vol. 18A, pp. 603-619.

12. R. Kiusalas and L. Backerud: "Influence of Production Parameters on Performance of Al-Ti--B Master Alloys", Solidification Processing 1987, Institute of Metals, London, 1988, pp. 137-140.

Precipitation hardening in A356 alloys

S. Shivkumar, C. Keller, M. Trazzera and D. Apelian
Aluminum Casting Research Laboratory, Department of Materials Engineering, Drexel University, Philadelphia, Pennsylvania 19104, U.S.A.

ABSTRACT

Cast aluminum alloys are generally heat treated to the T6 condition in order to obtain an optimum combination of strength and ductility. The improvement in strength properties after the heat treatment has been attributed primarily to the precipitation of Mg_2Si during the aging treatment. The precipitation sequence in A356 alloys is as follows: needle shaped GP zones →rod like precipitates → eqilibrium platelets of Mg_2Si. The precipitation characteristics are determined by several parameters including natural aging time, preaging time and temperature, heating rate to the final aging temperature and artificial aging time and temperature. The influence of these parameters on tensile properties of A356 alloys has been studied to determine optimum aging conditions necessary to obtain the desired property level. The data have been used to construct a TTT diagram for the aging process.

KEY WORDS

Aging, Precipitation Hardening, Aluminum Base Alloys, Casting, Heat Treatment

INTRODUCTION

Cast Al-Si-Mg components are generally subjected to a T6 temper in order to obtain an optimum combination of strength and ductility. The heat treatment consists of solution treatment at 540C for 4 to 8 hr, quenching in water and aging at temperatures between 150 to 200C for 4 to 8 hr. The primary purpose of the solution treatment is to place the precipitating phase, Mg_2Si, in solid solution and to alter the Si particle morphology [1,2]. The aging cycle enhances yield strength (YS) and ultimate tensile strength (UTS) substantially while there is a reduction in ductility. In most cases, the castings are used in the underaged condition to obtain an acceptable combination of strength and ductility.

The enhancement of strength properties obtained during the aging treatment is primarily due to the precipitation of metastable phases from the supersaturated solution. The precipitation sequence in Al-Si-Mg alloys is as follows:
Needle-shaped GP zones → rod-like β' precipitates → platelets of Mg_2Si
The decomposition of the supersaturated solution begins with the clustering of silicon atoms [3]. This clustering leads to the formation of coherent spherical GP zones that elongate along the cube matrix direction to assume a needle shape [4]. The zones are initially disordered with a large vacancy concentration and become ordered with longer aging [5]. This needle shaped GP zone corresponds to the monoclinic transition precipitate (β'') reported by other investigators [6]. GP zones are relatively

stable and may exist up to temperatures of about 260C [7]. With prolonged aging, the needle-shaped GP zones grow to form rods of an intermediate phase, β'. The β' particles are semi-coherent with the matrix and the rod axes are parallel to the cube matrix directions. While Jacobs [8] has shown that β' has a hexagonal structure, Livak [5] reports that β' has a face-centered tetragonal structure. The final equilibrium Mg_2Si phase (β) forms as an incoherent platelets on the aluminum matrix and has an ordered face-centered cubic structure that is anti-isomorphous with the flourite (CaF_2) structure with the Si forming an FCC sublattice and the Mg atoms at tetrahedral positions [9]. Peak hardness is achieved before the platelets form [10]. The maximum size of the particles before the hardness begins to decrease is of the order of 0.03 μm.

The mechanical properties in the casting after the aging treatment are significantly influenced by various process parameters such as natural aging, preaging, heating rate to the final aging temperature and artificial aging conditions. Apart from a few publications in the Japanese literature [11-13], most of the previous research on cast Al-Si-Mg alloys has focussed primarily on two parameters: artificial aging temperature and time. Consequently, there are no established guidelines for optimizing the aging parameters to obtain peak tensile properties. Extensive investigation of these parameters in wrought Al-Mg-Si alloys have led to the development of innovative heat treatment procedures [14]. It would be desirable to develop similar aging cycles in cast alloys to maximize tensile properties or perhaps reduce the duration of the aging treatment. In addition, molten metal treatments such as grain refinement and modification may influence the aging characteristics appreciably. Misra and Oswalt [15] report secondary elongation peaks in castings grain refined with Ti. The secondary elongation peaks coincide with the maximum in strength properties. Thus, a different heat treatment procedure may become necessary in grain refined castings. In this contribution, some of the results of an ongoing project on the aging characteristics of A356.2 alloys are presented.

EXPERIMENTAL PROCEDURE

The effects of aging parameters were evaluated for two different casting techniques: green sand casting and permanent mold casting. In both cases, grain refined and modified ASTM B-108 test bar samples were used to examine the influence of selected variables on tensile properties. Sand cast test bars were produced at Littlestown Hardware & Foundry Co., Inc. in Littlestown, Pennsylvania. Permanent mold test bars were cast at Stahl Specialty Co. in Kingsville, Missouri. Salient details regarding the production of test bars are given below. Other pertinent information pertaining to the casting procedure can be found in our previous publication [2,16].

The charge material consisted of 100% primary A356.2 alloy. The metal was thoroughly degassed with high purity nitrogen. Several chemical analysis, vacuum degassing and AluDelta samples were cast periodically to check chemical composition, gas content, grain size and eutectic undercooling. The metal was filtered with ceramic foam filters to reduce the incidence of inclusions in the casting. All the castings were grain refined with Al-5%Ti-1%B master alloy. Both unmodified and Sr-modified castings were produced from the same melt. Sr modification was achieved with a Al-10%Sr master alloy. The pouring temperature was maintained between 760±10C and 740±10C for permanent mold and sand castings respectively. All the castings were radiographed and test bars containing visible defects were rejected. The as-cast test bars were examined carefully by optical and scanning electron microscopy to determine the quality of the cast bars. Typical chemical compositions of permanent mold and sand cast test bars are shown in Table I.

All the heat treatments were conducted in our laboratory at Drexel University. The castings were heat treated in a resistance-heated air circulating box-type muffle (Lucifer) furnace. The T6 heat treatment cycle is summarized below [2, 16]:

- Solutionize permanent mold bars at 550±2C for 50 min
- Solutionize sand cast bars at 550±2C for 100 min
- Quench in water at 60C
- Natural age at room temperature for times varying from 0 min to 72 hr
- Age test bars at temperatures ranging from 145±2C to 201±2C for times

between 0 min to 100 hr

The quench interval was less than 7 s. At least 4 samples were heat treated under identical conditions. The heat treatment process was monitored carefully by positioning several thermocouples in the heat treatment basket. Heat treated samples were analysed to assess the microstructural changes occurring during heat treatment.

Extensive Rockwell hardness measurements were conducted to determine the precipitation characteristics. A 1.59 mm steel ball indentor and a load of 60 Kg (F scale) were used during the measurement. The heat treated samples were subjected to tensile tests in a computer-controlled Instron-1127 machine. A strain rate of 2 mm/min was used in all tests. The yield strength (0.2% proof strength), ultimate tensile strength and %elongation (gage length of 2.5 cm) were evaluated from the tensile tests.

RESULTS AND DISCUSSION

A typical natural aging curve is shown in Fig. 1. It is seen that precipitation does not begin for at least about 30 min at room temperature. Subsequently, the hardness begins to increase and a maximum value is observed after about 2 days. The total increase in hardness is of the order of 30 to 40%. No significant difference was observed between the aging behavior of permanent mold and sand cast alloys. Sr modification did not have any detectable influence on the precipitation kinetics. Aging curves for a temperature of 171C are also plotted in Fig. 1. Clearly, the rate of hardening increases with temperature. In this case, the total increase in hardness is of the order of 60 to 70%. When the temperature is increased above 250C , there is a sharp decrease in hardness after about 10 to 15 min. Hardness in samples aged at 171C and 300C for 15 min is of the order of 75 and 58 respectively. Furthermore, at times greater than about 30 min to 1hr, the T6 hardness at 300C is lower than the as cast value.

The data shown above indicate that precipitation in cast Al-Si-Mg alloys occurs very rapidly. Even at room temperature precipitation begins at relatively short times. Evidence of precipitation can be detected within about 30 min at room temperature. By comparison, in wrought Al-Mg-Si alloys containing the same amount of Mg_2Si, precipitation rates are very slow and at least 20 hr may elapse before there is any significant change in hardness at room temperature [17]. The rapid precipitation in cast alloys has been attributed to the presence of excess silicon in solid solution. The term "excess silicon" refers to the amount of silicon that is in excess of what is stochiometrically necessary for the formation of Mg_2Si. Even small amounts of excess silicon have a significant effect on the precipitation rate. In A356 alloys containing about 0.4% Mg, excess Si is typically of the order of 1.35%. The solid solubility of Mg_2Si in aluminum is reduced slightly by the presence of excess silicon. Also, excess silicon, leads to an increase in the solvus temperature at a given Mg_2Si level [5]. Thus at any given temperature a greater degree of supersaturation is achieved in alloys containing excess silicon and hence, a finer dispersion of precipitates is obtained [18].

The time to attain a particular value of hardness is plotted in Fig. 2 as a function of temperature. As can be expected, increasing the time or temperature improves the hardness of the sample. In general, it can be observed that increasing the temperature by 10C is equivalent to enhancing the aging time by a factor of two. Similar results have also been reported by Drouzy et al [19]. The hardness data can be used to construct a TTT curve shown in Fig. 3. It is evident that the highest precipitation rates are obtained in the temperature range 250 to 400C. Therefore, it is imperative that castings be cooled rapidly in this temperature range. The nose of the TTT curve is placed at about 45 s to 1 min. Thus, quench rates should be such that the residence time of the casting in the temperature range 250 to 400C is less than about 45 s. This result is agreement with the data of Tsukuda et al [12,13] who have reported that mechanical properties start to deteriorate when the quench interval is greater than about 30 to 45 s.

Tensile data for various test conditions are shown Tables II to V. The general trend is such that increasing the aging time or temperature improves strength properties and lowers elongation. At temperatures less than about 175C, yield strength increases gradually with aging time and a maximum value is attained after about 10 to 12 hr (Fig. 4). As the temperature increases beyond this value, the material hardens very rapidly and a high YS is observed even at short times. By comparison, variations in the ultimate tensile strength with aging conditions is relatively small.

The ductility of the sample exhibits an an interesting behavior. Initially, elongation decreases with aging time as the material hardens (Fig. 5). At aging times greater than about 10 to 12 hr, however, elongation begins to increase. No significant reductions in strength properties are observed at this point. In order to investigate this behavior further, several samples were aged for times of up to 100 hr. At aging temperatures less than about 180C, elongation continues to increase slowly, while there is a rapid decrease in strength properties. Note that in Table III, at a temperature of 155C, elongation increases from about 4.1 to 6.4% while UTS decreases from 325 to 164 MPa for aging times of 12 hr and 72 hr respectively. At temperatures greater than about 180C, there is a substantial improvement in ductility at long aging times. For example, elongations greater than about 25 to 30% were observed in samples aged at 200C for 100 hr (Fig. 6). In this case, the strength decreases rapidly to the as cast value.

Both sand cast and permanent mold test bars exhibit a similar behavior during the aging treatment. Hence, the cooling rate during solidification does not appear to have any significant effect on the precipitation characteristics. Similar results have been reported by Gustafsson et al [20]. An examination of the data also reveals that Sr modification does not affect the precipitation sequence. It should be noted, however, that Na modification has been reported to lower the age hardening rate [21]. This smaller age hardening rate has been attributed to the formation of NaAlSi ternary compounds.

Various events occurring during the aging treatment can be analysed by plotting the UTS as a function of elongation (Fig. 7). The curve begins at point A which represents the value in the solution treated and quenched (T4) condition. As precipitation begins to occur, the material begins to harden and the strength increases, while there is an overall reduction in ductility. After a sufficiently long period of aging, the material attains a peak strength value. This point corresponds to a minimum in the elongation. Subsequently, however, the strength remains fairly constant at the peak value while elongation increases. It is this portion of the curve that requires attention. A purely overaging phenomenon would suggest that increases in elongation should be accompanied by a reduction in the strength levels. But the strength is essentially constant at about 340 MPa. Note that the difference in UTS and YS at this point is relatively small (< 30 MPa) indicating that the material does not work harden appreciably. A similar behavior has been reported in Al-2.6%Li-0.09%Zr alloys where the minimum in elongation has been observed to precede the maximum in strength properties [22]. (In most precipitation hardenable systems the minimum in elongation generally coincides with the maximum in strength.) This unique variation of strength and ductility has been attributed to the intense planar deformation occurring in the sample. In cast aluminum alloys, it is possible that the observed trend may be related to the secondary elongation effects reported previously [15]. All the samples in this study were grain refined with Ti-B. $TiAl_3$ particles in the sample may have influenced the precipitation of Mg_2Si as indicated by Misra and Oswalt [15]. After sufficiently long times, the precipitates begin to coarsen and there is a substantial increase in elongation. It is interesting to note that at an aging temperature of 201C, elongation after about 100 hr is greater than the T4 value indicating the effectiveness of Orowan looping.

It appears that the improvements in ductility observed at long aging times are more significant in permanent mold samples than in sand cast bars. It can be seen from Tables IV and V that nearly all the permanent mold samples exhibit a good combination of yield strength and ductility at aging times of the order of 10 to 12 hr. Yield strengths of 305 to 340 MPa and elongations of 5 to 7% are observed at several temperatures. The overall increase in elongation at high aging times also depends on temperature. Note that at an aging temperature of 201C, an elongation of 9.6% is observed after 12 hr. The yield strength at this point is around 340 MPa. Such a combination of strength and ductility is extremely desirable in premium quality aluminum castings.

The influence of natural aging on T6 hardness is shown in Fig. 8. During the experiment, sectioned samples obtained from permanent mold bars were either directly aged at 171C for various times or stored at 25C for 24 hr and subsequently aged at 171C. In this case, the high temperature aging treatment was conducted in a salt bath furnace. A silicone heat transfer fluid (Syltherm 800 supplied by Dow Corning) was used in place of the salt bath. In the samples aged immediately after quenching, the hardness increases rapidly and a high value of hardness is obtained after about 3 to 4 hr. In naturally aged samples, however, precipitation occurs relatively slowly and the maximum value of hardness is attained after about 5 to 6 hr. Furthermore, the peak hardness in naturally aged castings is lower than in samples that were directly aged at the high temperature.

The effect of natural aging time on T6 tensile properties is shown in Table VI. Both YS and UTS decrease with increasing natural aging time. It appears that this decrease occurs within the first 4 to 6 hr of natural aging time. This behavior is observed in both unmodified and modified sand cast bars. Subsequently, the strength properties do not change appreciably. Similar results have been reported by Fujuki et al [12]. Natural aging, however, has a beneficial effect on elongation. For example, in modified sand cast alloys, elongation increases from 4.6% in samples aged immediately to about 6.9% in castings natural aged for 24 hr.

The effects of storage at room temperature on T6 properties have been studied by several investigators [4,23,24]. It is generally indicated that natural aging has a detrimental effect on strength properties. This behavior may be explained based on the model developed by Pashley et al [14]. During natural aging two factors need to be considered: (i) formation of clusters and (ii) supersaturation of the matrix. If the clusters formed during the natural aging process are stable (or have attained a critical radius), they can act as nuclei for the formation of GP zones when the alloy is aged at the high temperature. Consequently, a large nucleation density is obtained and the strength properties increase. Such behavior has been reported in Al-Zn-Mg alloys and in pseudo-binary Al-Mg_2Si alloys at certain concentrations of Mg_2Si [14,25]. However, if the clusters have not attained a critical radius, the supersaturation in the matrix diminishes and the size at which the clusters become stable during artificial aging increases. In this case, the clusters may dissolve during artificial aging until some of them are stabilized by the increased level of solute in the matrix. Thus large clusters grow at the expense of smaller ones leading to a fewer number of coarse precipitates. This phenomenon may explain the observed decrease in strength properties in A356 alloys when they are subjected to natural aging.

Alloying elements in solid solution may exert an influence on the extent of reduction in strength properties when the castings are stored at room temperature. In general, solute atoms which enhance the rate of clustering at room temperature increase the sensitivity to natural aging. Because of high rates of clustering, the effects of natural aging are enhanced in the presence of excess Si [26,27]. Ghate et al [28,29] have shown that the decrease in strength properties is proportional to the Mg concentration. This effect has been attributed to an increase in the rate of cluster formation. It should be noted that the rate of clustering increases at high Mg contents because of the greater supersaturation in the matrix. Hence, it would appear that 357 castings are more sensitive to room temperature storage than 356 specimens.

Two techniques have been suggested in the literature to overcome the detrimental effects of storage at room temperature. Pashley et al [14] propose innovative aging cycles to improve the mechanical properties. According to their model, a short high temperature treatment prior to artificial aging may be beneficial. This high temperature treatment may redissolve all the clusters formed during storage at room temperature and thus enhance the supersaturation in matrix. Consequently, the nucleation density increases and the tensile properties may be improved. Recently, there are indications that the effect of natural aging may be reduced considerably when trace elements such as In, Sn, Cd or Cu are present [12,28-30]. The concentration of these elements necessary to produce the desired effect is generally between 0.05 to 0.1%. It is suggested that these elements may form vacancy-solute complexes and thus lower the rate of cluster formation.

In many commercial applications, it may be necessary to heat treat castings which are already in the T6 condition, especially in components which undergo minor repairs. The influence of subjecting castings to the same heat treatment twice is shown in Table VII. In this case, the castings were solutionized at 550C for 50 (Sand cast) to 100 min (permanent mold), quenched in water at 60C, and held at room temperature for 24 hr. Sand cast samples were then aged at 155C for 4 hr and permanent mold bars were aged at 171C for 4 hr. Immediately after the artificial aging, castings were quenched in water and then subjected to the whole heat treatment cycle once again. It can be seen from Table VII that strength properties after the first and second heat treatment cycle are comparable. But the ductility after the second heat treatment is significantly lower than after the first cycle. This behavior is observed in sand cast and permanent mold samples which are unmodified or Sr modified.

CONCLUSIONS

Precipitation of Mg_2Si in cast aluminum alloys is a complex process and is influenced by several process parameters including alloy composition, natural aging time and preaging and artificial aging conditions. Because of the presence of about 1.4% excess silicon in solid solution, precipitation of Mg_2Si occurs much faster than in wrought alloys. TTT diagrams indicate that the rate of precipitation is extremely rapid in the temperature range 250 to 400C. At these temperatures, precipitation begins within about 30 to 60 seconds and hence, it is imperative that castings be cooled very rapidly in this temperature range during quenching.

An increase in elongation has been observed at aging times of about 10 to 12 hr. Interestingly, this increase is not accompanied by a reduction in tensile strength. Tensile properties as high as 305 to 340 MPa YS, 340 to 355 MPa UTS and about 10% elongation have been measured in sampled aged at 201C for 12 hr. At this temperature, elongations of about 25 to 30% have been estimated in samples aged for times greater than about 100 hr. The increase in elongation at large aging times may be very useful in the production of premium quality castings and needs further investigation.

Storing the castings at room temperature may lower the strength properties by about 10 to 20%. Maximum strength values may be obtained when the castings are aged immediately after quenching. It appears that the reduction in strength properties because of natural aging occurs within 4 to 6 hr of storage at room temperature. Natural aging, however, has a beneficial influence on ductility.

ACKNOWLEDGEMENTS

This research was conducted as a part of an ongoing research program at the Aluminum Casting Research Laboratory (ACRL). The authors would like to gratefully acknowledge the financial support of the consortium of companies supporting the Aluminum Casting Research Laboratory: Alcan International, Aluminum Company of America, CMI International, COMALCO Aluminum, Doehler-Jarvis, KB Alloys Inc., Littlestown Hardware and Foundry Co., Inc., Metallurgical Products and Technology, Metaullics Systems, Pechiney Corporation, Reading Foundry Products, Reynolds Metals Co., Selee Corporation,. Shieldalloy, and Stahl Specialty Co. The assistance of Mr. Len Potter of Littlestown Hardware and Foundry Co., Inc. and Mr. Ken Whaler, Mr. Richard Andriano and Mr. Frank DeHart of Stahl Specialty Co. in the production of castings is greatly appreciated.

REFERENCES

1. D. Apelian, S. Shivkumar and G. Sigworth, *AFS Transactions* , **97**, (1989), 727-742
2. S.Shivkumar, S. Ricci, Jr., B. Steenhoff, D.Apelian and G.Sigworth, *AFS Transactions* , **97**, (1989), 791-810
3. J. Lendvai, T. Ungar and J. Kovacs, *Mat. Sci. & Eng.* , **10**, (1963), 151-159
4. H. Westengen and N. Ryum, *Z. Metallkund* , **70**(8), (1979), 528-535
5. R.J. Livak, *Met. Trans. A* , **13A**, (1982), 1318-1321
6. T. Shchegoleva, *Physics of Metals and Metallography* , **25**(2), (1968), 56-64
7. A. Lutts, *Acta Met.* , **9**, (1961), 57-69
8. M. Jacobs, *Phil. Mag.* , **26**, (1972), 1-19
9. E. Ozava and H. Kimura, *Acta Met.*, **18**, (1970), 995-1006
10. J.F. Mondolfo: "Aluminum and its Alloys: Structure and Properties", Butterworth and Co., London and

Boston, 1976

11. S. Fujuki, M. Tsukuda, S. Koike and I. Fukui, *J of Japan Inst of Light Met.*, **33**(12), (1983), 712-718
12. S. Fujuki, M. Tsukuda, S. Koike and I. Fukui, Journal of Japan Institute of Light Metals, 1982, 32(6), 277-283
13. M. Tsukuda, S. Koike and K. Asano, *Journal of Japan Institute of Light Metals*, **28**(11), (1978), 531-540
14. D.W. Pashley, M.H. Jacobs and J.T. Vietz, *Phil. Mag.*, **16**, (1967), 51-76
15. M.S. Misra and K.J. Oswalt, *Trans AFS* , **90**, (1982), 1-10
16. S. Shivkumar, S. Ricci, Jr., C. Keller and D. Apelian, *J. of Heat Treating*, (1990), in press
17. R.C. Dorward, *Met Trans A* , **4**, (1973), 507-512
18. M. Kanno, H. Suzuki and Y. Shiraishi, *J. of Jpn. Inst. of Light Metals* , **28**(11), (1978), 553-557
19. M. Drouzy, S. Jacob and M. Richard, *AFS International Cast Metal Journal,* (1980), 43-50
20. G. Gustafsson, T. Thorvaldsson and G.L. Dunlop, <u>The Metallurgy of Light Alloys</u> [Proc. Conf.], Loughborough University, England, 24-26 Mar. 1983, 288-294
21. H. Tamada and T. Tanaka, *Shiga-Kenritsu Tanki Daigaku Gakujutsu Zasshi* , **16**, (1975), 43-47
22. J.E. Pegram, T.H. Sanders, Jr., C.J. Hartshorn, P.C. Mckeighan, M.G. Valentine and B.M. Hillberry, <u>Aluminum-Lithium Alloys - Vol. 1</u>: Ed. T.H. Sanders, Jr. and E.A. Starke Jr., 5th International Conference on Al-Li Alloys, Williamsburg, VA, March 27-31, 1989, 261-272
23. D.W. Pashley, J.W. Rhodes and A. Sendorek, *J. of Inst. of Metals* , **94**, (1966), 41-49
24. R.C. Harris, S. Lipson and H. Rosenthal, *Trans AFS* , **64**, (1956), 470-482
25. H. Nakamura, *J. of Japan Inst. of Light Metals,* **23**(9), (1973), 389-392
26. H. Suzuki and M. Kanno, *J. of Japan Inst Met.*, **29**, (1979), 197-208
27. M. Tsukuda, M. Harada, T. Suzuki and K. Susumu, *Journal of Japan Institute of Light Metals*, **28**(3), 1978, 109-115
28. G.P. Ghate, K.S. Raman and K.S.S. Murthy, <u>International Conference on Aluminum - 85</u> [Proc. Conf.], 30th oct-2 nov 1985, New Delhi, India, 485-494
29. G.P. Ghate, K.S.S. Murthy and K.S. Raman, *Aluminium* , **60**(1), (1984), 18-19
30. Gouthama and Kishore, *Trans. Indian Inst. of Metals* , **36**(4), Aug-Oct 1983, 283-289

Table I

Chemical composition (wt%) of test bars

Element	Sand Cast		Permanent Mold	
	Unmodified	Modified	Unmodified	Modified
Si	7.19	7.10	6.90	6.83
Mg	0.40	0.39	0.41	0.42
Fe	0.09	0.10	0.11	0.09
Ti	0.16	0.17	0.15	0.17
Sr	0.002	0.022	0.003	0.023
P	-	-	0.0013	0.0014
Sb	-	-	0.0059	0.0034
Cr	0.0012	0.0017	0.0022	0.0013
Ni	0.0028	0.0048	0.0053	0.0050
Cu	0.001	0.001	0.022	0.0149

Table II

Tensile Properties of green sand cast unmodified alloys for various aging conditions

Time (hr)	YS (MPa)	UTS (MPa)	% Elongation
		145C	
2	190.4 ± 20.0	278.6 ± 17.2	9.8 ± 2.3
4	226.9 ± 26.2	289.0 ± 26.9	6.6 ± 2.8
6	239.3 ± 24.8	294.5 ± 24.8	8.4 ± 5.4
10	228.3 ± 21.4	287.6 ± 12.4	4.1 ± 2.2
12	240.7 ± 31.7	296.6 ± 14.5	5.9 ± 1.1
		155C	
2	239.3 ± 12.4	289.7 ± 10.3	9.7 ± 1.7
4	229.7 ± 14.5	298.0 ± 21.4	5.4 ± 2.4
6	244.8 ± 29.0	305.5 ± 6.2	7.1 ± 1.6
10	251.7 ± 27.6	313.8 ± 22.8	1.4 ± 0.7
12	269.0 ± 17.2	287.6 ± 18.6	1.1 ± 0.1
		165C	
2	193.8 ± 13.1	277.3 ± 12.4	6.8 ± 1.5
4	235.2 ± 13.1	284.2 ± 13.1	5.5 ± 2.8
6	261.4 ± 5.5	302.8 ± 28.3	6.2 ± 2.9
10	297.3 ± 26.2	306.2 ± 29.0	1.9 ± 0.6
12	284.2 ± 20.7	328.3 ± 15.9	4.8 ± 0.7
		175C	
2	309.7 ± 17.2	338.6 ± 4.1	3.3 ± 0.9
4	313.8 ± 11.0	338.6 ± 11.7	1.7 ± 0.4
6	319.3 ± 15.9	348.3 ± 18.6	1.0 ± 0.1
10	327.6 ± 6.9	356.6 ± 5.5	1.4 ± 0.4
12	330.4 ± 2.8	356.6 ± 24.8	1.4 ± 0.5
		185C	
2	320.7 ± 22.8	341.4 ± 10.3	2.1 ± 1.1
4	316.6 ± 15.2	340.7 ± 10.3	2.7 ± 1.1
6	319.3 ± 25.5	350.4 ± 8.3	2.3 ± 1.3
10	329.0 ± 6.2	342.1 ± 6.9	1.4 ± 0.5
12	333.8 ± 13.8	352.4 ± 2.8	1.9 ± 0.2

Table III

Tensile properties of green sand cast Sr-modified alloys for various aging conditions

Time (hr)	YS (MPa)	UTS (MPa)	% Elongation
		145C	
2	215.2 ± 17.9	289.0 ± 12.4	7.8 ± 1.9
4	217.3 ± 16.6	304.8 ± 4.8	10.1 ± 3.2
6	212.4 ± 7.6	304.8 ± 13.8	9.6 ± 6.2
10	238.6 ± 26.2	284.8 ± 19.3	5.3 ± 2.2
12	247.6 ± 15.9	298.6 ± 10.3	7.8 ± 3.0
		155C	
2	224.2 ± 4.8	260.0 ± 19.3	8.1 ± 4.0
4	232.4 ± 17.2	280.0 ± 21.4	4.8 ± 2.7
6	231.0 ± 17.9	286.2 ± 21.4	3.5 ± 1.2
10	289.0 ± 8.3	329.0 ± 6.2	1.6 ± 0.8
12	318.6 ± 8.3	325.5 ± 4.1	4.7 ± 2.4
72	117.9 ± 17.2	161.4 ± 9.0	6.4 ± 1.8
		165C	
2	176.6 ± 2.1	283.5 ± 11.0	8.5 ± 1.3
4	200.7 ± 1.4	291.1 ± 10.3	9.8 ± 2.9
6	253.8 ± 16.6	304.2 ± 6.2	5.6 ± 1.4
10	332.4 ± 14.5	355.9 ± 8.3	2.1 ± 0.3
12	280.7 ± 26.2	327.6 ± 12.4	4.2 ± 0.9
		175C	
2	289.7 ± 15.9	332.4 ± 10.3	2.3 ± 0.8
4	318.6 ± 8.3	340.0 ± 9.7	3.5 ± 0.8
6	311.7 ± 9.0	351.1 ± 15.9	2.1 ± 0.8
10	347.6 ± 17.2	362.1 ± 9.0	1.8 ± 0.2
12	329.0 ± 30.3	361.4 ± 8.3	2.1 ± 0.4
		185C	
2	324.8 ± 6.2	342.1 ± 11.0	3.2 ± 1.9
4	318.0 ± 4.8	333.1 ± 11.7	1.9 ± 0.6
6	327.6 ± 17.2	349.7 ± 15.9	1.8 ± 0.5
10	345.5 ± 7.6	355.2 ± 4.1	2.3 ± 0.6
12	318.6 ± 21.4	333.8 ± 17.2	1.8 ± 0.2

Table IV

Tensile properties of permanent mold cast unmodified alloys for various aging conditions

Time (hr)	YS (MPa)	UTS (MPa)	% Elongation
		161C	
2	204.8 ± 11.7	323.5 ± 9.0	9.5 ± 0.7
6	320.7 ± 30.3	363.5 ± 9.0	5.3 ± 1.6
10	304.2 ± 9.0	369.0 ± 4.8	4.3 ± 1.2
12	295.2 ± 33.1	366.9 ± 1.4	3.7 ± 0.5
18	327.6 ± 25.5	371.1 ± 13.8	5.7 ± 3.1
		171C	
2	258.6 ± 13.1	345.5 ± 9.7	8.9 ± 1.6
4	297.3 ± 13.1	353.8 ± 18.6	4.1 ± 2.1
6	318.0 ± 24.1	368.3 ± 5.5	5.8 ± 2.5
10	333.1 ± 20.7	360.0 ± 4.1	4.9 ± 0.4
12	321.4 ± 35.9	358.0 ± 11.7	5.4 ± 2.2
		181C	
2	266.2 ± 17.2	315.2 ± 14.5	10.4 ± 2.4
4	311.7 ± 25.5	359.3 ± 7.6	8.4 ± 3.0
6	324.8 ± 31.0	354.5 ± 11.0	5.4 ± 2.4
10	304.2 ± 13.8	338.6 ± 5.5	6.7 ± 5.4
12	315.9 ± 6.9	361.4 ± 2.1	5.0 ± 1.2
		191C	
2	309.7 ± 17.2	338.6 ± 4.1	3.2 ± 1.1
4	345.5 ± 11.0	361.4 ± 15.9	5.8 ± 1.5
6	306.2 ± 4.8	338.6 ± 11.7	3.3 ± 1.0
10	303.5 ± 16.6	338.6 ± 14.5	3.1 ± 0.9
12	315.9 ± 30.3	346.9 ± 7.6	4.2 ± 0.8
		201C	
2	299.3 ± 26.2	331.1 ± 20.0	3.2 ± 1.0
4	304.2 ± 7.6	326.2 ± 17.9	4.9 ± 3.3
6	306.2 ± 30.3	337.3 ± 12.4	6.1 ± 1.1
10	329.0 ± 6.2	338.0 ± 6.9	1.4 ± 0.5
12	313.1 ± 17.2	343.5 ± 17.2	9.6 ± 3.5
100	128.3 ± 24.1	155.2 ± 20.0	25.6 ± 5.7

Table V

Tensile properties of permanent mold cast Sr-Modified alloys for various aging conditions

Time (hr)	YS (MPa)	UTS (MPa)	% Elongation
		161C	
2	222.8 ± 14.5	318.0 ± 11.0	7.8 ± 1.9
6	306.9 ± 8.3	371.7 ± 8.3	6.0 ± 2.7
10	320.0 ± 15.2	343.5 ± 13.1	1.5 ± 0.5
12	306.2 ± 28.3	366.2 ± 11.7	3.6 ± 1.0
18	338.6 ± 30.3	359.3 ± 6.2	2.6 ± 0.9
		171C	
2	253.1 ± 16.6	344.9 ± 2.1	8.6 ± 4.4
4	298.6 ± 35.2	353.8 ± 9.0	5.5 ± 1.5
6	298.0 ± 11.0	354.5 ± 5.5	5.4 ± 1.7
10	325.5 ± 29.0	371.7 ± 4.8	4.8 ± 0.9
12	315.2 ± 26.9	366.9 ± 2.1	5.3 ± 1.3
		181C	
2	262.8 ± 14.5	339.3 ± 17.9	13.6 ± 3.9
4	282.1 ± 11.0	358.0 ± 7.6	4.9 ± 1.9
6	315.2 ± 13.8	352.4 ± 5.5	3.9 ± 1.0
10	333.8 ± 15.2	363.5 ± 10.3	5.1 ± 1.6
12	340.0 ± 20.0	366.9 ± 2.1	6.5 ± 0.7
		191C	
2	290.4 ± 15.9	332.4 ± 10.3	2.7 ± 0.9
4	322.1 ± 33.1	371.7 ± 5.5	6.6 ± 1.4
6	344.9 ± 14.5	362.1 ± 7.6	4.4 ± 2.0
10	315.2 ± 17.2	355.9 ± 11.0	6.6 ± 2.1
12	305.5 ± 27.6	336.6 ± 11.0	6.6 ± 1.0
		201C	
2	306.9 ± 22.1	334.5 ± 21.4	4.7 ± 1.9
4	287.6 ± 12.4	310.4 ± 8.3	7.0 ± 2.3
6	298.6 ± 11.7	329.0 ± 6.9	8.3 ± 1.4
10	345.5 ± 7.6	355.2 ± 4.1	2.3 ± 0.6
12	346.2 ± 10.3	355.9 ± 9.0	9.6 ± 3.2
100	107.6 ± 7.6	149.0 ± 13.1	26.7 ± 2.6

Table VI

Variation of T6 properties with natural aging time in sand cast test bars

Natural Aging Time (hr)	Unmodified			Sr modified		
	YS (MPa)	UTS (ksi)	%Elon.	YS (MPa)	UTS (MPa)	%Elon.
0	246	303	3.7	255	311	4.6
4	211	279	4.5	239	287	5.6
16	232	280	5.3	214	281	5.2
24	243	287	5.1	222	282	6.9

Table VII

Tensile properties of samples subjected to a T6 heat treatment cycle twice.

Condition	After 1st cycle			After 2nd cycle		
	YS (MPa)	UTS (ksi)	%Elon.	YS (MPa)	UTS (MPa)	%Elon.
PM - U	296	352	4.1	302	352	2.0
PM - M	297	352	5.5	310	361	2.4
SC - U	246	314	4.9	254	310	1.8
SC - M	266	306	4.4	292	312	2.0

PM - U : Permanent Mold - Unmodified SC - U : Green Sand Cast - Unmodified
PM - M : Permanent Mold - Sr-Modified SC - M : Green Sand Cast - Sr-Modified

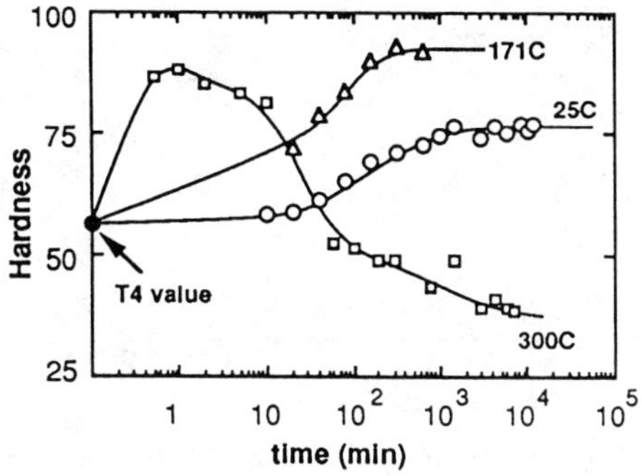

Fig. 1 Variation of hardness (Rockwell F) with aging temperature and time.

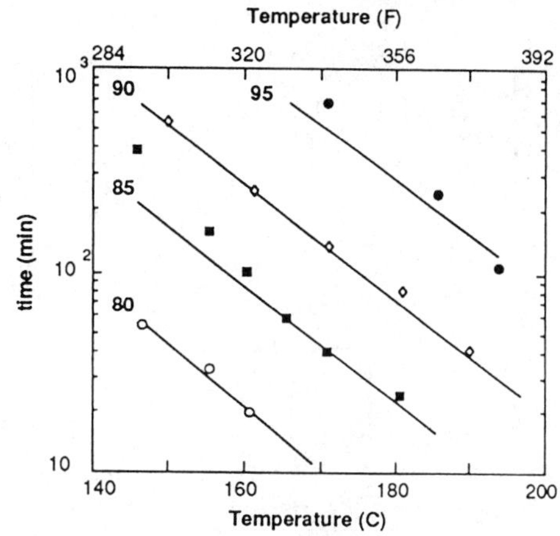

Fig. 2 Time required to attain a particular value of hardness (Rockwell F) as a function of temperature.

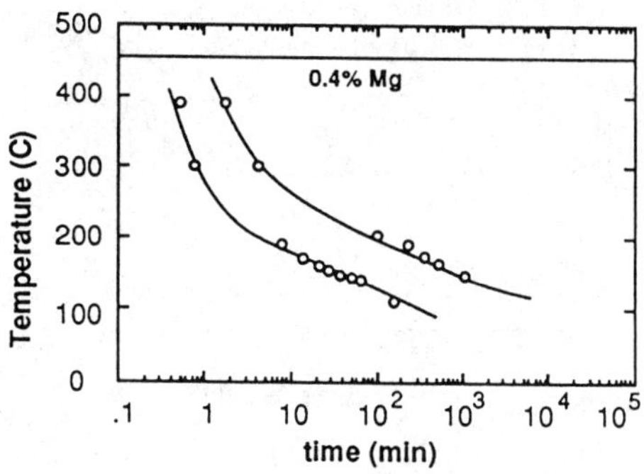

Fig. 3 TTT curves for A356 alloy containing 0.4% Mg.

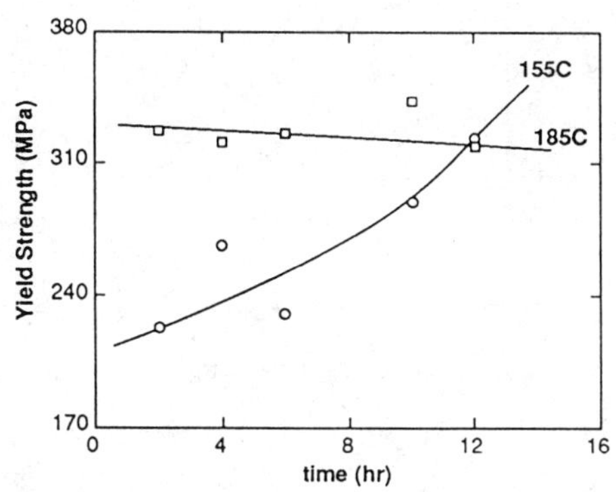

Fig. 4 Variation of T6 yield strength with aging time and temperature.

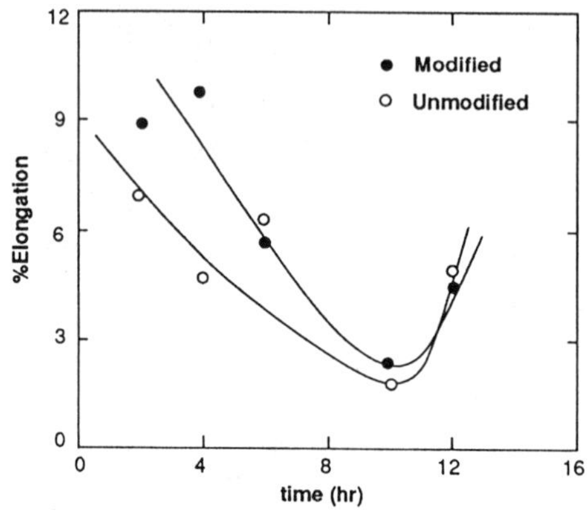

Fig. 5 Variation of T6 elongation in unmodified and modified alloys. The castings were aged at 155C.

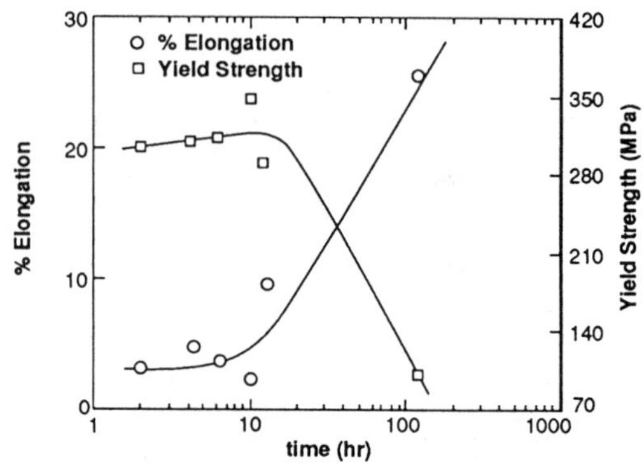

Fig. 6 Variation of T6 elongation and yield strength with aging time. The castings were aged at 201C.

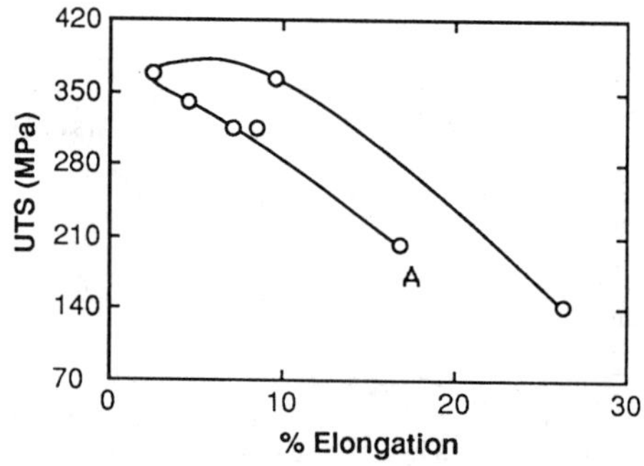

Fig. 7 Variation of ultimate tensile strength (UTS) and elongation in samples aged at 201C.

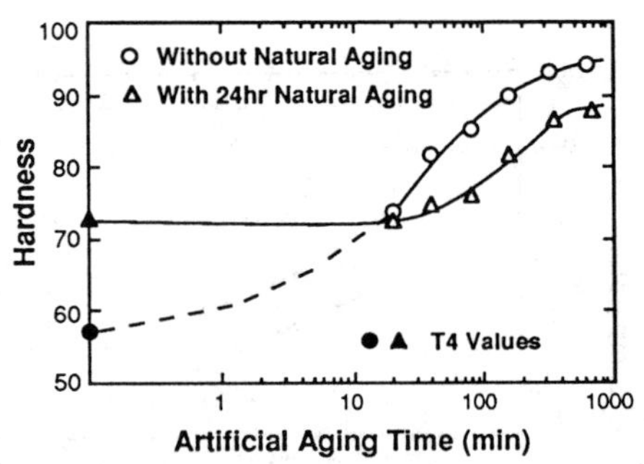

Fig. 8 Variation of T6 hardness (Rockwell F) in samples aged directly after quenching and in samples stored at room temperature for 24 hr before the artificial age.

Effect of Be on S phase precipitation in pseudobinary Al (α)-Al_2CuMg alloy

W. Fang and W.V. Youdelis
Engineering Materials Group, University of Windsor, Windsor, Ontario, Canada, N9B 3P4,

Abstract

The effect of microadditions of Be on the aging behaviour of Al-2.5%Cu-1.2%Mg alloy is investigated. It is shown that the addition of 0.15%Be significantly increases the peak hardness levels when the alloy is aged at various temperatures from room temperature to 300°C. Optical and scanning electron microscopy results show a significantly higher S phase (Al_2CuMg) precipitate density for the Be-containing alloy, indicating a Be-enhanced nucleation rate for its precursor phases, S' and GPB zones. The S precipitates are rod or lath-like in shape, and show an orthogonal orientation in their distribution.

Keywords

Age-hardening Al alloys, S phase precipitation, nucleation, precipitate morphlogy

Introduction

Aluminum alloys containing copper and magnesium in Cu:Mg weight ratios ranging from 2.2:1 to 7:1 form the basis of several commercially important age-hardening alloys. The strengthening of these alloys is associated with the presence of coherent (GP) zones and finely dispersed, partly-coherent precipitates: S' (Al_2CuMg) for the pseudobinary alloys (2.2:1), and both S' and θ' (Al_2Cu) for the alloys with the higher Cu:Mg weight ratio (1). The S phase is orthorhombic, and its partly-coherent precursor (S') tends to nucleate preferentially at dislocation lines and grow as lath-like particles with a {120} habit plane (2,3). The nucleation and refinement of the S' phase is enhanced by plastic deformation (which increases the dislocation density), which increases the peak hardness values of the aging curves (3). It has been shown that small additions of Ag (4) or Si (2) also increase the nucleation rate and refinement of the S' precipitate, with attendant increases in the peak hardnesses for the aging curves.

Previous investigations by one of the authors and co-workers have shown small additions of Be to be highly effective in accelerating precipitation in Al alloys through the nucleation mechanism. The Be-enhanced nucleation and refinement of θ' in Al-3.0%Cu[1] alloy (5,6) and β' (Mg_2Si) in the natural aging Al-0.75%Mg-0.5%Si alloy (7,8) with corresponding improvements in the age-hardening responses of the alloys, suggested that Be may also have a beneficial effect on precipitation in Al-Cu-Mg alloys. In this investigation, the effect of Be on the aging behaviour and microstructure of the pseudobinary Al-Al_2CuMg alloy is determined and the results compared with those of previous investigators.

---

1 All concentrations in wt%.

Experimental

Superpure Al and Mg (99.99%), electrical conductivity grade Cu (99.99%), and Al-5.23%Be master alloy were used in the preparation of an Al-2.5%Cu-1.2%Mg base alloy and the base alloys containing 0.15 and 0.25%Be. The alloys were prepared in graphite crucibles by induction melting in air, using a 15 kVA Ajax Inductotherm unit at a frequency of 10 kHz. The melts were heated to well above the liquidus temperature (~750°C), maintained for about 10 minutes to ensure complete homogenization, then stirred and poured into graphite molds 25 mm dia. x 70 mm length at room temperature.

For the age-hardening study, sections ~5 mm thick were cut from the central region of the ingots. The solution treatment for all samples consisted of a 60 h anneal at 575°C, followed by a quench in iced brine. The samples were then aged at five temperatures: room temperature (22°C), 130°C, 190°C, 240°C, and 300°C, in an electric furnace with the temperature controlled to ±1K. Microhardness measurements (HV0.05) to monitor the age-hardening process were carried out by constantly interrupting the aging treatments. At least five readings were randomly taken for each hardness determination to obtain a mean with a typical uncertainty of about 5%.

Specimens for optical and scanning electron microscopy (~5 mm thick) were prepared in the conventional manner, starting with 0.05 μm alumina and finishing with colloidal silica suspension (Buehler) as the polishing media. The etchant used was dilute hydrofluoric acid solution (0.5 ml HCl (48%) + 99.5 ml H_2O) and Keller's etchant.

Results and Discussion

Hardness Measurements

The age-hardening curves are shown in Fig. 1(a) to 1(e). The room temperature age-hardening results for the base alloy reported by Wilson et al (2) show good agreement with the present investigation. To determine if the independent precipitation of Be from the supersaturated alloy contributed to the age-hardening process, an Al-0.2%Be alloy was subjected to an aging treatment for two solution treatment temperatures, 540°C and 630°C, the latter to obtain maximum supersaturation. The room temperature aging curve for the Al-0.2%Be alloy (solution treated at 630°C) is given in Fig. 1(a). No significant precipitation hardening is associated with Be, and this was also confirmed for the higher aging temperatures (where a decrease in hardness results due to precipitation of the excess Be from solution). It is evident that the Be-containing alloys attain hardness maximums or peaks 15 to 35 points above those for the base alloy. The 190°C aging temperature gives the highest hardness peak (~HV 130) for the Be-containing alloys compared to the ~HV 110 for the base alloy after a 1 day age. The hardening rate is significantly increased when aging at 240°C, the Be-containing alloy reaching a maximum hardness of ~HV 120 after approximately 2 hours aging time, compared to ~ HV 82 for the base alloy. It is also evident that increasing the Be content to 0.25% in the alloy does not significantly change the age-hardening response of the alloy from that obtained with 0.15%Be.

The age-hardening curves show evidence of several stages for the precipitation process depending on the aging temperature. Three stages are readily distinguished for the base alloy and both Be-containing alloys for the 190°C aging treatment. The 300°C age-hardening curves show two stages for the Be-containing alloys and a single stage for the base alloy. The dependency of the number of stages for the age-hardening curves on aging temperature is well known. Hardy (9) reported that for the Al-Cu-Mg alloys with the Cu:Mg weight ratio of 7:1, three stages are evident in the curves at lower aging temperatures, decreasing to two and then one stage as the aging temperature increases above

220°C. The stages are generally associated with cluster and coherent zone formation, followed by their transformation (or dissolution and reformation) as partly-coherent and the non-coherent (stable) precipitate forms. Assuming the generally accepted sequential representations of the transformation stages for precipitation in age-hardening alloys, and in the absence of corroborative TEM results, the following stages might be associated with the 190°C aging curve (identified in Fig. 1(c)): stage (1) - GPB zone formation associated with the gradually rising curve to ~HV 110 (0.1 day), stage (2) - S' formation as the curve rises to ~HV 130 ($\lesssim$1 day), stage (3) - resolution of GPB zones to produce the initial fall in the curve to ~HV 120, and nucleation of additional S' phase to produce the second (lower) peak at ~HV 125 (~1 day) and finally overaging of S' to give the non-coherent S phase as the aging curve continues to fall.

It is noted that the initial (as-quenched) hardness values show a certain amount of variation, which is a complex function of several factors, including cooling rate (quench severity), alloy composition, as well as the elapsed time and temperature in preparing the alloy samples for the hardness measurements. However, the consistently higher initial hardness of the Be-containing alloy is in agreement with earlier work (10) which showed that Be had a strong solid solution hardening effect in Al.

Optical and Scanning Electron Microscopy

An optical and scanning electron microscopy study was carried out to determine the effect of Be on the distribution and density of the precipitating phase after aging. To obtain an optically resolvable precipitate required severely overaging the alloy, which was accomplished by a 300°C age for 7 days. Figures 2(a), (b) and (c) show the rod-like morphology of the S precipitate, which also appears round-like depending on the orientation of the grain. The large, grey, globular phase at the grain boundaries for the Be-containing alloys has the likeness of free Be that is observed in Al-Be alloys exceeding the maximum Be solubility in Al (10), which is approximately 0.1%Be at 644°C (11). The rod or lath-like morphology of the S phase is more readily evident in the SEM micrographs shown in Figures 3(a), (b) and (c). An orthogonal relationship in the orientation of the laths is evident, and is in agreement with the observations of other investigators that S' laths precipitate with a {120} habit plane and grow in <100> directions in the Al lattice (2,12). Both the optical and SEM micrographs clearly show a much higher density of precipitate particles in the Be-containing alloys, estimated to be 3 to 4 times that in the base alloy. The higher S precipitate density suggests a Be-enhanced nucleation rate for its precursor, i.e., the S' phase, and perhaps also the GPB zones, since Weatherly (13) reports strong evidence for the transformation of the GPB zones to S' precipitates for the pseudobinary alloy containing Si (Al-2.7%Cu-1.5%Mg-0.2%Si). An investigation is now underway to obtain TEM results for the earlier stages of the transformation (GPB zone and S' formation), and the kinetic parameters for the transformation using the resistivity method described in the investigation on precipitation in the Al-Cu (6) and Al-Mg-Si (7) alloys.

Conclusions

1. The addition of 0.15%Be to the pseudobinary Al-2.5%Cu-1.2%Mg alloy increases the S precipitate density, obtained after a solution and aging treatment, by a factor of 3 or 4 times, indicating a Be-enhanced nucleation rate for S'.
2. The S precipitates are lath or rod-shaped, and show an orthogonal orientation relationship in their distribution consistent with the reported {120} habit plane and <100> growth directions in the Al lattice.
3. The increased precipitate density for the Be-containing alloys is associated with significantly higher age-hardening rates and hardness levels.
4. The 0.15%Be and 0.25%Be alloys show no significant difference in precipi-

tate density or age-hardening behaviour, suggesting that the optimum effect is obtained with Be saturation in the Al lattice.

Acknowledgement

The authors wish to acknowledge the Natural Sciences and Engineering Research Council of Canada for financial support (grant OGP 0000836) and the Aluminum Co. of Canada, Kingston, Ontario, for providing the superpure aluminum.

References

1. Silcock, J.M. (1960-61). J. Inst. Met., 89, 203.
2. Wilson, R.N., D.M. Moore and P.J.E. Forsyth (1967). J. Inst. Met., 95, 177.
3. Sen, N. and D.R.F. West (1969). J. Inst. Met., 97, 87.
4. Polmear, I.J. (1964). Trans. Met. Soc. A.I.M.E., 230, 1331.
5. Karov, J., W.V. Youdelis and R. Herring (1986). Materials Science and Technology, 2, 547.
6. Karov, J., and W.V. Youdelis (1987). Materials Science and Technology, 3, 1.
7. Xiao, T. and W.V. Youdelis (1988). Proc. Inter. Symposium on Reduction and Casting of Aluminum, Ed. Christian Bikert, Pergamon Press, p. 37.
8. Xiao, T. and W.V. Youdelis (1989). Materials Science and Technology, 5, 991.
9. Hardy, H.K. (1954-55). J. Inst. Met., 83, 17.
10. Lowes, T.D., M.A.Sc. Thesis, University of Windsor, 1986.
11. Murray, J.L. and D.J. Kahan. Bulletin of Alloy Phase Diagrams, 4 (1), (1983).
12. Radmilovic, V., G. Thomas, G.J. Shiflet, and E.A. Starke, Jr., (1989). Scripta Metallurgica, 23, 1141.
13. Weatherly, G.C., Ph.D. Thesis, Univ. Cambridge (1966).

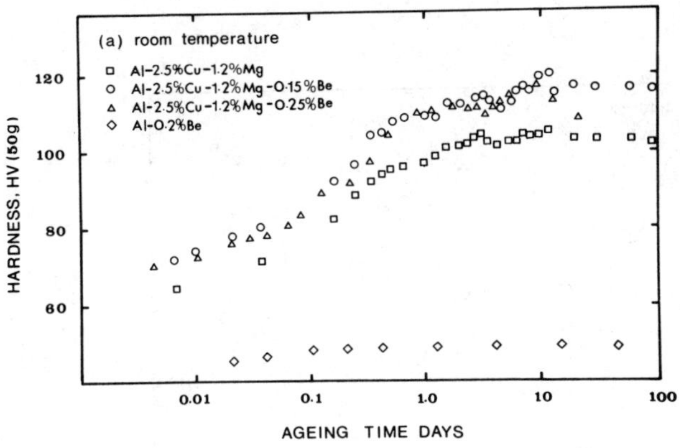

FIG. 1(a)

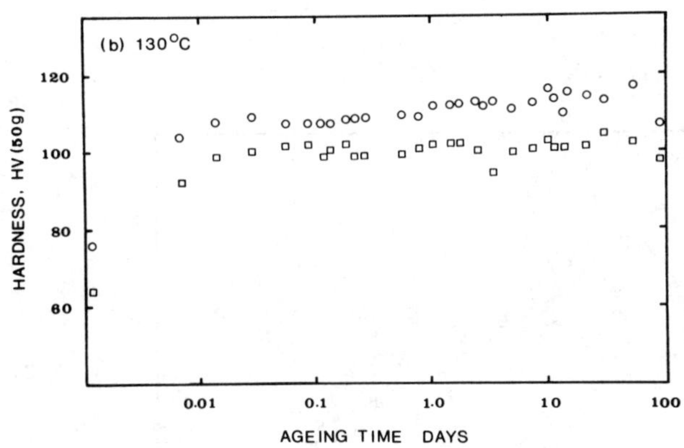

FIG. 1(b)

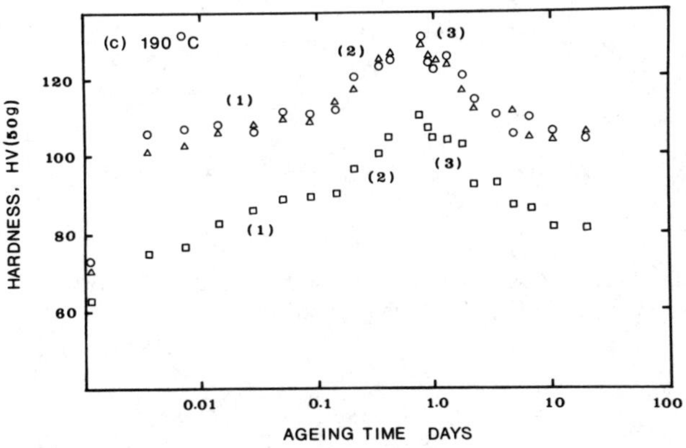

FIG. 1(c)

FIG. 1 (Continued).

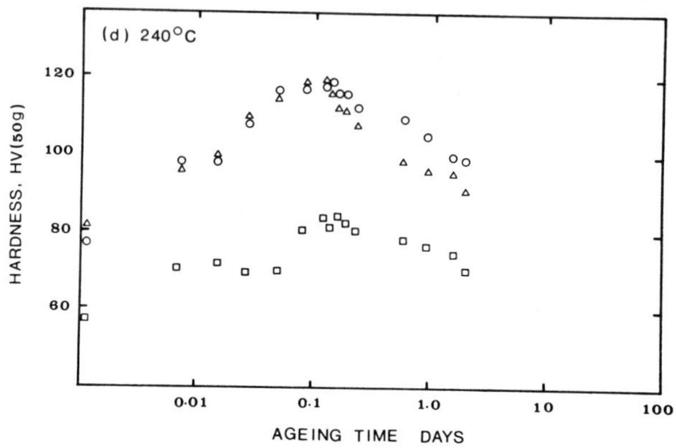

FIG. 1(d)

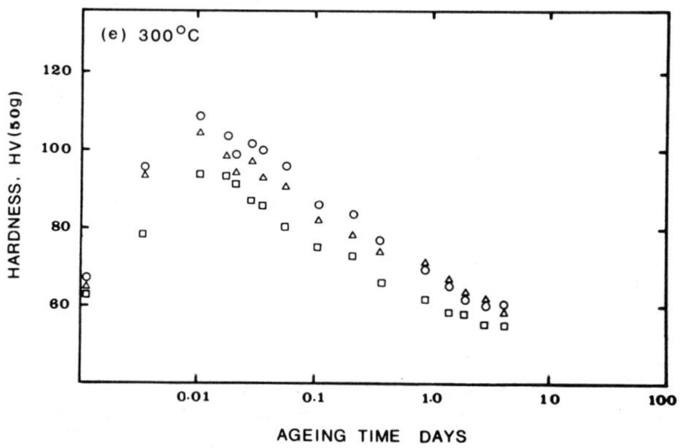

FIG. 1(e)

FIG. 1. Age-Hardening Curves for Alloys Aged At Various Temperatures (50 g Hardness Load).

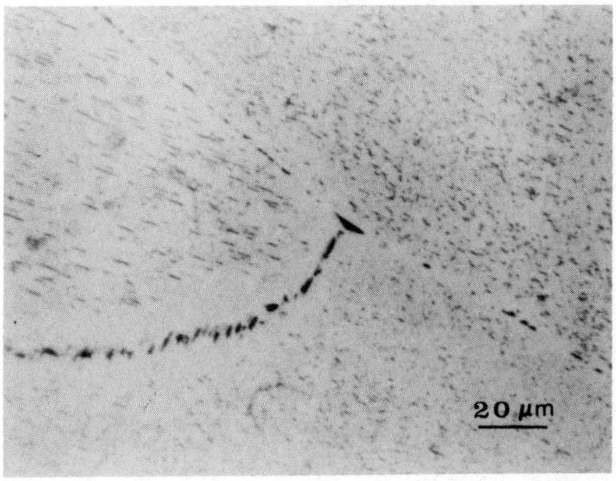

(a)

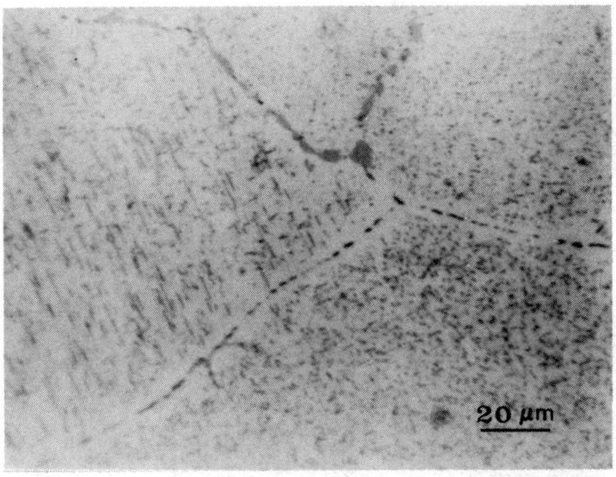

(b)

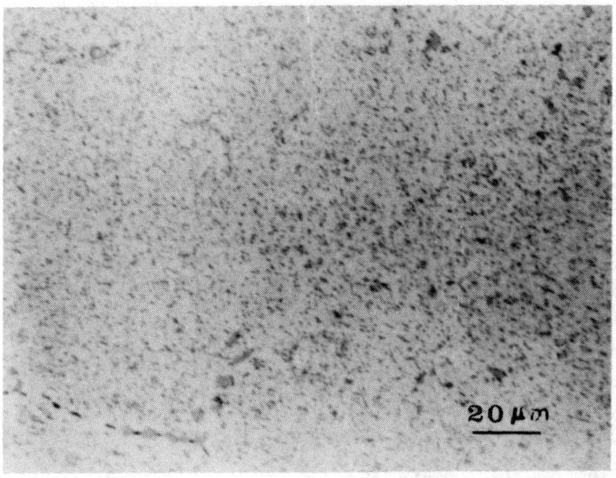

(c)

FIG. 2. Alloys Aged for 7 Days at 300°C (Optical). (a) Al-2.5%Cu-1.5%Mg; (b) Al-2.5%Cu-1.5%Mg-0.15%Be; (c) Al-2.5%Cu-1.5%Mg-0.25%Be.

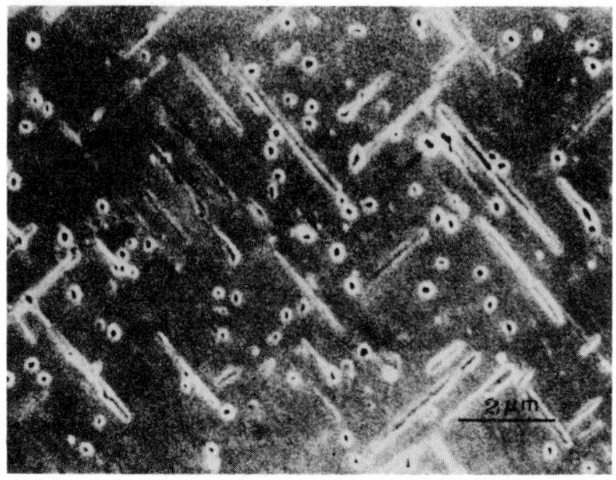

(a)

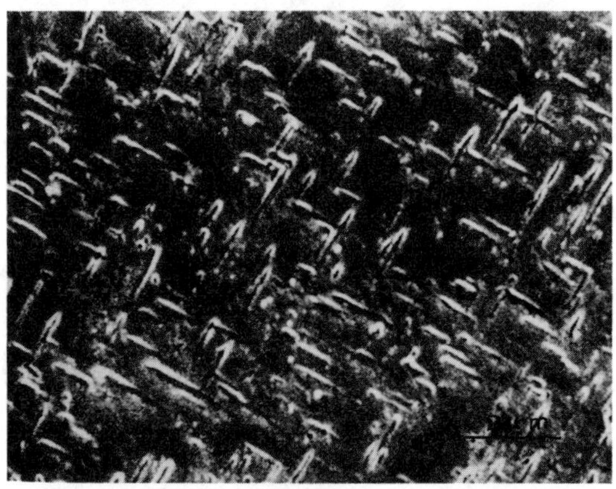

(b)

(c)

FIG. 3. Alloys Aged for 7 Days at 300°C (SEM). (a) Al-2.5%Cu-1.5%Mg; (b) Al-2.5%Cu-1.5%Mg-0.15%Be; (c) Al-2.5%Cu-1.5%Mg-0.25%Be.

Microstructural characterization of centrifugally atomized Al-Li-X powders: effect of Si and Co additions

F.H. Samuel
Department of Applied Sciences, Université du Québec à Chicoutimi, Chicoutimi, Quebec, Canada, G7H 2B1

G. Champier
Lab. de Physique du Solide, École des Mines, Parc Saurupt, 54042, Nancy, France

Abstract

The present observations made on the microstructure of centrifugally atomized Al-3 wt% Li-X powders, where X=Si and Co, can be summarized as follows:

i) Binary Al-Li alloy exhibits a dendritic microstructure over the entire range of selected particle size, 25 to 200μm.
ii) Addition of 3 wt% Li to the Al-Si system shifts the eutectic point to less than 4 wt% Si. The eutectic structure is a mixture of aluminum and AlLiSi phases. Increasing the cooling rate, through reducing the powder particle size, increases the extent of solid solubility.
iii) Addition of Co in quantities up to 0.8 wt% changes the dendritic structure of Al-3 wt% Li into complete cellular structure, regardless of the powder particle size.

Introduction

Rapid solidification of powders is an attempt to maximize the physical and chemical properties through control of the distribution of solute elements. The homogeneity of the powders can be explained in terms of extremely small casting of the order of 4μg. This mass is allowed to cool rapidly from a molten state at a rate of 10^4 to 10^6 K/s. These rates are many orders of magnitude greater than those employed in conventional casting (1).

In binary Al-Li alloys, the strengthening effect is due mainly to the long range order of the Al_3Li precipitates (2). Nevertheless, this strengthening is reduced when the particles are sheared by the dislocations and this produces a slip concentration leading to cracking in the extended slip bands and/or across grain boundaries. Furthermore, the precipitate free zones which appear along the grain boundaries are sites of preferential deformation leading to high stress concentration at grain boundaries: cracks can nucleate and propagate in an intergranular way within the precipitate free zone. These mechanisms reduce the ductility of the Al-Li binary alloys (3).

A method for making the deformation less non-homogeneous is to add elements that form incoherent dispersoids during solidification and/or heat treatment. Several types of additives have been studied, such as Cu, Mg, Mn, Zr, Ti (4).

In the present work we report on the effect of light elements like Si (with respect to the density of pure aluminum) and heavy metals like Co on the microstructure of as-solidified Al-Li powders. Table I shows the expected phases and their densities which ultimately affect the alloy density. The volume fractions of these phases could only be calculated for Co since it forms no compounds with Li.

TABLE I
Precipitate specifications

System	Precipitation	Structure	Density ($g\ cm^{-3}$)
Al-Li	Al_3Li	Cubic	2.33
Al-Li-Si	AlLiSi	Cubic	1.95
Al-Li-Co	Al_9Co_2	Monoclinic	3.62

Experimental Procedures

Alloys used in the present study were prepared from high purity metals (99.99%) under an inert atmosphere of argon. Rapidly solidified powders were produced by the conventional centrifugal atomizing process, with helium used as the cooling media to obtain high speed convective coolings. Ingots of 400g were melted in a graphite crucible having an orifice of 1mm diameter and coated with a thick layer of boron nitride refractory. The atomizing chamber was evacuated and filled with helium up to 1 bar. The liquid metal was ejected onto a rotating atomizer (a 43mm diameter graphite disc, coated with boron nitride, heated to 800°C, and running at 27300 rev./min.) with a helium overpressure of 0.4 bar. At the end of atomization, the temperature of the rotating atomizer had fallen between 600 and 650°C. The designation of the alloys and their chemical compositions, are given in Table II.

The microstructures of powder particles (with sizes ranging between 25 and 200μm) were examined by light microscopy, scanning electron microscopy (SEM operating at 14kV), and transmission electron microscopy (TEM operating at 200kV). The precipitates were identified by X-ray diffraction using a Co Kα source.

TABLE II
Alloy Specifications

Designation	Composition (wt%)			Initial Temperature (°C)
	Li	Si	Co	
L3	3	-	-	800
LS1	2.65	1	-	850
LS2	2.65	2	-	850
LS4	2.7	4.1	-	850
L88	2.7	7.4	-	850
LC4	-	-	0.4	800
LC8	-	-	0.8	800

Both Si and Co form a eutectic structure with Al (binary system). The eutectic temperature, eutectic composition and maximum solubility at room temperature are given in Table III.

TABLE III
Details of Al-X Systems

System	Eutectic Composition of X (wt%)	Eutectic Temperature (°C)	Maximum Solubility (wt%)
Al-Li	9.9	600	1.28
Al-Si	12.6	577	4.1
Al-Co	1.0	657	nil

Results and Discussion

Al-Li-Si system

Four Al-Li-Si alloys were prepared by induction melting of Al-Li alloy and Si in a crucible of graphite coated with boron nitride in open air; they were chill cast into graphite molds. The atomic concentrations of Li and of Si were, respectively, LS1: 8.97% and 0.98%; LS2: 8.9% and 1.79%; LS4: 8.74% and 3.78%; LS8: 8.47% and 7.28%.

The microstructures were examined by light microscopy. Fig. 1a shows the essential features of LS3 revealing the presence of primary α-Al crystals separated by the eutectic product (Al+AlLiSi phases). Increasing the silicon content up to 4 wt% (Fig. 1b) resulted in replacing the primary α-Al crystals with primary AlLiSi crystals surrounded by silicon depleted regions. The rest of the liquid decomposed into the eutectic mixture, constituting about 80% of the microstructure (based on quantitative metallographic measurements). These results, thus, enable us to say that the addition of 3 wt% Li to Al-Si alloys shifts the eutectic point from 12.6 wt% for the binary system to less than 4 wt% for the ternary one.

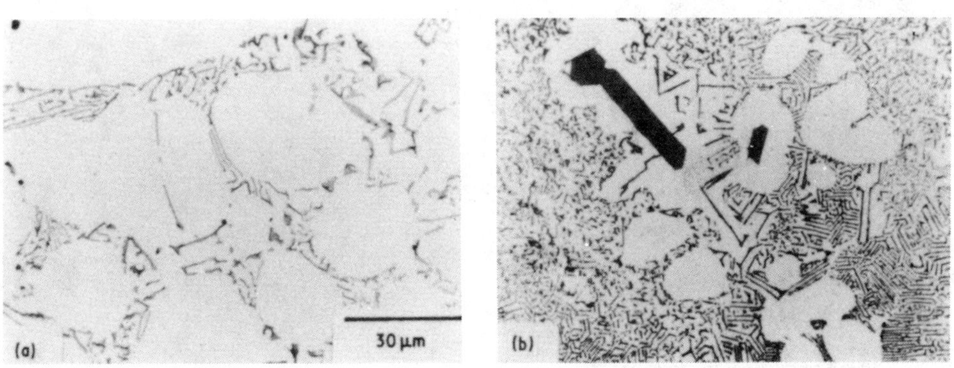

Figure 1 Optical micrographs of chill-cast Al-3 wt% Si alloys: (a) 2 wt% Si, (b) 4 wt% Si. Cooling rate $\dot{T} \sim 10^2$ K sec $^{-1}$

The microstructure of the powder particles is greatly influenced by the powder particle size. For sizes up to 50μm, the microstructure (Fig. 2a) reveals a dendritic structure with AlLiSi fine precipitates delineating the secondary dendrite arms. From measurements of the secondary dendrite arm spacing, the cooling rate was determined to be in the range 10^4 to 10^6 K/s.

Increasing the particle size up to 80µm (Fig. 2b) produces a microstructure similar to that presented in Fig. 1a for LS3 chill-cast alloy. Comparing these two figures highlights the refinement effect of rapid solidification. From the TEM micrograph (Fig. 3) representing a eutectic region, the distance between two AlLiSi lamellae is of the order of 0.1 to 0.2µm, i.e. one-tenth of the corresponding distance obtained from Fig. 1a. According to the Al-Li equilibrium diagram (5), the maximum solubility of Li in α-Al is 5 at%. The present alloy contains ~10 at% Li. Therefore, this alloy does not show any effective volume fraction of δ' (Al_3Li) precipitates which are responsible for strengthening the alloy on ageing.

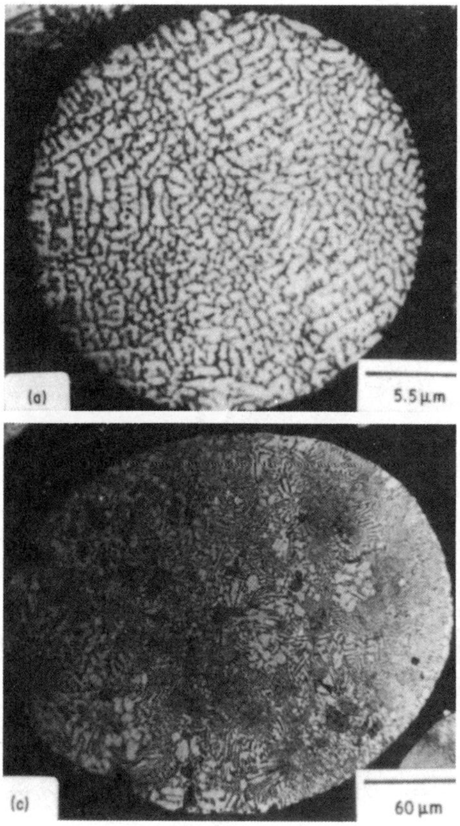

Figure 2 Optical micrographs of LS4 powder of different particle sizes: a) ~50µm, b) ~80µm ($\dot{T} \sim 10^4$ to 10^6 K sec^{-1}), c) ~200µm.

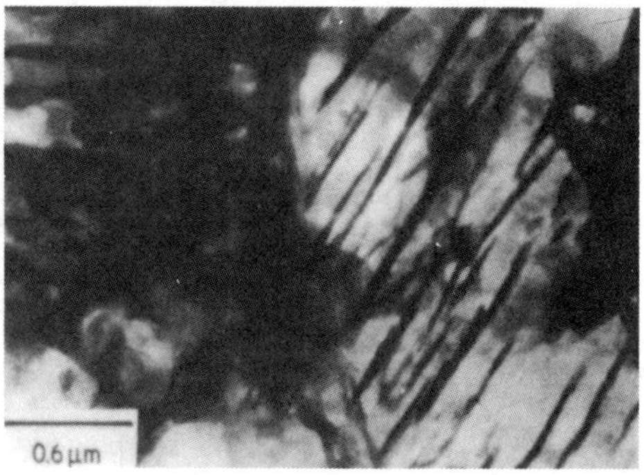

Figure 3 TEM micrograph of a eutectic area corresponding to Fig.2b.

Fig. 2c is the microstructure of powder particle size of the order of 200μm. The main features in this micrograph are essentially the same as those shown in Fig. 1b for LS4 chill-cast alloy. The effect of rapid solidification appears in reducing (i) the amount of primary AlLiSi crystals (~7% in Fig. 2b and 20% in Fig. 1b), and (ii) the size of AlLiSi primary crystals from 10 to 50μm in solid-state alloy to 1 to 3μm after rapid solidification. Comparing Figs. 2b and 2c emphasizes the extended solid solubility, enhanced by increasing the cooling rate via decreasing the powder particle size.

The above mentioned results allow us to conclude that if the Si/Li atomic concentration ratio is less than unity, all the silicon is included in the AlLiSi phase. If the Li concentration is less than 5 at%, Li remains in solid solution with no occurrence of $\delta'(Al_3Li)$ precipitation. Thus, for obtaining an appreciable volume fraction of Al_3Li precipitates in the presence of Si, the concentration of Li in the alloy should be sufficiently high. This, in turn, reduces further the alloy density (ρ). According to our density measurements, performed on loose powders containing various concentrations of Si and Li, the density of an Al-Li-Si alloy can be expressed by the relation,

$$\rho = 2.7 - 0.24\ C_{Si} - 2.008\ C_{Li}$$

where C_{Si} and C_{Li} are, respectively, the atomic concentrations of Si and Li. The expected density of the present alloy is 2.49 x 10^3 kg/m^3.

During annealing of the LS1 alloy, precipitates of the Al_3Li phase appeared inside the α-Al grains; they were 10-30nm in diameter. Some precipitate-free zones appeared around the initial AlLiSi phase particles. The size of the latter remained constant but new, very fine particles of AlLiSi phase precipitated inside the α-Al grains; they were 15nm in diameter and led to the formation of a narrow precipitate-free zone along the grain boundaries.

In the LS2 alloy, the precipitation of Al_3Li was again visible but the density of these precipitates was lower. The initial AlLiSi precipitates remained practically unchanged and very fine new ones appeared with greater density. In the LS4 and LS8 alloys no more Al_3Li phase precipitated and only a very small enlargement of the previous AlLiSi precipitates was the main observation.

AL-Li-Co system

The microstructure of L3 powder reveals well-defined colonies of dendrities for particles of 200μm size as shown in Fig. 4a. These dendrites are comprised of cylindrical arms probably elongating along the three equivalent <100> directions. In this case the solidification is presumably proceeding as an advancing front that radiates from the initial point of nucleation. Coarse particles >200μm exhibit a mixed dendritic/equiaxed type of structure. The equiaxed structure was characterized by the absence of directional growth in a polished section.

In the LC4 powder with a particle diameter larger than 25μm, a large proportion of the microstructure is cellular. Also, in the LC8 alloy quenched from 800°C, the cellular structure is predominant, irrespective of particle size (Fig. 4b). We are assuming here that all Co will precipitate in the form of Al_9Co_2 which has a density of 3.6 x 10^3 kg/m³. Under these conditions, the volume fraction of Al_9Co_2 precipitates is about 2.25%. What needs to be emphasized is that the partitioning of intermetallic Al_9Co_2 particles may reduce the stability of a planar liquid-solid interface, leading to cellular ondulations during cooling of a superheated liquid alloy droplet. For this the G_L/R ratio (G_L=thermal gradient in liquid, R=interface velocity) should be high enough to allow the transition from dendritic to cellular structure for a given cooling rate.

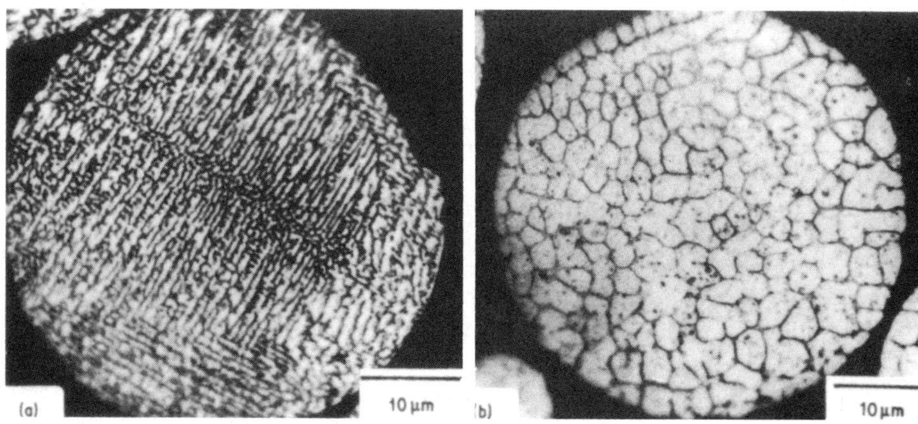

Figure 4 Optical micrographs of 80µm powder particle size of (a) Al-3 wt% Li, (b) Al-3 wt% Li-0.8 wt% Co.

A direct relation between SDAS and powder particle diameter is obtained for L3 alloy as shown in Fig. 5. In the atomized condition, Webster et al.(6) have reported that binary Al-Li alloys with Li content in the range of 3 to 5 wt% and particle dimensions ranging from 5 to 200µm, show a dendritic structure. Their powders were produced by argon gas atomization (GA method) yielding a cooling rate in the range of 10^2 to 10^4 K/s. The results of Webster et al. on Al-2.7wt% Li are superimposed in Fig. 5. As can be seen, their values are almost four times the present case. The difference between the present results and theirs is explicable in terms of the gas thermal conductivity (1.7 x 10^{-2} and 15 x 10^{-2} W/mK for argon and helium, respectively). From the plot of Matyja et al. (7), the average cooling rate for the present alloys is determined to be about 10^4- 10^6 K/s.

Figure 6a is a typical electron micrograph from L3 showing dendritic colonies of ~2-3µm diameter. The interdendritic network phase could be indexed on the basis of Al_3Li (δ') phase. Figure 6b shows a region consisting of images of a primary dendrite (1-1.5µm) with associated secondary dendrite arms (~0.5µm). The cooling rate estimated from Fig. 6b is about 10^6 K/s against 10^4 K/s as determined from Fig. 5.

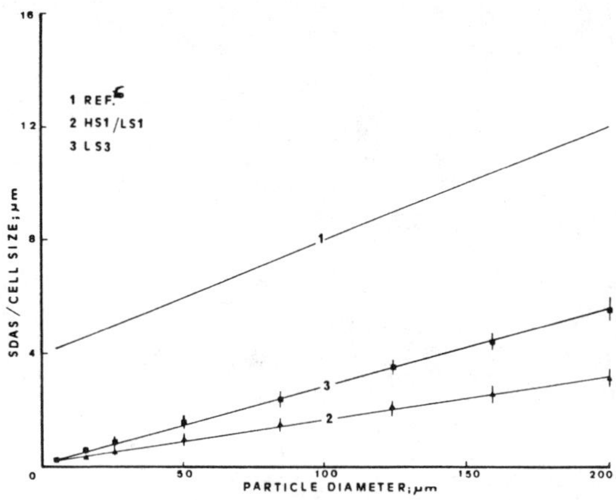

Figure 5 Variation in secondary dendrite arm spacing (SDAS) and cell width as a function of powder particle diameter deduced from light microscopy.

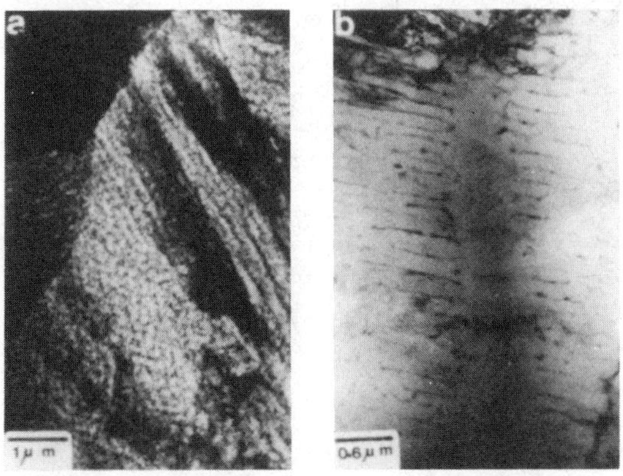

Figure 6 TEM micrographs of as-quenched powder in Al-3 wt% Li alloy.

The microstructure of LC8 reveals a cellular structure. Fig. 7 corresponds to this alloy and consists of two high-angle cells where the rows of low-angle cells are only seen in one of them (area A).

Figure 8 represents the light micrograph of $\sim 6\mu m$ particle diameter of LC8 showing three nucleation events in the form of a spherical cap occuring at the droplet surface. Each event is comprised of two parts, one being the featureless "white zone" without response to etching (marked H) and the other, a cellular zone characterized by radial growth. The microstructural details of the white zone are revealed in Figs. 8b and 8c. The former is a bright field micrograph exhibiting the presence of columnar grains of $\sim 0.2\mu m$ width and $\sim 1.5\mu m$ length. This structure is developed in two stages: first, the supersaturated solid solution of X-Al is formed from the melt by partitionless solidification, and second, the solid decomposes by precipitation during continuous cooling immediately after

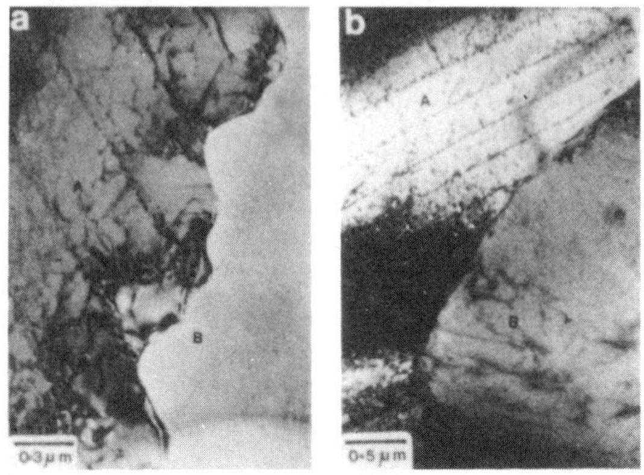

Figure 7 Two TEM micrographs of as-quenched powders of LC8 alloy

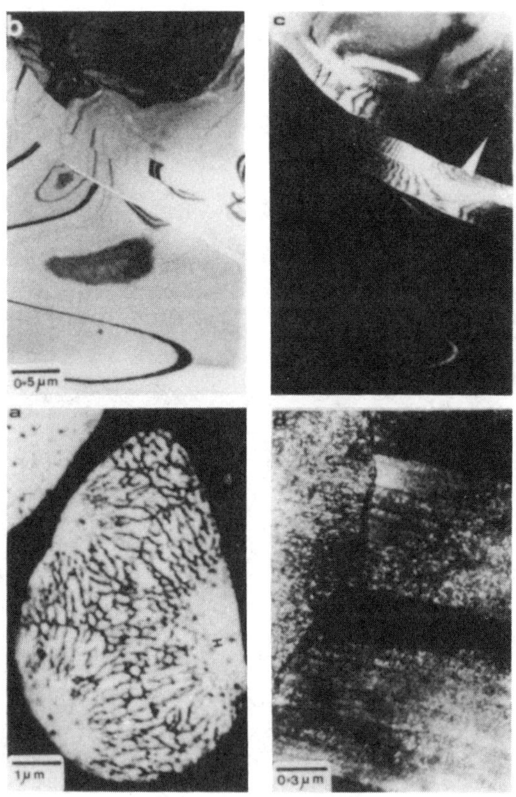

Figure 8 A series of micrographs of as-quenched 6μm powder particles in LC8 alloy, (a) light; (b), (c), (d): TEM

solidification. The achievable undercooling for this microstructure is about 360K. Immediately at the end of this micrograph, the interface is seen breaking down, leading to a cellular structure, Fig. 8d.

As the droplet recalesces, the interface velocity (R) and undercooling are reduced. When the interface temperature goes above the temperature required for supersaturation, solute is rejected ahead of the interface and favours cell formation. The solute in the intercellular regions has a higher supersaturation. This gives rise to nucleation of Al_9Co_2 in the liquid ahead of the interface. Now, for the trapping of an Al_9Co_2 particle by a moving interface, the nucleated particles must possess a certain critical size for a given interfacial velocity (8). Based on the present TEM investigations, these are, respectively, 30 to 40 nm and 5 x 10^4m/s. Below the critical size, the particles will be pushed by the moving interface. For example, the size of particles in the Al_3Li phase is about 4nm and they are segregated into the secondary dendrite arms.

The homogeneous δ' formation during the quenching of Al-3wt% Li is followed by its coarsening throughout the matrix during artificial ageing at 473 K. Fig. 9a shows the persistence of the dendrite structure after an ageing time as long as 100 h. Since the precipitation reaction in LS4 and LC8 was found to be almost the same, only observations on the latter alloy have been reported here to bring out the comparison between the dendritic and cellular reactions after ageing at 473 K.

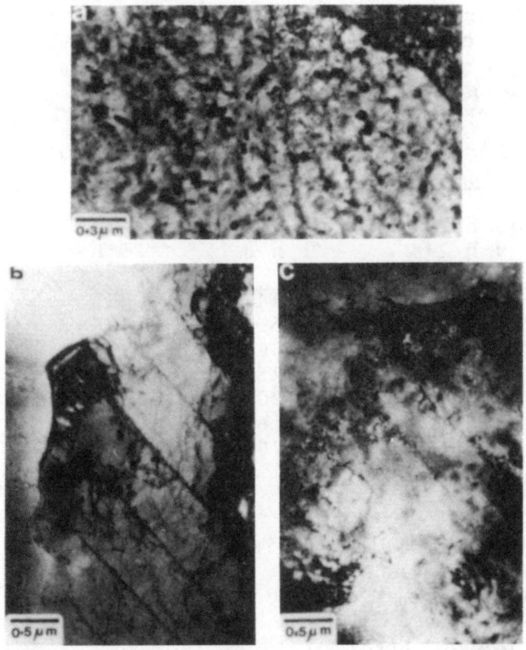

Figure 9 TEM micrographs of quenched powders following ageing at 473 K: (a)Al-3 wt% Li, 100 h, (b)Al-3 wt% Li-0.9 wt% Co, 16 h, (c) Same as (b), 100 h.

A heavy precipitation of the δ'-phase with a nearly uniform particle size is occurring on the cell walls as well as within their interiors. After 16 h annealing at 473 K the essentially observed process consists of a continuous dissolution of the Al_9Co_2 phase. This process is accompanied by the emergence of a new grain. The emerging grain grows into the cellular structure by boundary-boundary migration, Fig. 9b. This type of dissolution mechanism is described as cellular dissolution (9). Considering the simultaneous precipitation and coarsening of δ'(Al_3Li) particles, this leads to a slowing down of the movement of the newly formed grain boundary and, in turn, reduces the rate of cellular dissolution as shown in Fig. 9c, following ageing at 473 K for 100 h.

Pawlowski and Truszkowski (9) have reported that the degree of dissolution can be determined from X-ray measurements of the angle Θ_x according to the formula

$$X = \frac{\Theta_x - \Theta_{min}}{\Theta_{max} - \Theta_{min}}$$

where Θ_{min} denotes the value of Θ in equilibrium at a given ageing temperature, Θ_{max} is the respective value in equilibrium at the dissolution temperature. The change in the Angle 2Θ in the course of ageing LC8 for times between 1 h and 100 h is shown in Table IV. As can be seen, the maximum deviation in 2Θ for (200) α-Al is obtained for the overaged condition (473 K/100 h as determined from microhardness measurements) and is about 0.86. On the other hand, the increase in the angle 2Θ for (200) α-Al (as an example) is found to be associated with an increase in the relative intensity of (200) α-Al from 44 % to 71 % for the same conditions of treatment.

TABLE IV
Variation of 2Θ for (hkl) α-Al as a Function of Ageing Treatment

Condition (hkl) -Al	As-quenched 0 h	Under-aged 1 h	Peak-aged 16 h	Over-aged 100 h	$\Delta\Theta^*_{max}$
	2Θ (deg)				
(111)	45.109	45.034	45.035	45.082	+0.043
(200)	52.432	52.450	52.484	52.518	+0.086
(022)	77.364	77.389	77.392	77.435	+0.071
(113)	94.247	94.275	94.271	94.298	+0.051
(222)	99.888	99.920	99.897	99.935	+0.047
(004)	124.198	124.216	124.211	124.241	+0.043
(133)	148.686	148.770	148.779	148.850	+0.164

$\Delta\Theta^*_{max}$ = 2Θ (100h) - 2Θ (0 h).

Based on our metallographic observations, the phase transformation occurring during the ageing of Al-3wt% Li - 0.8wt% Co is the outcome of three successive processes: 1) the dissolution of the Al_9Co_2 phase and the formation of grains of solid solution, 2) homogenization of the solid solution and then migration of subgrain boundaries of the newly formed grains, 3) formation of a stable solid and its grain growth.

Conclusions

The present work was performed on two types of powders, having the compositions Al-2.7% Li-X, where X=Si (ranging between 1 and 7.5 wt%) and Co (0.48 and 0.8wt%). The important micro-structural changes as viewed by light and SEM microscopy are:

A) Al-Li-Si alloys

1) Addition of 3wt% Li to the Al-Si system shifts the eutectic point to less than 4wt%. The eutectic structure is a mixture of aluminum and AlLiSi phases.
2) If the Si/Li ratio is less than unity, all the silicon is included in the AlLiSi particles. If the Li concentration is too low, it remains in solid solution with no δ' (Al_3Li) precipitation.
3) For a given volume fraction of Al_3Li, addition of Si permits an increase in the concentration of Li in the alloy, which produces a remarkable decrease in the alloy density.

B) Al-Li-Co alloys

1) The presence of four different structures, a) dendrite, b) cellular without secondary arms, c) equiaxed-type structure characterized by the absence of directional growth, and d) featureless zones with no response to etching.
2) Addition of Co, as also increasing the powder particle diameter, favors the transition from dendritic to cellular microstructure.
3) From the linear relation between secondary dendrite arm spacing and powder particle diameter, the cooling rate was determined to lie in the range of 10^4-10^6 K/s.
4) Ageing of quenched powders at 200°C for times up to 100 h shows a dissolution of the cellular structure.

References

1. F.H. Samuel, J. Hinojosa-Torres amd G. Champier, 11th Conf. on Applied Crystallography, September 10-14; 1984, Kozubnik, Poland.
2. T.H. Sanders Jr., E.A. Ludwiczak and R.R. Santell, Mater. Sci. Eng. 43, 247 (1980).
3. T.H. Sanders Jr., E.A. Starke Jr., Acta Metall. 30, 927 (1980).
4. W.X. Feng, F.S. Lin and E.A. Starke Jr., Metall. Trans. 15A, 1209, (1984).
5. D.B. Williams, Aluminum-Lithium I, Stone Mountain, May 1980, (TMS-AÎME, 1981), p. 89.
6. D. Webster, G. Wold and W.S. Cremens, Metall. Trans. 112, 1495, 1981.
7. H. Matyja, B.C. Gissen and N.J. Grant, J. Inst. Met. 96, 30, (1968).
8. F.H. Samuel, Metall. Trans. 17A, 73 (1986).
9. A. Pawlowski and W. Truszkowski, Acta. Metall. 30, 37 (1972).

Aluminum and magnesium recycling

Economics of recycling magnesium

J.C. Agarwal, A.R. Frydenlund and F.E. Katrak

Magnesium is not a mature metal as compared to steel, copper, and aluminum. Moreover, its use as a structural metal has been limited in the past because of its high impurity content which historically caused unacceptably high corrosion rates in magnesium-based alloys. These factors have contributed to the lack of an organized and well-established magnesium scrap market. However, with recent improvements in the quality of magnesium metal, all the primary producers are vigorously pursuing the structural markets for magnesium, especially in the automobile industry.

Historically, most of the magnesium has been used for nonstructural use. For example (see Table 1), more than 50 percent of primary magnesium is used in the aluminum industry for alloying. The magnesium content of the aluminum-based end products is too small to be recovered and recycled as magnesium metal. However, some of the magnesium is recycled as part of the campaign to recycle used beverage cans. Since this magnesium never enters the marketplace, it will not be discussed here.

TABLE 1

Use Pattern for Primary Magnesium in the Noncommunist World

Aluminum alloying	131,000	(tonnes/yr)
Chemical uses	72,000	
Structural	37,300	
Other	6,900	
TOTAL	247,200	(tonnes/yr)

SOURCE: Based on International Magnesium Association data, 1989.

Similarly, the chemical uses of magnesium consume 35 percent of the primary magnesium in chemical reactions and is lost forever.

Only 15 percent of the primary magnesium produced, or about 37,000-40,000 mt/yr, are currently employed for structural uses. This is the pool from which essentially all magnesium scrap will continue to come.

Availability of Scrap Magnesium

Scrap magnesium is available in three forms:

- *Old scrap* -- magnesium that has had a previous use, such as reclaimed engine blocks from automobiles;
- *New scrap* -- scrap from magnesium casting and machining operations, such as defective castings, flash, and turnings; and
- *Slag and dross* -- waste streams from the production of primary magnesium.

The availability of each of these forms is discussed below.

Old Scrap

Old scrap is reclaimed through scrapyards and brokers. The scrap generally is cleaned, remelted (which allows the bolts, clips, and other nonmagnesium components to be separated from the magnesium), and cast prior to use. Some magnesium scrap that is free of other components is cleaned and ground directly into powder for reagents; the melt step is omitted in order to reduce powder cost. In the melt step, as a result of imperfect scrap separation, the magnesium is usually contaminated with aluminum and/or copper. If the end use of the melt is ingots for casting, the contaminants must be diluted to a specified level through addition of primary magnesium to the melt. If the end use is desulfurization, the melt is cast into grinding bars without refinement.

The U.S. Bureau of Mines (BOM) places annual collection of old scrap in the United States in the range of 4,000 to 5,000 tonnes from 1984 to 1988. As is the case with other metals, the amount of scrap collected is price-driven, accelerating in times of tight supply and elevated prices. CRA expects that current and pending primary magnesium capacity in North America will be sufficient to meet domestic and export demand in the short term (through 1993). This will keep overall old scrap availability at around 4,000 to 5,000 tonnes per year.

The scrap metal collection industry in the United States has a well-established and efficient infrastructure. The scrap remelters and reagent producers have long-standing relationships with scrap sources for both old and new scrap, and there is little competition among reagent producers for sources of supply. Competition does exist between reagent producers and casters for scrap supply, however.

New Scrap

New scrap is acquired from casters and fabricators, and takes the form of defective castings, flash trimmed from castings, and turnings from machining. Clean scrap may be directly ground to powder or remelted; dirty scrap must first be degreased and cleaned.

The U.S. Bureau of Mines data place annual collection of new scrap in the United States in the range of 1,000 to 3,000 tonnes from 1984 to 1988. CRA estimates that from 20 to 30 percent of the magnesium used in fabrication of structural products becomes scrap through the manufacturing process, depending on the weight of the part produced and the extent of machining. At this rate, annual new scrap generation may have been as high as 4,000 to 6,000 tonnes from 1984 to 1988, or about twice the BOM estimate. (Consumption of primary magnesium for production of structural products was in the range of 13,000 to 20,000 tonnes from 1984 to 1988.) The BOM estimate may not comprehend all scrap generated due to company reporting requirements; some of this scrap is recycled directly by the fabricator by remelting. The portion that is remelted for structural parts versus that which is made into desulfurization reagents depends on relative margins for each fabricator.

For this paper, CRA has assumed that fully 30 percent of the primary magnesium used in production of structural products becomes scrap in the manufacturing process and is available at a suitable price for use in reagents. This figure may overstate the scrap generation rate for some fabricators or products. However, the amount of magnesium reported as used in production of structural products is understated by perhaps 10 percent, since it includes only the primary magnesium used, and none of the secondary. Thus, CRA believes that a 30-percent average new scrap availability is a reasonable best estimate for the short term (through 1993). Note that as the volume of new scrap generated per facility increases, remelting of scrap becomes economically viable and attractive for more fabricators. In the long term this trend may decrease the new scrap availability from 30 percent to 20 percent or less.

The International Magnesium Association (IMA) projects that magnesium consumption in castings will grow by about 65 percent from 1988 to 1993. Based on an equal growth in new scrap generation, the United States will produce approximately 10,000 tonnes of new scrap in 1993.

Slag and Dross

The slag and dross produced in the manufacture of primary magnesium contain some high-purity magnesium metal along with oxides, chlorides, and trace amounts of other materials such as silica and iron. If the slag and dross are removed from melting pots, alloying agents such as aluminum and zinc are also present. Northwest Alloys, Alcoa's primary magnesium subsidiary, estimates the magnesium content of reclaimed slag and drosses at 5 to 10 percent of its total primary magnesium production.

The Northwest slag is granular, and it is crushed, treated, screened, and sold without remelting. Northwest sells all of its slag to a reclaimer, who sells the reclaimed magnesium to a major desulfurization reagent supplier. The reclamation process for slag and dross from the other U.S. primary magnesium producers, Dow and Magcorp, and from Norsk Hydro in Canada, involves salt removal, then remelting and casting of the remaining metal. This metal enters the product stream along with other primary and secondary magnesium, and may be used for various products, including desulfurization.

CRA estimates the production capacity of Northwest at 36,000 tonnes annually. If an additional 5 percent of this capacity is available in the form of magnesium reclaimed from slag and dross, approximately 2,000 tons per year of magnesium scrap may be available from Northwest. CRA expects this level to remain constant through 1993.

Timminco in Canada has a process similar to that of Northwest, but it has a small annual production of primary metal. Its scrap and dross generation of about 600 tonnes per year is not included in this U.S. analysis because it originates in Canada.

Uses of Scrap Magnesium

There are two major uses of scrap magnesium:

1. Desulfurization of iron in the steel industry; and

2. For casting.

Overview: Magnesium Scrap Availability and Use in Desulfurization

All of the categories of magnesium scrap mentioned above are used by reagent producers. Steel companies are aware of the presence of contaminants in the reagent through regular in-house analysis and the certified test reports that reagent suppliers are required to provide. Steelmaker specifications require minimum amounts of magnesium and lime and sometimes aluminum and fluorspar, but do not currently stipulate maximum levels for contaminants such as zinc. Pricing of reagents takes contaminants into account: discounts are given for off-spec batches. However, the steelmakers' major concern is effectiveness of the reagent and how much they are paying per tonne of steel for sulfur removal. As a reflection of this order of priorities, the prices of at least one reagent supplier are based on sulfur removal effectiveness, rather than pounds of reagent.

Contaminants are not a concern for steelmakers, because contaminant levels in the reagent are low (less than 5 percent), the "contaminants" generally are benign or even helpful (aluminum aids in desulfurization), and the amount of reagent used per ton of hot metal is low (about 0.6 kg of contained Mg per tonne iron). Any dirt or grease present on old or new scrap is removed through degreasing, heating, or other cleaning processes.

CRA expects reagent contaminant levels to become a concern for steel desulfurization and not for iron desulfurization. The steel desulfurization market will require high-purity reagents. The bulk of reagent sales (i.e., to iron desulfurization markets) will probably not be affected in the short or long term by concern over contaminants.

CRA estimates 1988 magnesium scrap availability in the United States to be about 12,000 tonnes. This amount may grow to 15,000 or 16,000 tonnes by 1993, fed by growth in new scrap from fabrication of structural products (see Figures 1 and 2). All forms of magnesium scrap are used in the manufacture of desulfurization reagents, and CRA expects no restrictions on scrap use for the bulk of iron desulfurization in the long and short terms.

Economics of Scrap Magnesium

The scrap processor, whether a melter or a desulfurization reagent producer, generally buys scrap at about U.S. 70 cents/lb of contained magnesium. The melting and casting costs are generally 25-30 cents/lb. Including the yield losses and capital charges, the total costs are about $1.10 to $1.20/lb. The selling price of secondary ingot is about $1.30 to $1.50, giving a before-tax profit margin of about 20-30 cents per lb of magnesium.

If the scrap is reprocessed as a desulfurization reagent, additional costs are incurred in producing the magnesium powder of about 20-25 cents/lb. The selling price is generally U.S. $1.80-1.85 per lb of contained magnesium, which results in similar profit margins.

The largest future growth in the magnesium industry will be in the structural sector. It is expected to grow to more than 100,000 mt/yr of primary metal consumption by the year 2005. The scrap magnesium business will grow proportionally.

Figure 1

AVAILABILITY OF MAGNESIUM SCRAP IN THE UNITED STATES; OLD, NEW, SLAG: 1984-1988

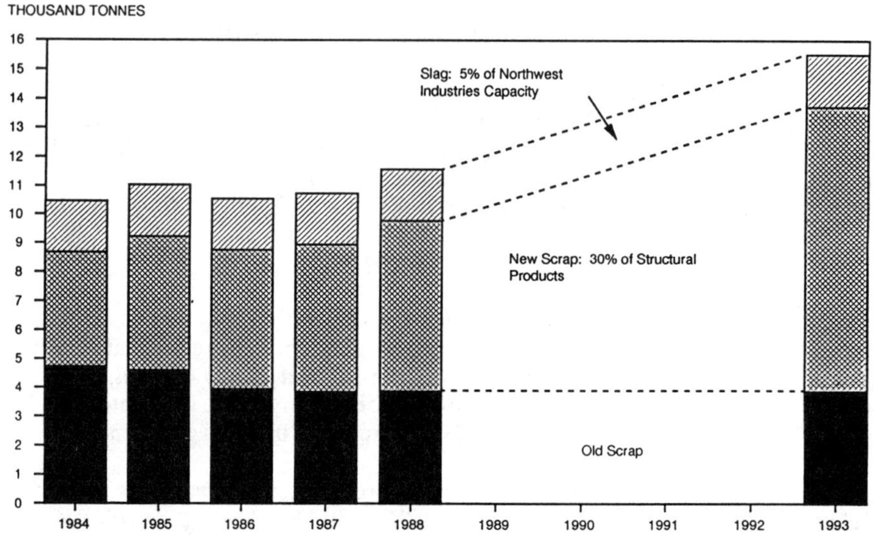

SOURCE: Charles River Associates, 1990.

Figure 2

AVAILABILITY OF NEW MAGNESIUM SCRAP IN THE UNITED STATES: 1984-1988

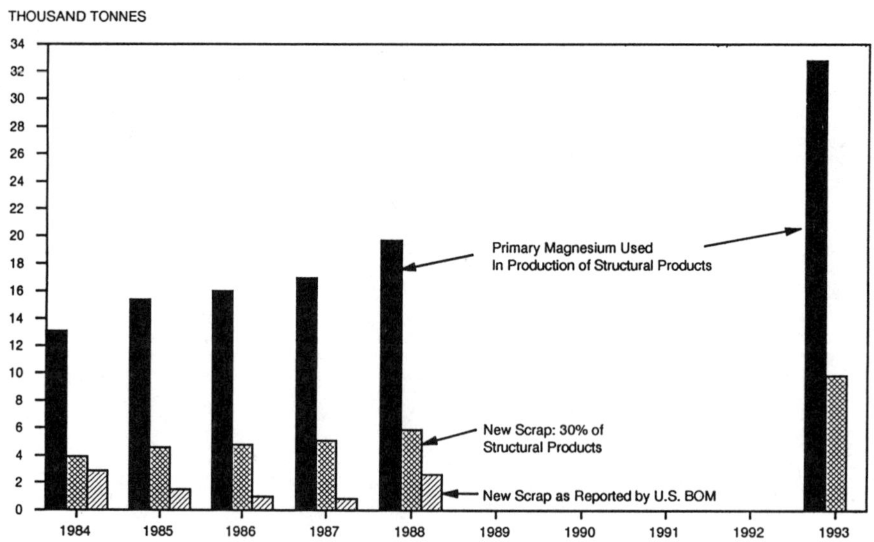

SOURCE: Charles River Associates, 1990.

Aluminum recycling in the 90s

R. Yank
Alcan International Limited, Montreal, Quebec, Canada

Introduction

As we enter the 90's, already being called the "decade of the environment" or the "green decade", a growing awareness and concern over the impact of our industrial societies on the health of the planet has generated a high level of interest in recycling. Marching under the recently unfurled banner of "sustainable development", individual consumers, governments, and industry have identified recycling as one area where positive action can be taken now to reduce man's impact on the environment. The aluminum industry has for many years had a large and viable recycling sector which, as a result of this widespread and intensified attention, will certainly witness a period of rapid change. All facets of the industry are likely to be affected during this period, from process technologies and product engineering to materials management and corporate integration.

Against this backdrop, let us examine more closely the secondary aluminum industry of today and explore the major challenges and opportunities which it faces between now and the end of the century.

Importance of the Secondary Aluminum Sector

During 1988, the latest year for which complete data is available, more than 6 million tonnes of aluminum were recovered in the form of "new" (generated during manufacturing, after the semis stage) and "old" (post-consumer) scrap. This figure does not include "prompt" or "runaround" scrap, generated during the production of semi-fabricated shapes, which is typically recycled internally and not reported for statistical purposes. Scrap, therefore, furnished almost one-third (32%) of the 19 million tonnes of aluminum consumed in total in the Western World during the same year. Considering that growth over several decades (Figure 1) and that many applications for aluminum, such as transmission lines and residential construction, have long service lives, a more useful measure of the maturity of the secondary sector is the ratio of recovered scrap to available scrap. This ratio is not easily calculated due to lack of data on available scrap — available scrap being the total of new scrap generated and of old aluminum that reaches the end of its useful life and is discarded during the year. The most recent analysis to be published on this question was done for West Germany, based on data for the year 1987. This study arrived at an average service life of 14 years, and a "true recycling rate" of 72%(1).

Using this average service life for all of the Western World, and taking into account new scrap generation and recovery losses, would indicate that less than 12 million tonnes of aluminum became available as new and old scrap in 1988. This would suggest that the more than 6 million tonnes of scrap usage represents a true recycling rate in excess of 50%. This figure is obviously quite sensitive to the initial service life assumption. Let us now look at the principal sources of recovered scrap, and at the main products of the secondary aluminum industry that make use of this scrap.

Alcan estimates that "new" scrap currently accounts for over 45% of all recovered scrap. This scrap is a more or less fixed proportion of total aluminum consumption, equal to 15-16% of semi-fab shipments. Post-consumer scrap makes up the balance, with only two products responsible for nearly 70% of recovered "old" scrap: vehicles and beverage cans (Figure 2). Similarly, scrap consumption is concentrated into very few end-uses. The production of secondary alloys for the foundry industry, the traditional business of secondary smelters, still consumes about 55% of all recovered scrap: not only old scrap recovered from retired vehicles, but also a large percentage of the available new scrap and other old industrial scrap. Overall, 85% of aluminum foundry alloys are produced from scrap. With the growth in collection of used beverage cans (UBC) over the past decade, approximately 58% of rigid container sheet produced for the manufacture of new cans is now based on scrap. This market now consumes almost one-quarter of all recovered scrap, and has been almost totally responsible for the increase in secondary versus primary aluminum demand in recent years. Consequently, foundry alloys and can sheet together consume more than 75% of recovered scrap. Remaining secondary aluminum products include de-oxidizer for the steel industry and secondary sheet used mainly for residential construction.

Future Outlook — Cars and Cans

Growth in aluminum demand is expected to continue in the 90's, mainly as a result of further penetration of aluminum in cars — particularly in cast parts — and beverage containers. It is significant for the secondary aluminum sector that these two end-use markets are the ones that employ the highest proportion of scrap as a source of metal. So let us examine these applications a bit more closely.

As stated above, foundry alloys used in the production of cast parts are already made to a large degree (85%) from recovered scrap. Consequently, growth in this market will come mainly from more extensive use of lightweight aluminum castings in automobile engines, transmission components and suspension parts, rather than from an increased proportion of secondary versus primary metal usage. In fact, it is not unlikely that some of the more critical applications will be somewhat less tolerant of impurities and compositional variations inherent with secondary foundry alloys. There is a growing role for science here, in an industry traditionally based on art.

While a well-established network of dismantlers and auto-shredders already ensures that most cars are recycled, changes in material mix, including the increasing aluminum content of cars, are forcing the salvage industry to focus more attention on the efficient recovery of non-ferrous scrap from retired cars. Also, while large scale production of aluminum structured vehicles, made from aluminum sheet and extrusions, is not anticipated in the near future, designers are already concerned with the ultimate recycling of the aluminum. Car designers and aluminum producers must work together to ensure that these structures can eventually be recovered and re-used to manufacture wrought-alloy sheet and extrusions.

The situation is a little different for the aluminum beverage can. The creation of an efficient and viable recycling system for UBC's (used beverage cans) is by now a well known success story. In 1989, the industry collected 60.8% of all cans sold in the U.S.A., representing over 760,000 tonnes of aluminum that was recycled back into new canning sheet for the packaging sector (Figure 3). From a demand standpoint, the can has also been the success story of the 80's for the aluminum industry, and particularly in Europe and Japan, further growth is expected during the 90's. Why has can recycling been so successful? Firstly, because of the can's unique properties: short service life, ease of identification and segregation, and standard composition. Secondly, recyclers have adopted rigid quality standards for UBC's and have worked with collectors to ensure that these standards are maintained. Thirdly, the aluminum industry has developed process technologies tailored to the demands of UBC recycling. Cans are now largely recycled directly back into undifferentiated rolling ingot in dedicated, high-volume plants. Consequently, there is no technical reason to prevent higher collection rates in the future and the anticipated growth in demand for aluminum cans will rest on the industry's ability to achieve even higher UBC collection rates.

Alcan expects that by the year 2000 scrap will supply at least 40% of the Western World's aluminum needs (Figure 2). Recovered new and old scrap will exceed 10 million tonnes per year. This will result in large part from the increasing importance of the car and the can. Cars that will be recycled in the year 2000 are on the roads today, and those currently made in North America already contain an average of 157 pounds of aluminum per car. According to a recent survey of the U.S. automotive industry, forecasts for the year 2000 ranged from 170 to 200 pounds per car (2).

Within total scrap utilization, new scrap will become less important and may account for only one-third of the total by the end of the decade. While new scrap availability will remain a more or less fixed percentage of total aluminum semi-fab shipments, old scrap will be recovered to a greater extent. Not only will dismantled cars contain more aluminum and UBC collection rates rise above the 60% range, as described above, but we will see more old scrap arise from the building and construction boom of the 1950's and 60's, from the replacement of early transmission lines, from the increased use of aluminum in recyclable products (e.g. lithographic plates), and from the growing interest in recycling municipal waste.

Alcan in the Secondary Aluminum Sector

Alcan is well known as one of the world's largest producers of primary aluminum: production totalled 1.6 million tonnes in 1989, representing 12% of the Western World total. Less well known, however, is the fact that Alcan operates 214,000 tonnes of can recycling capacity in North America, and is presently building a similar 50,000 tonne-per-year plant in the U.K. In addition, Alcan and its related companies annually produce approximately 280,000 tonnes of secondary foundry alloys in five countries around the world: Canada, U.S.A., Italy, U.K. and Japan.

Challenges for the Industry

The inherent recyclability of aluminum will certainly fuel impressive growth in the secondary industry during the 90's. As we have seen, demand for already well-established secondary aluminum products is expected to grow faster than the overall industry. This growth will create both opportunities and challenges for the industry.

Higher demand for foundry alloys will require new investment in a sector long plagued with over-capacity. But, as their dependance on secondary aluminum increases, end-users such as the automotive companies will become increasingly concerned about security of supply which will lead them to seek stronger ties to their suppliers. Scrap purchasing will likely become more competitive, and more demanding alloy specifications will impact on the quality of scrap required by secondary smelters.

Tighter environmental standards are already playing an important role in reshaping the industry. Many of the smaller secondary producers have been unable to invest the capital needed to upgrade their facilities and have had to close down. The main issues facing smelter operators are the treatment of fume emissions from scrap decoaters and melting furnaces, the recycling of oxides (dross) and salt slags from the furnaces, and the recycling or safe disposal of solid wastes generated by the processing of contaminated scraps. In many cases, technological advances have been made in recent years. Implementation of these technologies will, however, require a significant outlay of capital by the industry.

The drive by society to reduce waste will place even more pressure on the aluminum industry to recycle. Even with current recycling efforts, which are substantial, more than 500,000 tonnes of beverage cans — equivalent to 32 billion containers ! — are still lost to the waste stream each year. And other aluminum packaging is recycled to even less an extent. Successful curbside collection or "Blue Box" programs in Ontario and elsewhere have shown that the consumer is ready to do his share to reduce municipal waste. Our industry must respond with the required processing technologies and product designs. With the exception of beverage containers and castings, few products currently use a high proportion of recycled metal. Wrought alloys, in particular, show a low tolerance for impurities and mixed scrap utilization.

Aluminum recycling must be approached at all levels: from the product designer through to the dismantler, the recycler and the aluminum producer. "Product design for recyclability" is becoming an increasingly important requirement. This offers a challenge not only to design engineers, but to process engineers and metallurgist as well. Improved technologies are required for scrap identification, sorting and segregation, and molten metal treatment. Alloy development is also needed.

References

(1) Recyclingrate bei Aluminium uber 70%., St-Glimm (1989), Aluminum 65(11): 1088-1094.

(2) Delphi V: "Forecast and Analysis of the U.S. Automotive Industry Through the Year 2000", Office for the Study of Automotive Transportation, University of Michigan, December 1989.

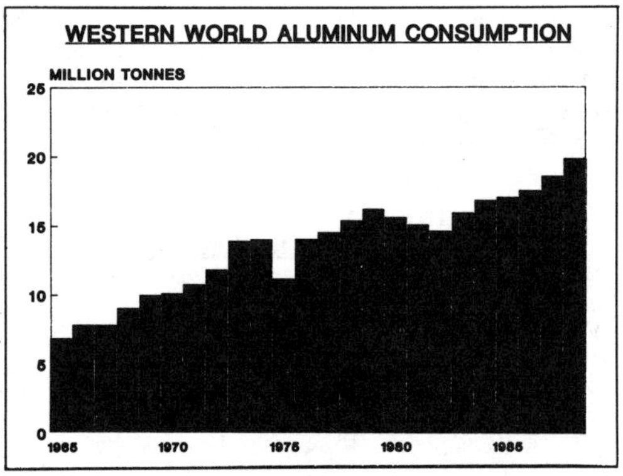

Figure 1

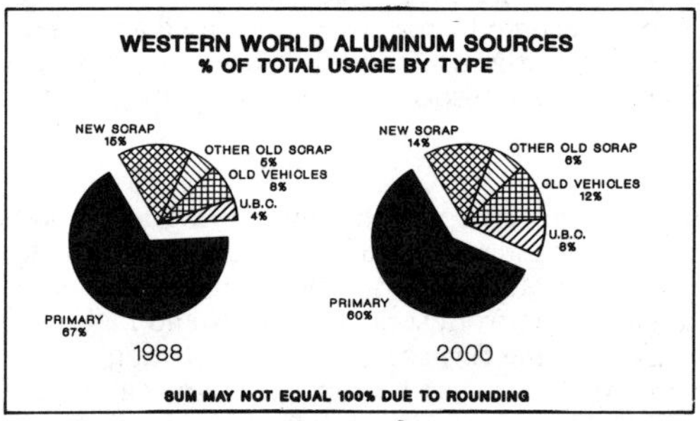

Figure 2

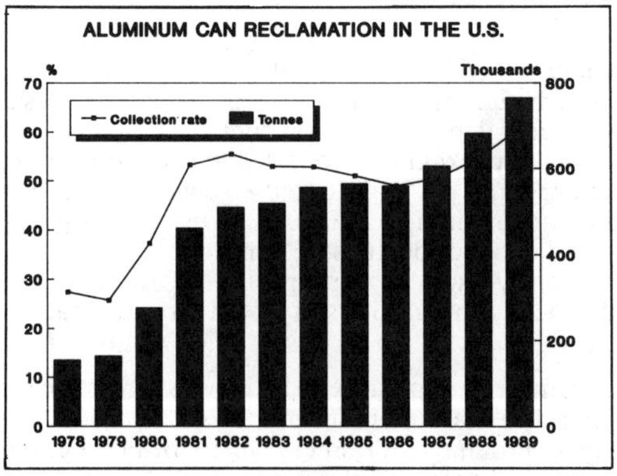

Figure 3

Fabrication and application of light metals

Ceramic components for light metal casting

R.H. Brown and G.E. Holling
Thor Ceramics Limited, Clydebank, Scotland

ABSTRACT

The use of specific ceramic materials for indirect heating of non-ferrous metals is described. The process is based on the use of isostatically pressed Silicon Nitride bonded Silicon Carbide tubular shapes, immersed in the molten metal bath and offering a choice of energy sources. This process offers important benefits to the user and this paper covers heating, melting and holding at temperature, non-ferrous metals for a variety of casting processes. The benefits, in terms of improved metal quality and energy management are included.

Keywords

Immersion Heating, Heater Tubes, Isostatic Pressing, Nitride Bond, Silicon Carbide, Riser Tubes.

Introduction

During the last decade there has been an increasing amount of ceramics used in the light metal industry. This paper deals with two aspects of the use of ceramics in the light metal industry, the first is with the handling, pouring and the general purification of the aluminium. The second is with the heating of aluminium melting and holding furnaces, and purification systems. The use of material such as graphite and steel, usually in the form of grey cast iron, is still a common practice but with the recent trends towards aluminium with more ductile properties, there has been a move away from the use of iron.

More recently, such materials as Silicon Nitride, Silicon Carbide, Sialons and Boron Nitride have become available in shapes suitable for use in aluminium processing. In the cases of some of these materials they are only available at considerable price premiums. This has tended to exclude them from general use.

Background

In theory the most suitable ceramic materials for handling molten aluminium are based on dense, fine grained Nitrides and Carbides which have been processed to a virtual zero porosity. The combination of aluminium's inability to wet the surface of these materials and the very low porosity enables them to have almost infinite lives in molten aluminium. They do tend however, still to be very brittle and are very susceptible to mechanical damage. In all cases they are only available in a few limited shapes and sizes, because of the linear shrinkages which occur during the densification process, and the high temperatures required to cause this densification to take place. For general use in the light metal industries,these advanced ceramic materials can be discounted.

It is therefore towards the top end of the refractory grades of Nitrides and Carbides that the industry must look for its more arduous requirements. Silicon Nitride of the reaction-bonded type will shortly be available in large sizes, such as 2 metre long Riser Tubes; or Tundish Boxes large enough for low pressure die casting of aluminium components for the automobile industry.

This general class of material has a low open porosity which ranges from about 12 to about 16%. They still have a tendancy to be highly non-wetting towards molten light alloys, have excellent mechanical properties at the operational temperatures, and are still highly resistant to corrosion from the slags associated with light alloys. The products which will be subsequently described are based on a refined version of Silicon Nitride bonded Silicon Carbide refractories which have been available for a considerable time. The general properties of Silicon Nitride bonded Silicon Carbide are shown below in Table 1.

TABLE 1

Typical Properties of Silicon Nitride Bonded Silicon Carbide

Physical:	App. Porosity		13 - 16%
	Bulk Density		2.55-2.65 g.cm
	A.S.D.		3.01 g.cm
	Thermal Conductivity		111 B.T.U./ins/hr/ft/^{o}F
	Coeff. Exp.		4.6 x 10 per ^{o}C
	MoR at 20°C		45 MN/m
	MoR at 1300°C		40 MN/m
Chemical:	SiC	76%	
	Si_3N_4	22%	
	Al_2O_3	1.0%	
	Balance	1.0%	

This material has excellent thermal shock properties due to its high strength, modest thermal expansion, and very high thermal conductivity. It is virtually non-wetting to slags and most metals and has exceptionally good mechanical properties at high temperatures. It does, however, begin to oxidise significantly above 1500°C, although in protected atmospheres, can operate up to nearly 1700°C. The introduction of isostatic compaction techniques to the fabrication of large tubular shapes for Immersion Heater Tubes, Riser Tubes and Pump Canals has advanced the technology and quality of aluminium melting and handling. The introduction of a reliable Immersion Heater Tube which could sustain long life when being heated by gas, has served to give the user confidence in the viability of the process which can yield beneficial and substantial fuel economies and metal quality improvement. During the last 5 years there has been a significant increase in the number of Immersion Heater Units, both gas and electrically powered, which have been introduced into the light alloys industries. The growth in the galvanising industry has been quite considerable, and furnaces with 18 tubes are already in operation, with larger furnaces being currently planned.

Immersion Heater Tubes

Figure 1 shows a typical Immersion Heater in an electrically powered aluminium melting application. A typical tube for such a furnace would be approximately 1 metre long overall, have a 20mm wall thickness and an internal diameter of 165mm.

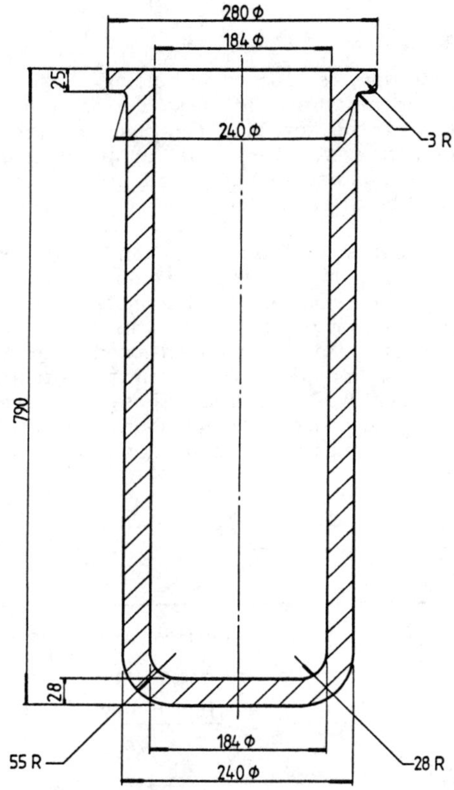

FIGURE 1: Typical immersion heater

The isostatic fabrication process requires a slight taper in the bore of such tubes.

Power Levels

The power output from such a tube when gas fired is considerably greater than when it is electrically powered. It is also possible to get more power from the same size tube in a zinc application than in an aluminium application. The reason for this is fairly simple: the temperature gradient which is the driving force for heat flux is much greater in the case of zinc where the operational temperatures are well under 500°C in the molten metal. In the case of zinc a heat input of up to 65 kilo watts can be obtained from one of the tubes shown in Figure 1. In the case of aluminium, 50 kilo watts heat input can be obtained. In the case of electrical heating which is only used in aluminium, a maximum heat output of 30 kilo watts is available. The limitation here is due to the size of the electric heating element that can be placed inside a tube of the given diameter. Two types of electric heating elements are in common use, the first is a Silicon Carbide heater element, which enables a high operational element temperature to be achieved, and hence a slightly higher output, and the second a spirally wound refractory metal element which is supported on a ceramic carrier. The advantage of electrically powered systems is the cleanliness and ease of control.

Temperature Profiles

Metal temperatures when melting, as opposed to holding temperatures, are slightly higher, as it is the flow of hot metal around the cooler solid metal that gives the heat transfer essential to melting of the metal. In the case of a melting furnace used for making zinc alloys for die-casting, a molten metal temperature of between 460 and 500°C is typical. In a straight forward holding application during galvanising operation, the metal temperature would be at about 460°C.

In an aluminium holding application then, the metal temperature would be in the region of 730°C, but in the case of a melting furnace, it would be slightly higher, at 750°C. In trials carried out at an aluminium smelter on a Fumeless Inline Degassing Unit (FILD), which was powered by two gas-fired recuperative burners with two Silicon Nitride bonded Silicon Carbide tubes of the previously mentioned dimensions, the temperature profile on the inside wall of the Immersion Tube from a position above metal level to a position well below metal level was measured. These measurements were carried out over a period of several days and ended when the thermocouples burned out. Table 2 shows the typical high and low values obtained when the burners were either high or low firing during modulation.

TABLE 2

Typical Temperatures taken from an Immersion Heater Tube in a FILD Unit

Position in Tube	Temperature High Flame	Temperature Low Flame
Molten Metal	715°C	750°C
Above Metal Level	887°C	820°C
Below Metal Level	1115°C	845°C

It shows that with a metal temperature of 750°C, the bottom of the tube, below metal level, tended to run hotter than the top of the tube above metal level. It should be remembered that the flame is directed through a Firing Tube towards the bottom of the tube, emerges and is re-circulated between the Firing Tube and the inner wall of the Immersion Tube. A typical lower tube temperature would be 1100°C on high fire and 850°C on low fire. The upper tube fluctuated between 941°C and 774°C. This temperature profile is to be expected, because the higher temperature below metal level is due to the fact that it is measured at the point nearest to the flame exit point from the Firing Tube. The table shows that the bottom of the tube is nearly always at a higher temperature than the top end of the tube and typical of tube temperatures in a gas environment.

Tube Life

It is very difficult to measure and assess tube lives in an objective manner. Immersion Tube lives vary from plant to plant and from application to application. In general, however, zinc applications have been yielding the longest lives, with typical tube lives of 6 months in zinc applications and slightly less in aluminium applications. Individual tube lives of up to 12 months are quite common in zinc, but less common in aluminium melting.

There has been, however, an elimination of catastrophic or early failures in tubes that have been made by the isostatic compaction process. In the more arduous aluminium melting processes and holding furnaces where higher tube internal temperatures prevail, Silicon Nitride bonded Silicon Carbide tubes have been found to give the longest lives. Tube lives can also be affected by the metal compositions. It has been noticed that in aluminium alloys containing sodium up to 50 or 60 parts per million, shorter tube lives have been obtained than in applications where sodium levels less than 10 parts per million occur. This difference in sodium levels has been reported to virtually halve the lives of the tubes.

Figures 2 and 3 show cut-away drawings of a gas-fired Heater Tube and an electrically powered Heater Tube; the electrical heating element and inner re-circulation Firing Tube can easily be seen.

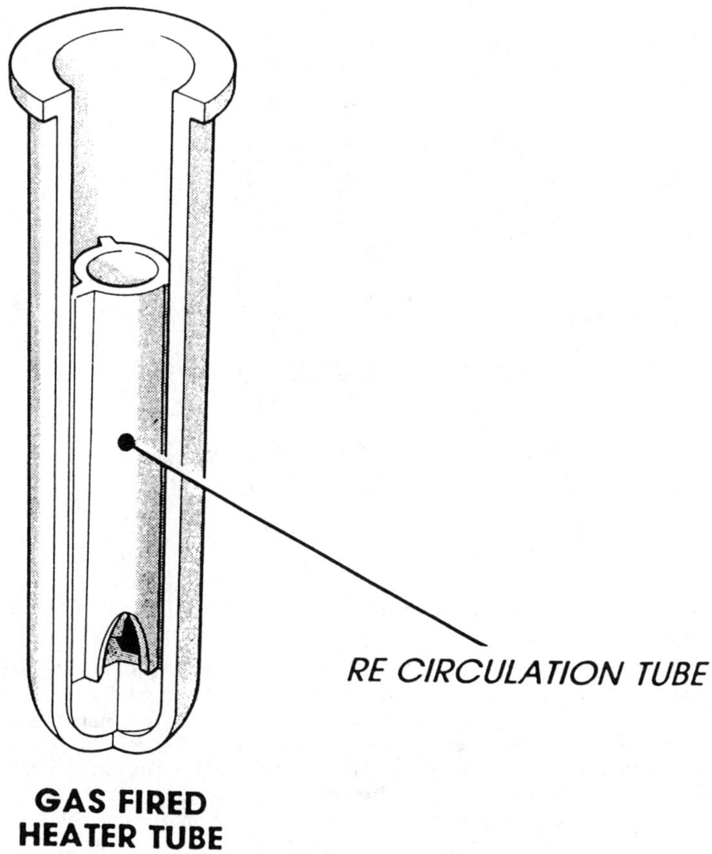

FIGURE 2: Cut-away drawing of a gas-fired heater tube.

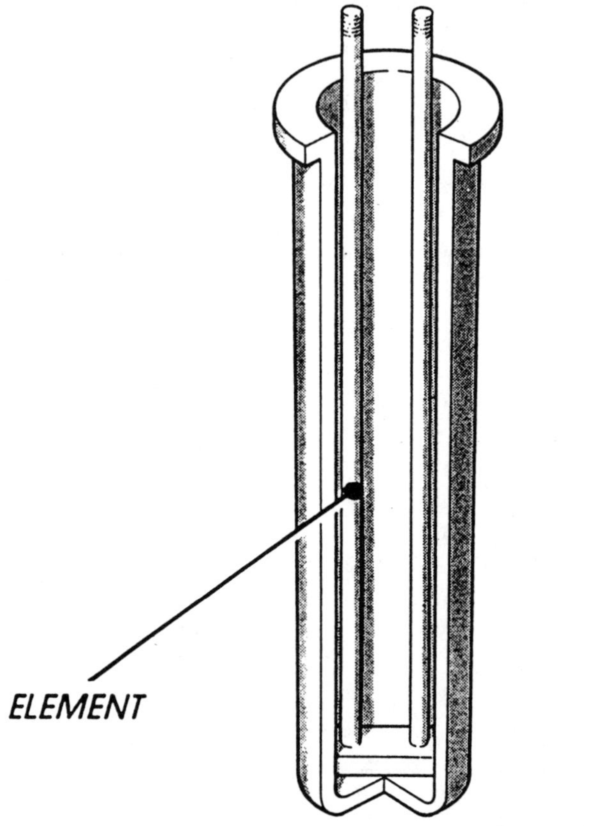

FIGURE 3: Cut-away drawing of an electrically powered heater tube.

The lives of Immersion Tubes also depend greatly upon the "Fluxing" practice and in applications where high sodium fluxes are used, relatively short lives of under a month are found and accepted. This tends to be the exception rather than the rule. Ceramic Immersion Tubes are now a well established technology and over 200 Ceramic Immersion Tubes have been sold in the UK to the British Gas design. The benefits are an economic fuel cost and an upgrading of local environmental conditions. In the case of zinc melting and galvanising, there are considerable technical benefits derived from the ability to get away from the use of steel baths which were externally heated, using gas or oil. The pick-up of iron in the zinc considerably affected the quality of the galvanising process. The use of external heating was particularly inefficient. Using refractory brick lined or cast refractory linings inside zinc baths, heat losses were minimised, iron pick-up was eliminated and an efficient heat input was available.

In the case of aluminium, a major use of Immersion Tubes has been in the filter box applications at aluminium smelters, and in holding furnaces. The input of heat into small furnaces is highly inefficient using open flames, is environmentally unattractive and conducive to metal contamination, due to re-oxidation and access for soluble gases. The use of Immersion Tubes overcomes all these disadvantages, with a net saving in fuel costs, a more pleasant environmental surrounding and superior metal quality. A novel development of the Immersion Heater Tube Furnace has been conversion of the furnace to facilitate pressurisation and dispension of molten metal from a totally sealed furnace. It is not known whether an inert gas is used above metal level, but it is thought highly probable.

Riser Tubes

The increase in the use of light alloy auto wheels has seen a shift from metallic Riser Tubes in the low pressure die-casting systems towards Ceramic Riser Tubes. In the low pressure die-casting systems, a reservoir of molten metal is sealed and subjected to sufficient excess air pressure to force the molten aluminium up a centrally located tube, as shown in Figure 4, to a steel mould.

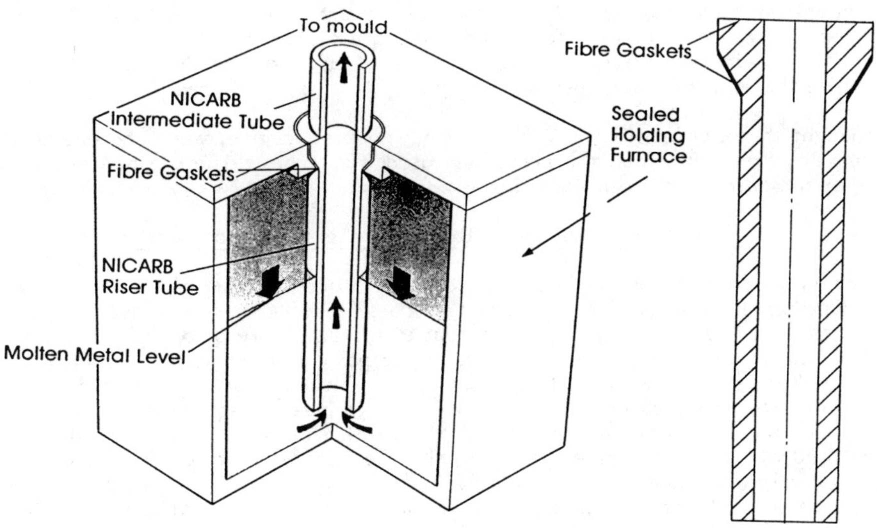

FIGURE 4 : Cut-away drawing of a low pressure die-casting system with ceramic riser tube.

The ductility of aluminium is adversely affected as the iron content increases, and it is therefore highly desireable to keep the iron content to a minimum. The use of Ceramic Riser Tubes instead of cast iron Riser Tubes helps greatly in this matter. The most suitable tubes have been found to be made from either Aluminium Titanate or Silicon Carbide derivatives. There is also a handling advantage due to the significant reduction in weight when changing from cast iron to the much less dense ceramic materials. However, ceramics are much more susceptible to mechanical damage during installation. It has been possible to reduce iron levels by changing from cast iron to Ceramic Riser Tubes. When "alloying" takes place within the foundry this helps to reduce costs, because of the scope of using cheaper alloys and still keep below 0.3% iron content. The typical lives of Ceramic Riser Tubes, which are in use for 24 hours a day for 5 consecutive days, with maintenance and cleaning taking place over the weekends, is between 2-3 weeks for a Carbon bonded Silicon Carbide tube, to 10-13 weeks for a Silicon Nitride bonded Silicon Carbide tube. Aluminium Titanate tubes have also been used successfully for applications where smaller tubes can be used. There is inherent difficulty in the fabrication of Aluminium Titanate and a tendency to be attacked when moderate levels of sodium are present in the molten aluminium. Long life Nitride bonded Silicon Carbide Riser Tubes also minimise production down-time due to less non-scheduled tube failures and changes.

The major problem using Ceramic Riser Tubes has been associated with mounting them and sealing them in the base of the die-casting machine, which also consitutes the upper structure of the furnace. This is seen as an engineering problem which will be overcome. In the case of ultra high pressure die-casting, such as the Cray process, where operating pressures above 400 pounds per square inch are used, the designers have been able to engineer a system which not only seals the furnace against this high pressure, but seems to have eliminated breakage of tubes during the mounting process. It is known that the tube flanges must be ground to fine tolerances.

It is thought that Ceramic Riser Tubes will be more attractive, when the engineering problems are overcome, than the previously used Iron Tubes. The use of Ceramic Riser Tubes has increased significantly during the last two years.

Isostatic Manufacturing Route for Ceramic Tube Shapes

The fabrication route being considered here is for Silicon Nitride bonded Silicon Carbide tubes made using the isostatic compaction process. The isostatic compaction route involves the use of moulds which have a central metallic core or pressing mandrel and an outer rubber membrane, in between which the powder which is to be compacted, is placed and hermetically sealed. The whole is then placed in a hydraulic pressure vessel and subjected to pressures between 10 and 15,000 pounds per square inch. The formulation grading and pre-mixing of the powders ensures homogeneity of the finished pressed product. After the pressing cycle, the tooling is dismantled from around the compacted ceramic shape. Because of the inherent hardness of most ceramic materials, it is desireable to design the tooling so that the ceramic component is pressed to near, if not exact, nett shape. Limited machining can be carried out before firing, but this is undesirable. After drying, the compacted but "green" shape, must be fired. In the case of Silicon Nitride bonded Silicon Carbide this takes place is a special atmosphere.

At Thor Ceramics, a high integrity Nitriding Furnace is used. This furnace is shown schematically in Figure 5.

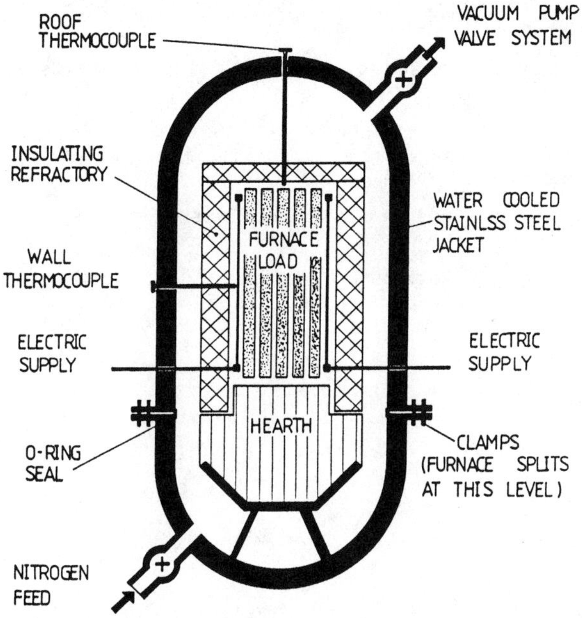

FIGURE 5: Sketch of the nitriding furnace used at Thor Ceramics.

The furnace is essentially a large stainless steel water cooled shell. The shell is split at the bottom, which separates and allows the load to be placed on the bottom half of the furnace.

The furnace is heated by an electric resistance element located around the load. A typical firing cycle subjects the furnace interior to a high vacuum followed by several purges of nitrogen. The furnace is then raised in temperature in an atmosphere of nitrogen and a reaction occurs between free silicon metal in the powder compacts and the nitrogen in the furnace atmosphere at temperatures above 1250°C. The firing cycle is considered complete at 1500°C. The resultant nitride bond is free from crystalline forms of quartz associated with less sophisticated nitriding regimes. The main bonding components are Silicon Nitride and Silicon Oxy-Nitride. The advantages of such a fabrication route is that porosities as low as 13% are obtainable, and the iso-pressed shapes are free from defects which are associated with other fabrication processes.

Conclusions

1. Immersion Heater Tubes are now well established technology for melting and holding light alloys.
2. They provide an efficient, economic and environmentally upgraded source of heating for light metals.
3. They provide accurate temperature control and higher levels of purity in the resultant molten alloys.
4. The high integrity nitriding combined with the isostatic compaction route produces superior quality ceramic tubulet shapes for Immersion Heater and Riser Tube applications.
5. The superior quality means longer service lives for the ceramic components, and reduces down-time on plant.
6. The use of Silicon Nitride bonded Silicon Carbide tube shapes such as Immersion Heater Tubes and Riser Tubes will increase significantly in the future.

Microporosity formation in A356.2 castings

J. Zou, K. Tynelius, S. Shivkumar and D. Apelian
Aluminum Casting Research Laboratory, Department of Materials Engineering
Drexel University, Philadelphia, Pennsylvania 19104, U.S.A.

ABSTRACT

Cast aluminum alloys are susceptible to extensive microporosity formation during solidification. This microporosity originates from the liberation of dissolved hydrogen upon freezing and from the volume contraction resulting from the liquid-solid transformation. The fundamental mechanisms associated with microporosity formation during solidification of A356.2 alloy have been investigated. The effects of initial hydrogen concentration, cooling rate and other process variables on pore size and distribution have been studied with simple test castings. The amount and distribution of porosity have been evaluated by image analysis techniques and density measurements. The results indicate that the pore density is essentially constant for various hydrogen concentrations while the pore size and amount of porosity increases with the initial hydrogen content in the liquid metal. The experimental data have been used to formulate a mathematical model to simulate microstructure evolution and pore formation in equiaxed structures. The model yields data on microstructural parameters such as dendrite arm spacing, grain size, amount of porosity and pore size. It has been shown that the predicted values compare well with experimental results.

Key Words: Aluminum base alloys, Casting, Computer modeling, Porosity, Solidification

INTRODUCTION

Microporosity defects are widely observed in aluminum alloy castings. Although porosity severely impairs mechanical properties and service life of the casting, fundamental mechanisms of microporosity formation are still not clearly established. Recently, mathematical modeling techniques are being applied extensively to illustrate some of the physical phenomena associated with pore nucleation and growth [1-10]. Mathematical modeling of solidification process is a simple and cost-effective technique for implementing process control and quality assurance in the cast shop. Computer models can be utilized to ascertain the influence of various process variables on product quality and thus may eliminate costly plant trials. The availability of inexpensive microprocessors over the last decade has promoted the application of computer models in the cast metals industry and various CAD packages are now available to simulate mold filling and solidification.

Macro-modeling methods, which utilize principles of heat transfer and fluid flow, have been used extensively to study the solidification of cast aluminum alloys. In recent years, attention has been focused on micro-models which incorporate nucleation and growth equations to predict microstructure evolution. A micromodel to simulate microstructure evolution and porosity formation in A356.2 alloy was developed at our laboratory [1]. This model incorporates the effects of hydrogen liberation at the metal front and interdendritic fluid flow. In this contribution, the above model has been used to highlight the principal physical concepts associated with pore formation in A356.2 alloy. The results of the model are compared with experimental data obtained under carefully controlled conditions. This work is part of an ongoing program on feeding of aluminum alloys being conducted at the Aluminum Casting Research Laboratory.

MICROSTRUCTURE EVOLUTION

A typical cooling curve obtained during the solidification of alloy A356.2 is shown in Fig.1. After the superheat is dissipated, the liquid metal undercools slightly below the liquidus temperature. The extent of this primary undercooling, ΔT_1, will determine grain size in the casting. Subsequently, a slight recalescence is observed because of the release of latent heat from the growing nuclei. During recalescence, the nuclei grow into dendritic grains and impinge on each other. This impingement for A356 alloys occurs at relatively low solid fraction (≈12-32 %) [11]. The dendrites then begin to coarsen until the liquid metal attains the eutectic temperature. At this point, the solid fraction is of the order of 55%. The interdendritic liquid is enriched in solute elements and solidifies with a secondary undercooling, ΔT_2. This solidification leads to the precipitation of Si, Mg_2Si, and other Fe-rich phases in the interdendritic regions. Our model can be used to describe the sequence of events occurring during solidification [1]. Thus, the model can be used to generate data on microstructure parameters such as grain size, dendrite arm spacing and eutectic spacing. Typical results obtained from our model are shown in Fig. 2. It can be seen that the predicted values compare well with experimental data [11-13].

MICROPOROSITY FORMATION

During solidification, the interdendritic liquid is gradually enriched with hydrogen as the fraction solid increases, since most of the hydrogen is rejected at the solid-liquid interface. Solubility of hydrogen in liquid and solid aluminum is in the order of 0.65 and 0.035 ml/100 g alloy respectively [14]. The variation of hydrogen concentration in the liquid during solidification as calculated by the model is shown in Fig. 3. As solidification progresses, the hydrogen content in the liquid increases and eventually exceeds the solubility limit. Ideally, a gas pore should nucleate at this point. However, the creation of a new pore requires the establishment of a new surface [15]. Because of this surface barrier, the hydrogen concentration in the liquid will continue to increase above the solubility limit until it reaches a maximum value at which pores can form, stage I in Fig. 3. At this point, pores begin to nucleate, stage II. This nucleation occurs predominantly at the root of dendrites or at other heterogeneous sites such as inclusions [1]. Once the pore is nucleated, there is rapid growth during the initial stages and the hydrogen concentration in the liquid decreases, stage III. Subsequently the pore detaches from the dendrite arm and grows at a relatively slow rate, stage IV. This process is schematically illustrated in Fig. 4.

EXPERIMENTAL PROCEDURE

Experimental data on microstructural parameters and on pore characteristics were obtained with simple test castings. The test casting design is shown in Fig. 5. The castings were produced in top poured cast iron molds, which were coated with boron nitride. The molds were equipped with a silica alumina fiberboard riser. This mold was designed to ensure proper risering and to minimize the formation of macroscopic defects. Prior to the experiment, the molds and the risers were held overnight in an oven at 70 °C, but were allowed to cool to room temperature before the casting operation took place. Two Chromel-Alumel thermocouples (0.032 in. diameter) connected to a data acquisition system, were positioned in each mold to record the cooling curve during solidification, Fig. 5.

Primary A356.2 ingots were melted in an induction furnace and the temperature increased to 710 °C. The melt was subsequently degassed at 710 °C with 99.998% pure nitrogen gas purged through a graphite rod. The temperature of the metal was then increased to 810 °C and a small quantity of metal was transferred into a preheated ladle. The casting was then poured at 760 °C into the mold. Immediately prior to pouring each casting, a sample for chemical analysis and a Ransley sample for hydrogen analysis were taken from the melt. Several castings were then produced at various stages during the degassing process in order to obtain castings at various hydrogen contents. The chemical analysis of the alloy is shown in Table 1. The hydrogen contents for which castings were produced were determined by LECO analysis and are given in Table 2. Several samples were taken from each casting for microstructural observation and density measurement as shown in Fig. 5.

The cooling rate, defined as the average cooling rate between the liquidus and solidus temperatures, was measured to be 2.5 °C/s. The same cooling rate was recorded at both thermocouples in the casting and the temperature gradients were negligible. The dendrite arm spacing was measured to be of the order of 31 μm. Density measurements were conducted according to ASTM C693. Image analysis was performed on a Macintosh II computer using the Automatix software system. Each sample was evaluated for area percentage porosity, pore size and pore density. In general, a good correlation was found between data obtained from image analysis and density measurements, Fig. 6.

RESULTS AND DISCUSSION

The measured and calculated amount of porosity as a function of initial hydrogen content is shown in Fig. 7. As can be expected, the amount of porosity increases with initial hydrogen concentration. The calculated results show good agreement with measured data. Furthermore, it appears that there is a linear variation between the amount of porosity and the initial hydrogen concentration. The amount of porosity increases from 0.25% to 1.7% when the initial hydrogen content increases from 0.183 to 0.445 ml/100 g. The experimental data in the literature have also been compared with the computed values, Fig. 7. This comparison further strengthens the validity of our model. Since the amount of porosity depends primarily on the initial hydrogen content rather than on the cooling rate [1], our experimental results are expected to be close to the published data although the casting conditions are different. Thomas and Gruzleski's data [16] (Al-6%Si) were obtained in sand-cast bars with a diameter of 2.5 cm. Deoras and Kondic's data [17] (Al-8%Si) were obtained in a water cooled permanent mold with a diameter of 10 cm. In both cases, porosity data were obtained by density measurement and the cooling rate of castings has not been mentioned. Since no thermocouples were used in their experiments, it is difficult to estimate the value of the cooling rate.

A typical pore size distribution obtained from our experiments is shown in Fig. 8a for a sample containing a hydrogen content of 0.183 ml/100 g alloy. The pore size is given as the equivalent diameter measured on the image analysis system. During the analysis it was observed that the pore size distribution follows a normal distribution. It is seen from Fig. 8a that a majority of the pores are less than about 50 μm in size. In this case, the maximum pore size is less than 200 μm. The pore size distribution for a sample containing 0.445 ml/100 g of hydrogen is shown in Fig. 8b. In this case, the pore sizes are more evenly distributed than in Fig. 8a, indicating that a majority of pores are less than 125 μm in size. The maximum pore size is now of the order of 325 μm. The corresponding micrographs for data shown in Fig. 8a,b are shown in Fig. 9a,b. It is evident that bigger pores are present at higher initial hydrogen concentrations. It should be noted that the value of pore sizes, (given as area or diameter) measured on the surface of the sample, cannot be directly used as pore size in the volume. Quite complex relationship between the values measured on the surface and in the volume emerges when we attempt to transform one to another [18]. It can be expected, however, that the pore volume fraction varies linearly with area percent porosity since experimental data suggests a linear relationship between density and area percent porosity, Fig. 6.

The non-uniform pore size distribution in the sample comes from non-uniform heterogeneous nucleation of pores in the castings. Since distribution of inclusions and other heterogeneous pore nucleation sites in the castings is different from one location to another, the possibility of pore nucleation depends upon the local heterogeneous pore nucleation sites and resistance of the dendritic network to fluid flow. This non-uniform pore size distribution is an important factor that affects final mechanical properties. It has been observed that tensile, fracture and fatigue properties in A356 alloys depend on the maximum pore size [19]. Therefore, a diverse pore size distribution may be more detrimental than a uniform distribution for the same value of average pore size and pore volume fraction.

The variation of calculated and measured pore sizes with initial hydrogen content is shown in Fig. 10. The measured values correspond to the average pore sizes. It is evident that the pore size increases with initial hydrogen content. Although the experimental data do exhibit some scatter, there is general agreement with calculated values. It should be noted that although pore size increases with initial hydrogen content, the density of pores is essentially independent of initial hydrogen content, Fig. 11. The density of pores is determined primarily by cooling rate.

Our model indicates that during solidification pores begin to form after the fraction solid has reached a critical value, Fig. 3. This value is termed as the "threshold fraction solid" and depends on factors such as cooling rate and initial hydrogen content. It is seen from Fig. 12 that the threshold fraction solid generally increases with the cooling rate. As indicated previously, during solidification, the hydrogen concentration in the interdendritic liquid increases to a value greater than the solubility limit, Fig. 3. The concentration of hydrogen in the interdendritic liquid at which porosity nucleates, $[H_n]$, is also plotted in Fig. 12. This critical hydrogen concentration increases with cooling rate. The variation of the threshold fraction solid and the critical hydrogen concentration with initial hydrogen content is shown in Fig. 13. It can be seen that the threshold fraction solid decreases with increasing initial hydrogen content indicating that pores begin to form at an earlier stage of the freezing process for higher initial hydrogen contents. Consequently, these pores will be bigger than those formed at a later stage. This is reflected in the pore size distributions, Fig. 8a,b. Furthermore, note that the critical hydrogen concentration at which pores nucleate is essentially independent of initial hydrogen content. As indicated earlier, our experimental results suggest that pore density is essentially independent of initial hydrogen content, Fig. 11. These data suggest that there may exist a relationship between the density of pores and the critical hydrogen concentration in the liquid, $[H_n]$. Thus, the number of pores in the casting may be proportional to the critical hydrogen concentration in the interdendritic liquid, which in turn is a function of the cooling rate.

CONCLUSIONS

A mathematical model has been developed to predict porosity formation in A356.2 alloy. This model has been used to elucidate various mechanisms associated with pore formation. Experimental results have also been acquired to test the validity of the computed results. The experimental data such as amount of porosity and pore size compare well with the predicted values and are in accordance with published experimental data. The principal observations from this work can be summarized as follows:

- During solidification the hydrogen concentration in the interdendritic liquid increases above the solubility limit because of surface restrictions associated with nucleation of the gas bubble. At a critical value of hydrogen concentration, pores begin to nucleate and grow. This critical value depends primarily on cooling rate.
- Density of pores does not show sensibility to initial hydrogen content.
- There may exist a relation between the critical hydrogen concentration in the liquid at which pores form and the pore density.
- Pore size increases with initial hydrogen content.
- Distribution of pore sizes depends on the initial hydrogen content.
- Amount of porosity increases with initial hydrogen content.
- There may exist a critical fraction solid at which porosity begins to form. This critical value is greater than about 50% and depends on cooling rate and initial hydrogen content.

In this work, the effect of inclusions or filtration, cooling rate, grain refinement and modification has not been presented. The influence of these process parameters is currently being examined. Other factors influencing porosity formation such as inclusion concentration, alloy composition and thermal gradient in the casting are also being studied.

ACKNOWLEDGEMENT

This research was conducted as a part of an ongoing research project at the Aluminum Casting Research Laboratory (ACRL). The authors would like to gratefully acknowledge the financial support of the consortium of companies supporting the Aluminum Casting Research Laboratory: Alcan International Ltd., Aluminum Company of America, CMI International, Inc., COMALCO Aluminum, Consolidated Metco., Doehler-Jarvis, Hitchcock Industries, KB Alloys, Inc., Littlestown Hardware and Foundry Co., Inc., Metallurgical Products and Technology, Metaullics Systems, Pechiney Corporation, Reading Foundry Products, Reynolds Metals Co., Selee Corporation, Shieldalloy, and Stahl Specialty Co.

REFERENCES

1. J. Zou, S. Shivkumar and D. Apelian: *AFS Trans.*, **98**, 1990, in press
2. J. Zou, K. Tynelius, D. K. Kim, S. Shivkumar and D. Apelian: *28th Annual Conference of Metallurgists of CIM*, 1989, 87-99
3. D. R. Poirier: *Metall. Trans.*, **18B**, 1987, 245-255
4. D. R. Poirier and S. Ganesan: Prepared for meeting at GE-Lynn, June 1989
5. K. Murakami, A. Shiraishi and T. Okamoto: *Acta Metall.*, **32**(9), 1984, 1423-1428
6. K. Murakami and T. Okamoto: *Acta Metall.*, **32**, 1984, 1741-1744
7. Q. T. Fang and D. A. Granger: *AFS Trans.*, 1989
8. Q. T. Fang and D. A. Granger: *Light Metals*, 1989, 927-935
9. K. Kubo and R. D. Pehlke: *Metall. Trans.*, **16B**, 1985, 359-366
10. D. R. Poirier, K. Yeum and A. L. Maples: *Metall. Trans.*, **18A**, 1987, 1979-1987
11. J. Tamminen : "*Thermal Analysis for Investigation of Solidification Mechanisms in Metals and Alloys*", Ph. D. Thesis, University of Stockholm, Sweden, 1988, A-21
12. L. M. Hogan and H. Song: *Metall. Trans.*, **18A**, 1987, 707
13. D. A. Granger and E. Ting: "*Solidification Processing of Eutectic Alloys*", Eds. D. M. Stefanescu, G. J. Abbaschian and R. J. Bayuzick, TMS, 1988, 105-118
14. D. E. Talbot: *Inter. Metall. Rev.*, **20**, 1975, 166-183
15. J. Campbell: "*The Solidification of Metals*", ISI Report No.110, 1967, 18-26
16. P. M. Thomas and J. E. Gruzleski: *Metall. Trans.*, **9B**, 1978, 139-141
17. B. R. Deoras and V. Kondic: *Foundry Trade J.*, **100**, 1956, 361
18. E. E. Underwood: "*Quantitative Stereology*", Reading, Massachusetts, Addison-Wesley Pub., 1970
19. J. C. Jaquet: "*2nd International Conference on Molten Aluminum Processing*", Orlando, Fla., Nov. 1989

Table 1. Chemical Composition of A356.2

Chemical Analysis	
Si	6.884
Mg	0.344
Cu	0.035
Fe	0.119
Ti	0.008
Sr	0.0000
Ca	0.0000
P	0.0005
Mn	0.0015
Cr	0.000
Zn	0.175

Table 2. Hydrogen Contents for Which Castings Were Produced

ml H_2/100 g alloy
0.183
0.184
0.223
0.236
0.304
0.355
0.445

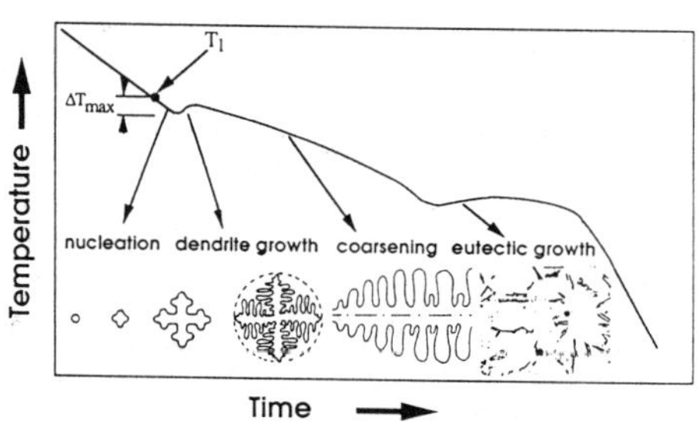

Fig. 1 Typical cooling curve for A356.2 alloy.

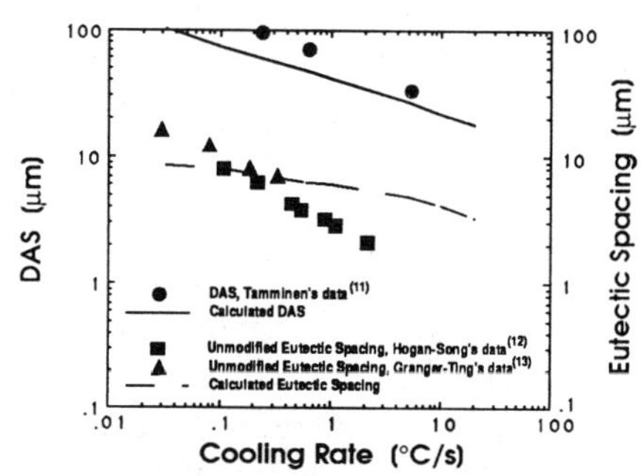

Fig. 2 Variation of secondary dendrite arm spacing and eutectic spacing with cooling rate. Experimental data of various investigators (11-13) are also shown for comparison.

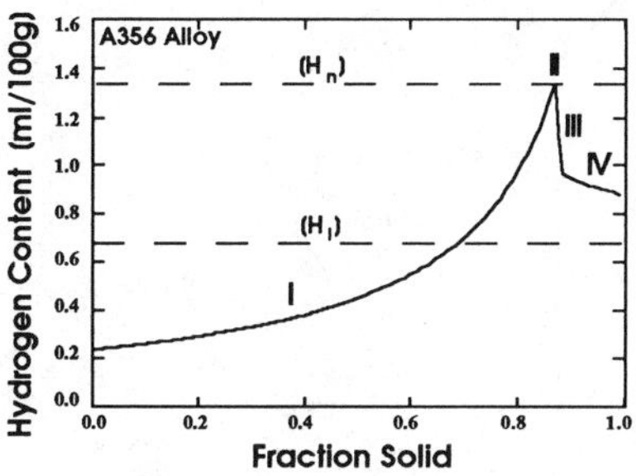

Fig. 3 Hydrogen concentration in the interdendritic liquid as a function of fraction solid.

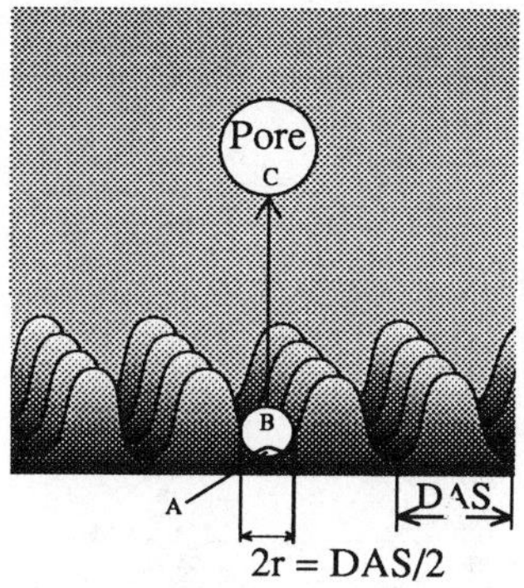

Fig. 4 Schematic illustration of pore nucleation at the root of a dendrite.
A: Nucleation of a pore at the dendrite arm.
B: Growth of a spherical pore from the dendrite arm.
C: Detachment of the spherical pore from the dendrite arm.

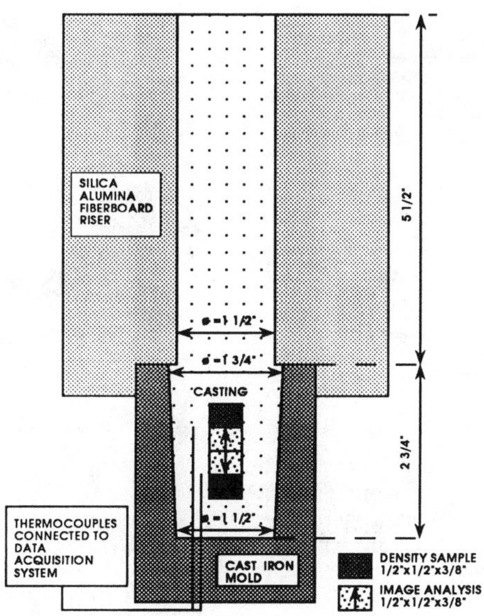

Fig. 5 Mold design and location of test samples in the casting.

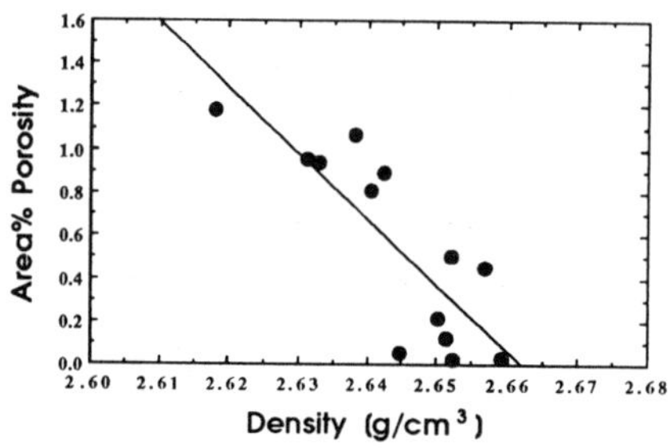

Fig. 6 Relationship between the amount of porosity obtained from density measurements and image analysis.

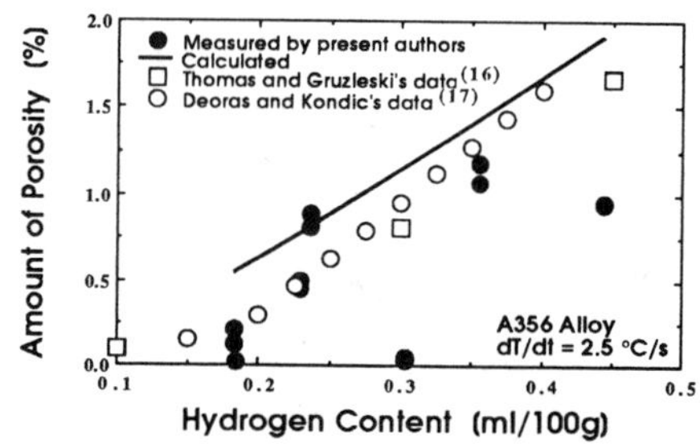

Fig. 7 Measured and calculated amount of porosity as a function of initial hydrogen content.

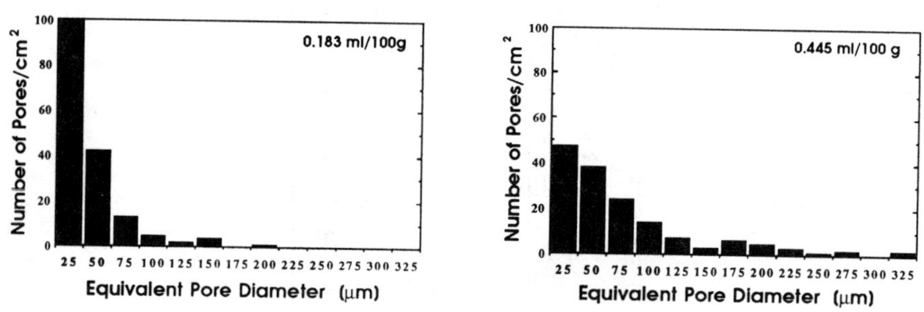

a)

b)

Fig. 8 Pore size distribution obtained by image analysis.
a) Hydrogen content 0.183 ml/100 g.
b) Hydrogen content 0.445 ml/100 g.

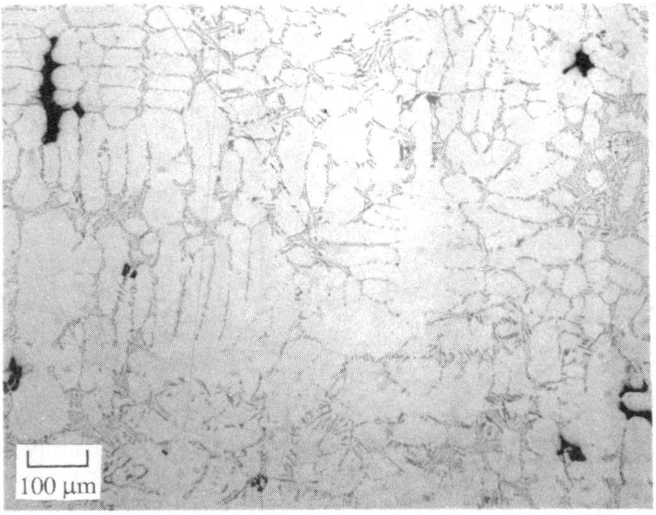

a)

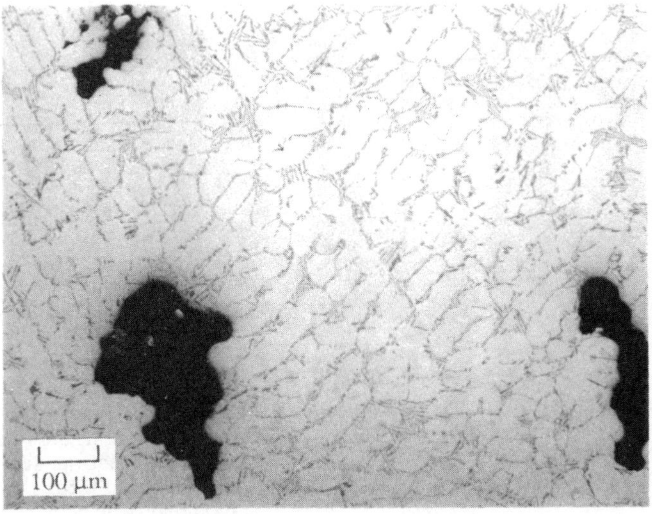

b)

Fig. 9 Micrographs illustrating pore morphology.
a) Hydrogen content 0.183 ml/100 g.
b) Hydrogen content 0.445 ml/100 g.

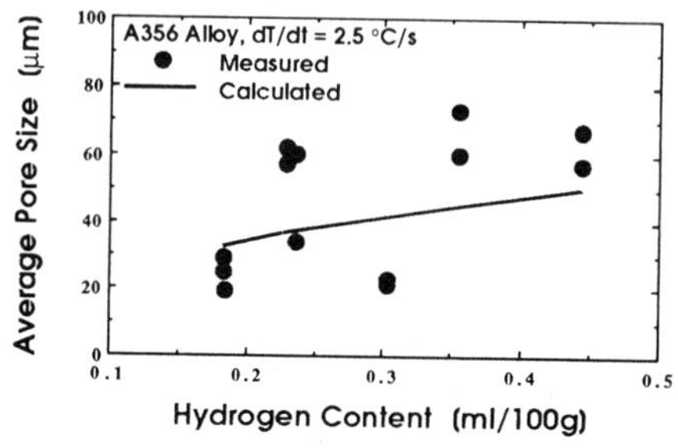

Fig.10 Measured and calculated equivalent pore diameter as a function of hydrogen content.

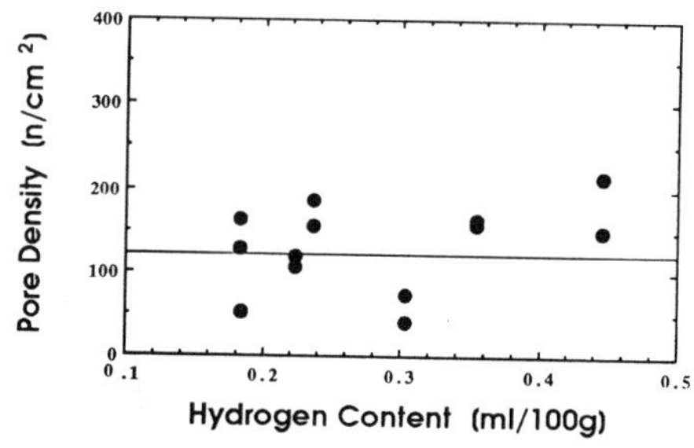

Fig. 11 Measured pore density as a function of initial hydrogen content.

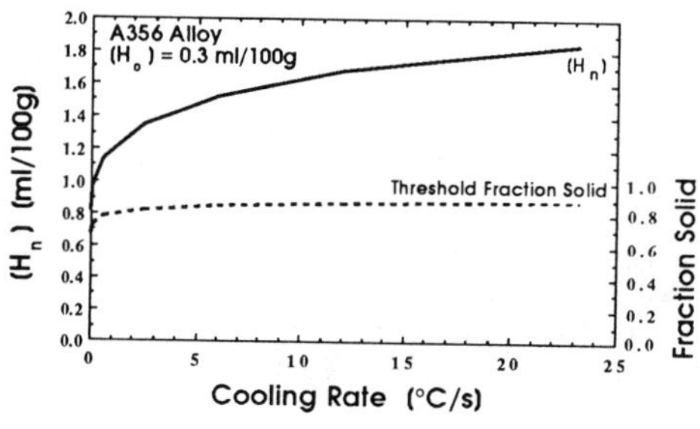

Fig. 12 Calculated maximum hydrogen concentration in the liquid, $[H_n]$, and threshold fraction solid at which pores begin to form as a function of cooling rate.

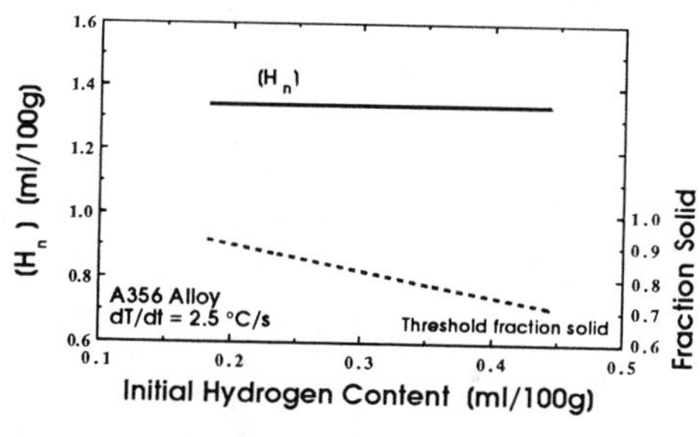

Fig. 13 Calculated maximum hydrogen concentration in the liquid, H_n, and threshold fraction solid at which pores begin to form as a function of initial hydrogen content.

Globularization of Al-8% Cu dendritic structures for rheocast slurries production

A. Damasco and M.H. Robert
Mechanical Engineering Faculty, State University of Campinas, DEF/FEM/UNICAMP, SP, Brazil

Abstract

The author's previously developed method to produce rheocast slurries by heat treatment above solidus temperature, of deformed dendritic structures, was applied to Al-8% Cu alloy.

The material was isothermally treated in different temperature and time conditions, from as cast or deformed structures.

Results show that roundness and average globular solid phase diameter increase as treatment time increases; high previous deformation leads to more perfect rheocast structures. Temperature influences kinetics of involved mechanisms but not their nature.

It was verified that dendrite-to-rounded phase transition occurs by either recrystallization followed by the newly formed grains detachment to the neighbour liquid phase, or natural dendritic coarsening, depending on the previous deformation degree.

Evolution of microsegregation profiles was also analysed. Homogeneous Cu distribution within rounded solid phase and increase in this element content in the liquid phase were verified.

1. Introduction

Rheocasting of metals has been studied since the 70's when M.I.T. group published some promising results, (1,2,3) pointing out the possibility of the production of metallic slurries with particular viscous behaviour, which they named rheocast material. The production process developed then required severe stirring of liquid metal during specific stages of its solidification, in specially built equipments (rheocaster).

The so obtained slurry has globular solid phases, instead of the conventional dendritic structure, immersed in liquid. This particular solid phase morphology is the one responsible for the thixotropic, non Newtonian viscous flow behaviour of the rheocast slurry.

M.I.T. results were followed by others form different research groups (4,5) which tried to improve slurry production procedure besides explaining the mechanisms involved in the slurry structure formation.

Relating to the production process, viscometer-type rheocasters were developed and successfully utilized for even the continuous slurry production (6).

However, the stirring method to produce rheocast semi-solid metals suffers from inherent limitations: the severe shear required in the metal being stirred, necessary to its structure globularization, leads to early and easy wearing out of rheocaster material, besides the difficult process parameters control.

Robert and Kirkwood (7) show that it is possible to produce rheocast structures obtained by an alternative procedure, meaning the isothermal heat treatment of previously deformed dendritic structures, at temperatures above liquid formation. By controlling process parameters it is possible to control secondary phase melting and globularization of solid phases surrounded by liquid.

Relating to the mechanisms involved in the particular slurry structure formation there is still some controversy and a definitive explanation is still to be given; however, most authors agree that the rounded primary phase is a consequence of dendritic growing crystals degeneration. Vogel (8) suggests that stirring causes dendrite arms bending followed by recrystallization, wetting of grain boundaries and detachment of such grain to the liquid. Kattamis (9) on the other hand, suggests that severe stirring improves mass transportation, accelerating coarsening phenomena.

This work deals with the globularization of Al-8% wt. Cu alloy as cast, and cold deformed structures by isothermal treatment within solidification range. Discussions on the involved phenomena are made to bring contribution to their clarification.

2. Experimental Procedure

Initially it was produced an equiaxial grained ingot from which samples for heat treatment were cut off. Slight stirring was promoted during ingot solidification for grain refinement and macrosegregation minimization.

Part of obtained samples was submitted to 80% height reduction (true strain) by cold squeezing before heat treatment, which was done at 560°C and 600°C, corresponding to a liquid fraction of 0.2 and 0.4 respectively. Treatment times from 5 to 60 minutes were utilized.

After isothermal treatment, samples were water quenched, sectioned and prepared for metallographic observations and microanalysis.

3. Results and Discussions

Obtained results are shown in the following micrographics. Figure 1 shows the evolution of as cast dendritic structure, as treatment time increases, when treated at 600°C.

It can be already observed for 5 minutes of treatment, homogenization within the structure and liquid formation at dendrite and grain boundaries taking place: dendrite arms start to fade away and small pools of liquid replace the original dendrite boundaries. Increasing treatment time makes it clearer the internal homogenization of original grains: dendrite arms can no longer be observed. The structure now is formed by irregular blocks resembling original grains, containing pools of entrapped liquid and surrounded by liquid phase (eutectic phase when solidified).

These isolated unshaped blocks tend to get rounded form as time increases, probably by coarsening and agglomeration plus coalescence mechanisms. The driving force for such transformations is the need of internal surfaces energy reduction.

The structure obtained by isothermally keeping dendritic as cast Al-8% Cu, at temperatures above secondary phase melting temperature, can be characterized as rheocast structure, though the big and irregular size of solid phase besides entrapped liquid, can compromise its rheological properties.

Reduction of original grain size and increase in eutectic phase amount in dendrite boundary should probably lead to smaller and more rounded globular solid phase in the rheocast slurry.

Results show therefore how easy it is to get reasonable rheocast slurries by simply heat treating dendritic structures. Involved mechanisms are secondary phase melting, grain coarsening and homogenization.

Figure 2 shows results obtained for samples previously cold deformed and then treated at 600°C for different times.

It can be clearly observed that the mechanisms involved in the structure modification is quite different from that discussed for the former situation. The resulted structure is now formed by small, rounded and almost homogeneously sized primary phase. It can also be observed no entrapped liquid within solid phase.

From the micrographics it is possible to observe recrystallization and liquid penetration in the newly formed, free of microsegregation grain boundaries. It occurs when $\gamma gb > 2\gamma SL$, which is probably the situation when 80% true strain is imposed to the material.

The resulting structure is quite homogeneous, relating to shape and size of primary solid phase. Rare liquid-entrapped pools can be observed, probably due, in this case, to agglomeration of two or more rounded particles.

The transition from dendritic to globular structure occurs in a direct way, there are no unshaped intermediate morphologies. Small equiaxed particles surrounded by liquid phase are rapidly obtained.

Increasing holding time leads to undesirable solid phase growth in the slurry; therefore excess of holding must be avoided.

Concerning the treatment temperature, it was observed the acceleration of globularization phenomena, since diffusion kinetics and liquid amount in the slurry are increased. No influence of temperature on the nature of the involved mechanisms could be observed.

It could be observed qualitatively, however, the influence of cooling rate on the rounded particles surface: it can be degenerated by small dendrites formation when the rheocast slurry is not quenched fast enough. This degeneration can mean the slurry has no longer suitable viscous behaviour.

The microsegregation profiles obtained from different times treated samples clearly show internal homogenization of original grains turned to spherical particles in the rheocast structure, and increase of Cu content in the particles boundary.

Figure 3 shows the evolution of Cu and Al profiles within the structure for different conditions.

It can be observed the homogeneous Cu and Al distribution inside rounded particles of rheocast structure (Fig.3-b1,c) as well as the boundary enrichment on Cu. This result is obtained more rapidly for the previously deformed material, once the structure is also more rapidly changed by recrystallization mechanism.

In a conventional homogenization treatment of Al-8% Cu alloy, Cu diffusion occurs from outer to inner layers of dendrites and grains. In case of additional liquid formation, as treated here, it can have an extra and opposite Cu diffusion to the liquid region, where its solubility is higher, resulting in the observed Cu content increasing in the inter-solid phase region. Such phenomena are the subject of our current studies and we hope in the near future to bring some more light on them.

4. Conclusions

1. Al-8% Cu rheocast slurries can be easily obtained by simply heat treating at temperatures above solidus, originally dendritic material. Treatment time can vary from 5 to 30 min depending mainly on whether the structure was previously deformed or not.

2. Obtained rheocast structures have their quality depending on process parameters as follows:

   * previous deformation - determines the responsible mechanisms for dendrite-to-rheocast evolution. At high deformation, such mechanisms are recrystallization followed by isolation of recrystallized grains which had their boundaries wetted by liquid. The result is small globular solid phase without entrapped liquid. For as cast, non deformed structures, their globularization is due to coarsening phenomena. The result is globular solid phase with dimensions similar to original grains, with regions of entrapped liquid within.

   * Temperature - defines kinetics of transformations. Higher the temperature, faster the globularization.

   * Time - must be carefully determined, once leads to undesirable solid phase growth.

   * Slurry cooling rate - must be small enough to prevent rounded phase surface degeneration.

3. Obtained rheocast structures have homogeneous Cu and Al content distributions within the globular solid phase, while Cu concentration is increased in grain boundary. Homogenization is faster when recrystallization instead of coarsening is the acting mechanisms.

5. Acknowledgments

The authors would like to thank FINEP - Financiadora de Estudos e Projetos - for its financial support.

6. References

1. Metz, S.A., Flemings, M.C. - "AFS Trans. vol 78, 1970, p.453-460.

2. Spencer, D.B., Mehrabian, R., Flemings, M.C. - "Metall. Trans. vol 3, July 1972, p.1925-1931.
3. Flemings, M.C., Rick, R.G., Young, K.P. - Mat. Science and Engineering, vol 25, 1976, p.103-117.
4. Mehrabian, R. et. al. - Metals Abstract, June 1976, p.710-715.
5. Vogel, A. - Metals and Materials, Feb. 1979, p.30-32.
6. Riek, R.G. et. al. - AFS Trans. vol 83, 1975, p.25-30.
7. Robert, M.H., Kirkwood, D.H. - Proceedings of Solidification Processing Conference, Sheffield, September 1987, p.373-376.
8. Vogel, A. - Metals Science, December 1978, p.576-578.
9. Kattamis, T.Z. et.al. - Journ. Mat. Science, 1972, p.888-894.

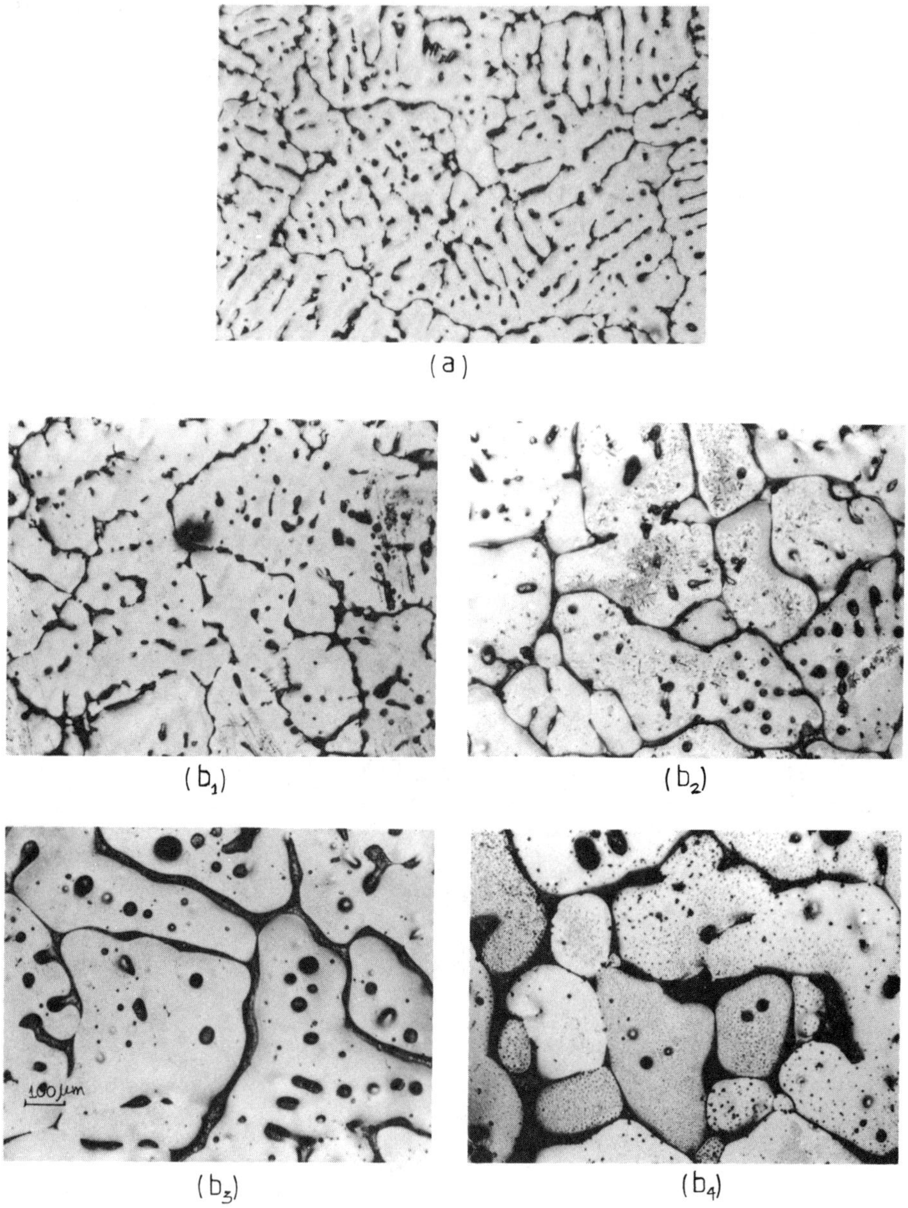

Fig.1. Microstructures of Al-8% Cu (a) as cast and treated at 600°C for (b_1) 5 min, (b_2)15 min , (b_3) 30 min, (b_4) 60 min. Magnification 65x.

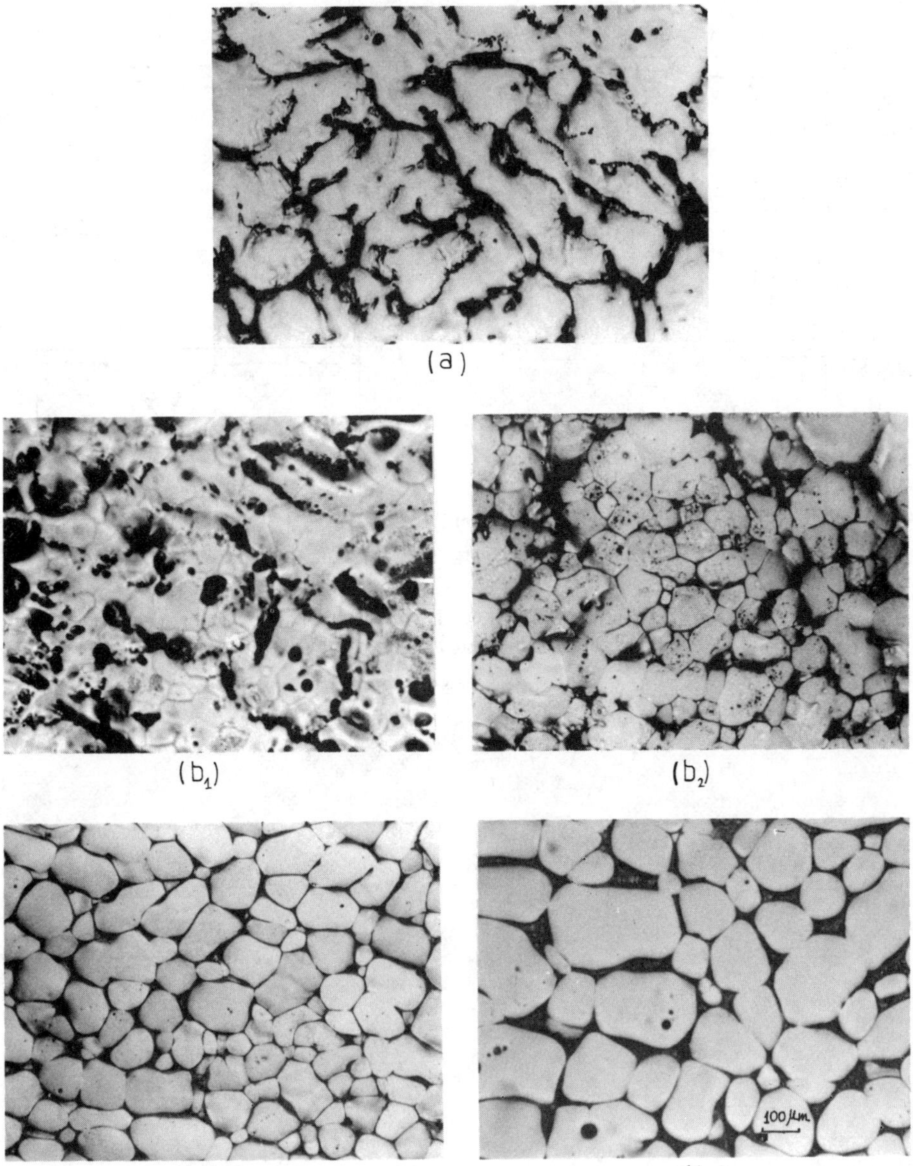

Fig.2. Microstructures of Al-8% Cu (a) cold deformed and treated at 600°C for (b_1) 5 min, (b_2) 10 min, (b_3) 15 min, (b_4) 60 min. Magnification 65x.

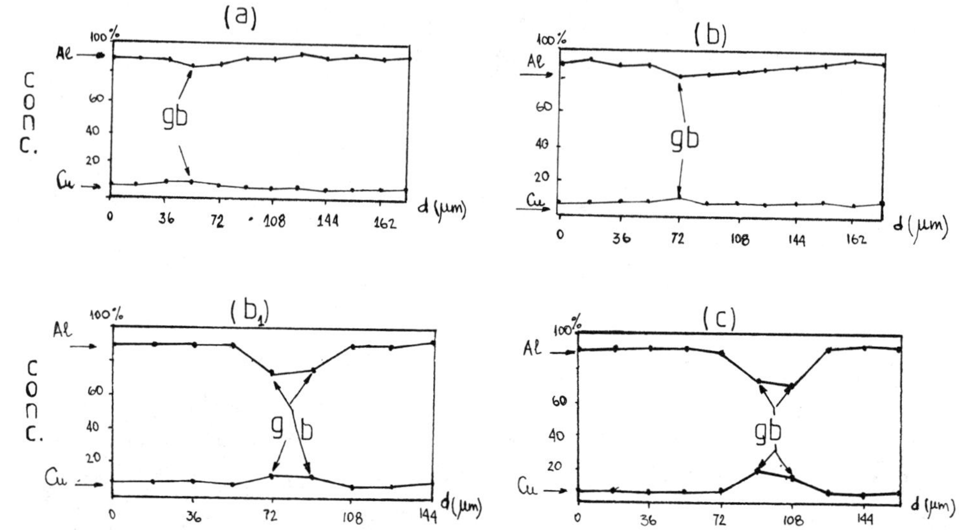

Fig.3. Al and Cu content profiles in (a) as cast dendritic structure, (b) after 10 min at 600°C; (b_1) after 60 min at 600°C; (c) after 10 min at 600°C from previously 80% deformed structure.

Structural design trends for magnesium die casting

T.J. Ruden
Norsk Hydro Magnesium, Southfield, Michigan, U.S.A.

Abstract

Structural uses of magnesium have traditionally been based on magnesium's low weight relative to its strength and rigidity properties. Historically, the development of structural applications has been driven primarily by necessity in wartime or by simple cost comparisons during peacetime.

In recent years, the status of magnesium as a structural material had advanced dramatically due to technological advances in both production and processing and from increases in production capacity. Today, magnesium die castings are considered "design efficient" compared to many other materials. New and enhanced principles now consolidate the various methods previously used by the aluminum, composites (plastic and metal based) and even steel industries resulting in components that are mechanically appropriate and cost effective for a wide variety of applications.

Keywords

Magnesium, die casting, design, light metals, fluxless melting, high purity alloys, high ductility alloys, structural materials, component consolidation, cold chamber, hot chamber, Gravimetric Metering Technology, fasteners, Norsk Hydro.

Historical Review

Since the early 1800's, magnesium has been known for its light weight and dimensional stability. The metal's use, however, was not unlike that of other materials for lightweight structural applications. It was not until World War II that the use of magnesium for structural applications increased dramatically (Fig. 1). There was a great demand for lightweight structural materials, so producing magnesium and fabricating it for high priority military equipment was more important than long-term performance considerations. As a result, little effort was made in magnesium research and development.

At the end of World War II, there was no solid knowledge base to support an effective marketing program for magnesium (1). Magnesium was known by a relatively small group of people and many considered its use limited in the post-war era. What's more, as production of magnesium diminished sharply, the cost became prohibitive except for special applications.

Not only did magnesium have "exposure" problems, research had not addressed two important issues: atmospheric and galvanic corrosion. The potential for galvanic corrosion when two dissimilar metals are bolted together had not been resolved. General corrosion from either salts or continuous weathering compounded the problem. As a result, magnesium for structural use was gradually abandoned and was considered only when weight savings was of primary importance (2).

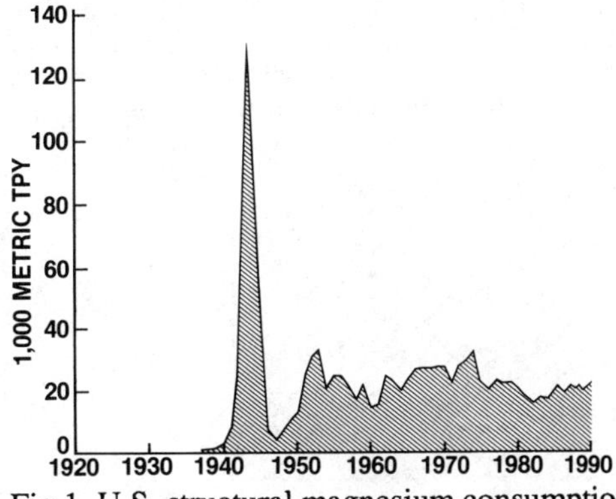

Fig.1. U.S. structural magnesium consumption.

Recent Developments

Fluxless Melting

In the molten state, magnesium alloys react with oxygen in the surrounding air and will burn unless some type of protective cover is maintained over the material. This leads to castings of poor quality and also promotes excessive melt losses (3). Protective covers provide the dual function of preventing oxidization and reducing vaporization, the immediate cause of burning.

The most common type of flux previously used was a salt-based flux cover. While providing adequate surface protection, flux has a tendency to mix with the melt, and unless extreme care is taken, may become trapped into the casting. Flux salts are hydroscopic and acidic. Any flux at the surface of the casting, or exposed by subsequent machining, creates active corrosion sites. Flux salt vapors also create a corrosive atmosphere for shop equipment.

A developed alternative to flux cover is a protective gas atmosphere which acts to modify the oxide film on the melt surface. Sulfur dioxide (SO_2) acts in this way, but is also toxic and corrosive in the shop environment. The most effective cover gas, and the one now universally adopted by die casters (Fig. 2), is a dilute mixture of sulfur hexaflouride (SF_6) in dry air or air and carbon dioxide (CO_2) (4).

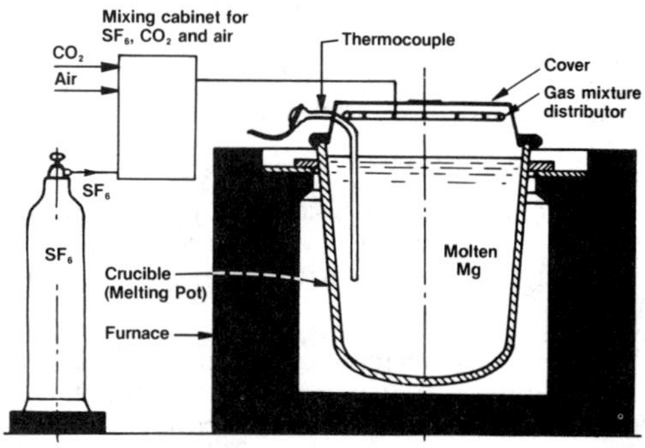

Fig. 2. Typical fluxless melting setup.

High Purity Alloys.

Complementing the fluxless melting process is the commercial development in the early 1980's of the high purity Mg-Al-Zn series alloys, the foundation for which was provided by J. D. Hanawalt (5). The basis for these alloys is the direct limitation of metallic impurity content (Fe, Ni, Cu) along with a controlled content of manganese.

These alloys have superior resistance to salt water corrosion. Although the potential for galvanic corrosion still exists, in many cases it may be limited and controlled simply by careful design of the assembly.

High Ductility Alloys

A more noted recent example of magnesium innovation is the development of alloys with higher ductility properties. These new alloys, AM20 and AM50, were developed to be utilized in die cast components that require a high degree of bending and flexibility during a failure mode while in operation. The primary components driving the development of these alloys came from the automotive industry. Integral front seats, such as those provided on the Mercedes-Benz 300/500SL Roadsters, and structural instrument panels, like those in production on several Audi passenger cars, are just two examples of components that require both high strength and more flexibility than other comparable materials. Mechanical property data for these and the other most common alloys is provided in Figures 3 and 4. Note that for AM20 and AM50, the strength of the material is lower than that of the most common alloy, AZ91D. This can be attributed directly to the aluminum content in the alloy. As a rule and general trend, the lower the aluminum content, the lower the strength but the higher the elongation (6).

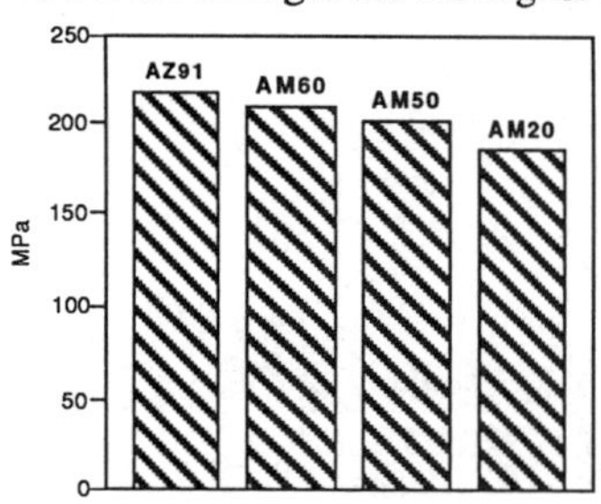

Fig. 3. Tensile strength comparison.

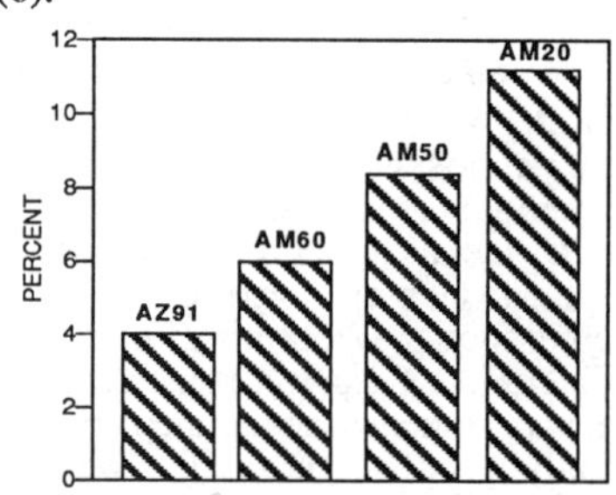

Fig. 4. Elongation comparison.

Die Casting Processes

In the magnesium pressure die casting industry there are basically two methods of producing components. The processes are hot chamber and cold chamber die casting. Magnesium alloys are considered flexible in that they may be run in either type of machine. In fact, today there are an even number of machines within the industry utilizing each process. In comparison, aluminum alloys are restricted to the cold chamber process and zinc based alloys may only be effectively cast in hot chamber machines.

Cold Chamber Developments

Since 1986, Gravimetric Metering Technology has proven to be successful in enhancing the cold chamber die casting process. Previously, hand ladling was the common practice. It was slow and led to a great deal of inconsistency, both in metal flow and cycle time. There were significant temperature changes between the furnace and actual casting phase of the process.

The development of Gravimetric Metering Technology removed virtually all inconsistencies within the cold chamber process. Consistent cycle times were established and placed in accord with the process controllers and SPC programs. Temperatures were held at a more consistent level, and uniformity of the die castings was improved. This technology leads to less overall scrap in the process, improves the efficiency of die casting and provides for more consistent and reliable castings (7) . Figures 5 and 6 illustrate the differences between conventional and gravimetric metal transfer system, respectively.

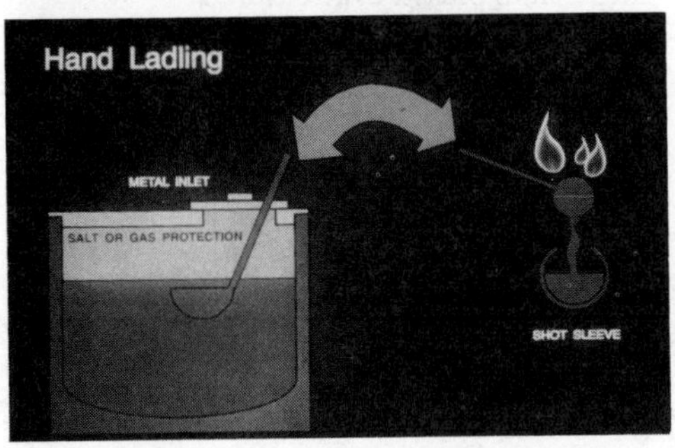

Fig. 5. Conventional metal transfer method.

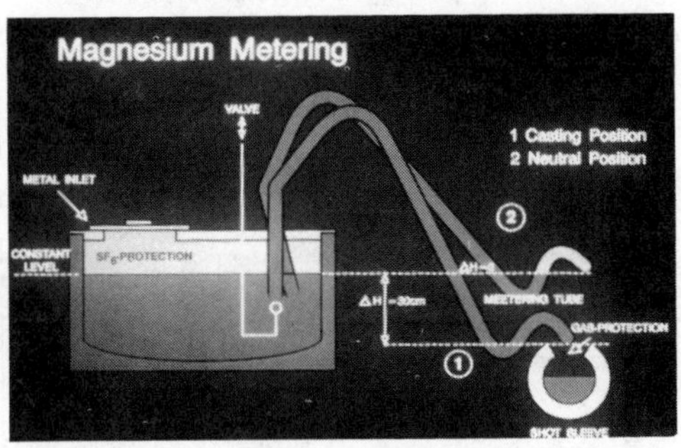

Fig. 6. Gravimetric metal transfer system.

Design Advantages of Magnesium Die Casting

Table 1 lists generally accepted design parameters for several materials commonly used in structural applications. The data are not intended to be used on all components, but rather, as a comparative guide prior to the design process.

TABLE 1
Design Parameters of Common Structural Materials
All Values in Millimeters

	MAGNESIUM	PLASTIC	ALUMINUM
Dimensional	±.06	.12	.12
Draft Angle	0-1.5°	1.5 - 3.0	2.0 - 3.0
Wall Thickness	1.5 - 2.5	2.0 - 3.0	2.5 - 3.5

Tolerances

Magnesium die castings may be easily designed with closer tolerances than many other competitive materials. The typical dimensional tolerance ("rule of thumb") of a magnesium die casting may be expressed as +/-.064mm. Or, in another way, .001mm/mm (8). In some cases, die castings can achieve tighter tolerances up to +/-.005mm without the requirement for machining or other secondary operations (9).

Draft Angles

Magnesium does not have the affinity for Fe in the tool steel (does not stick to the tool upon release) and provides for consistent and uniform shrinkage rates. Consequently, it is very common to produce many shapes including holes, ribs and walls of die castings with either near-zero or zero degree draft.

Wall Thickness

To fully utilize the many benefits of magnesium, die castings are generally cast with much thinner walls than plastics, composites (either metal or plastic based) or other non-ferrous alloys. In some types of designs, magnesium has been cast with wall sections as thin as .635mm thick without needing special equipment or tooling. Other materials may require more section material because of either casting/flow requirements or to uphold certain limits of mechanical and functional requirements. More detailed information on design is provided for in the following figures. The information presented, again, is intended to serve as an initial step in the design process (10).

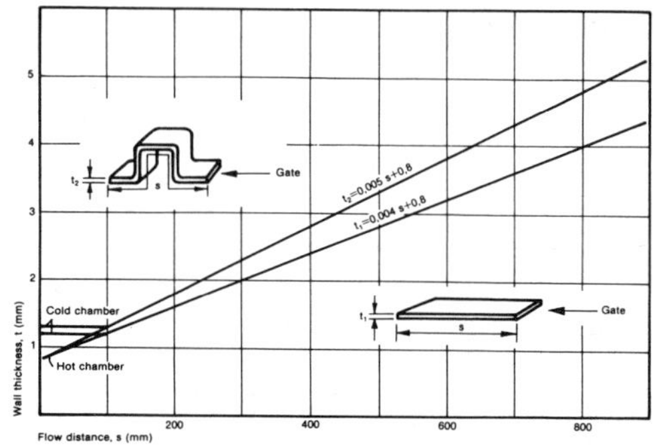

Fig. 7. Minimum section thickness for magnesium pressure die castings.

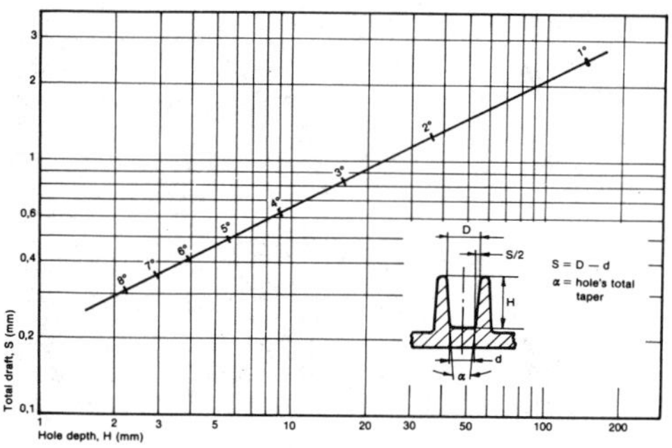

Fig. 8. Diagram of draft in cored holes for magnesium pressure die castings.

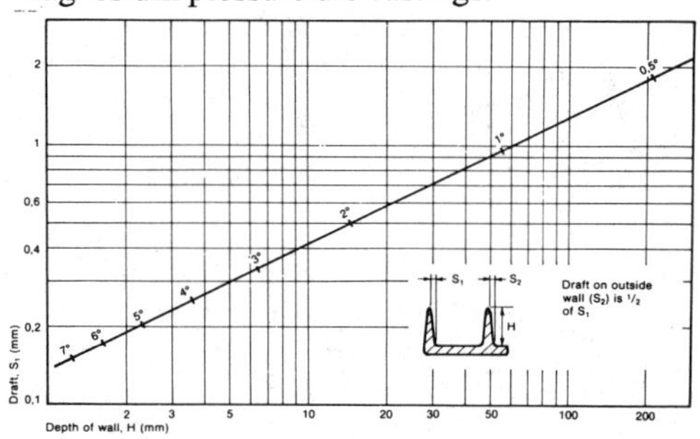

Fig. 9. Diagram of draft on inside walls (S_1) for magnesium pressure die castings.

Design Features

Component Consolidation

One of magnesium's most interesting features is the ability of the material to be cast into a single component, thereby replacing an alternate design made from several parts. The economic value of this type of designing is self evident and should be practiced more often by designers and engineers. The potential benefits to be realized include (a) improved reliability, (b) lower assembly costs, (c) less dunnage and scrap, (d) lower intangible costs, (e) improved warehousing, and (f) homogeneous designs.

Figure 10 outlines how a basic two-piece assembly of steel and plastic (an automotive ashtray door) can be designed for a one piece magnesium die casting.

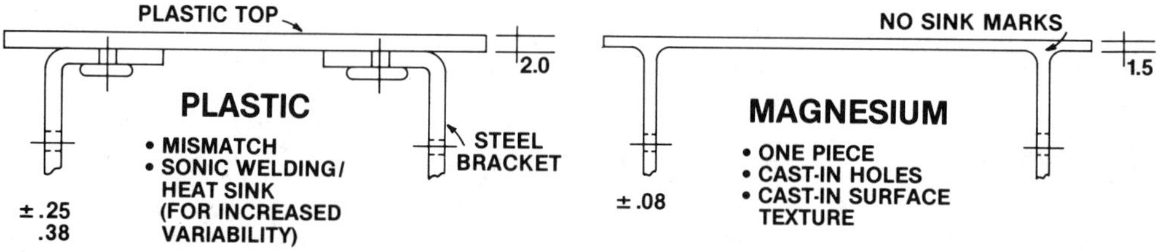

Fig. 10. Designs for automotive ashtray door.

Effective Use of Fasteners

Fastener and connection designs are easily fabricated with magnesium die castings. Threads may be easily tapped into bosses with cored holes to accommodate traditional fasteners. Self threading fasteners have also seen an increased use in recent years. The primary benefit in using self threading fasteners is their ability to reduce secondary machining and chip removal operations while still providing for a sound, bolted connection.

Several figures are presented to familiarize the designer with characteristics of fastener design. It should be noted that much credit for fastener design should go to T. B. Cameron who, in 1986, published the first major work in this area (11).

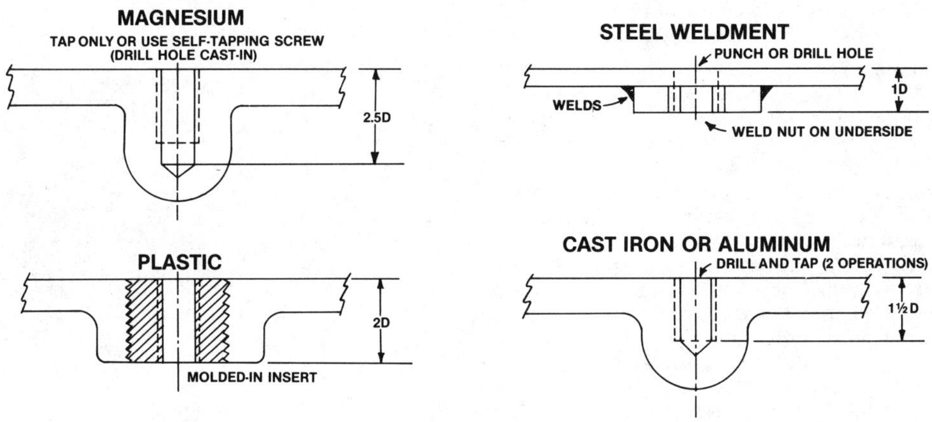

Fig. 11. Comparison of fastener designs.

Typical Sections

A wide variety of shapes and sizes of magnesium die castings exist today. To cover all detail in regard to this area of information would be time consuming and unproductive. In a general sense, then, it is recommended that minimum values of mechanical data be strictly adhered to. It is also recommended that each design be warranted under its own merits, and further, the proper design factors be incorporated within the component.

To illustrate some basic characteristics of magnesium die castings, the following diagrams have been developed. Again, these diagrams should be treated as an aid, not as definitive design criteria.

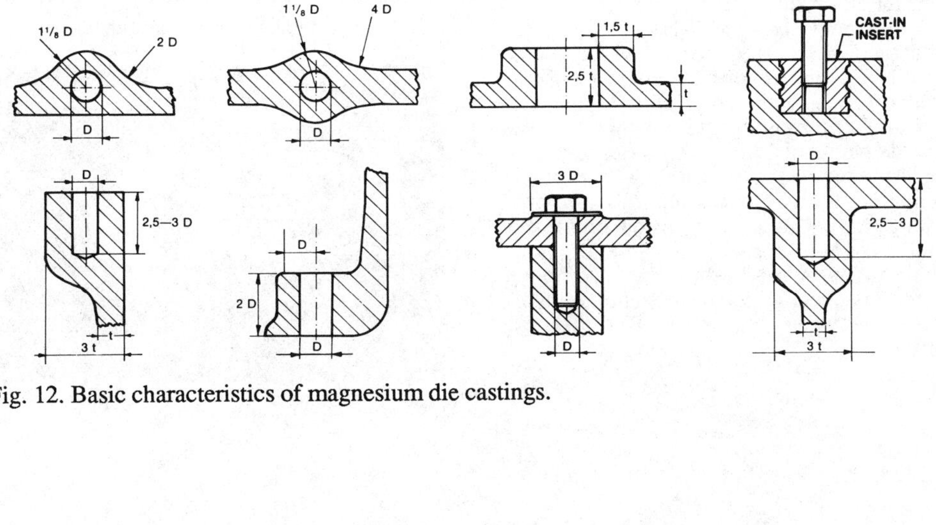

Fig. 12. Basic characteristics of magnesium die castings.

Case Studies

Automotive Seating

As stated earlier, Mercedes-Benz now utilizes high ductility alloys, AM20 and AM50, in its latest seating application (Fig. 13). Customer safety and comfort were the key determinants in developing this highly visible application. Magnesium alloys also provide for low weight (8.4 kg) and component consolidation with only five die castings per seat. A comparable steel seat design would weigh an estimated 35 kg and require between 20 to 30 stampings and weldments.

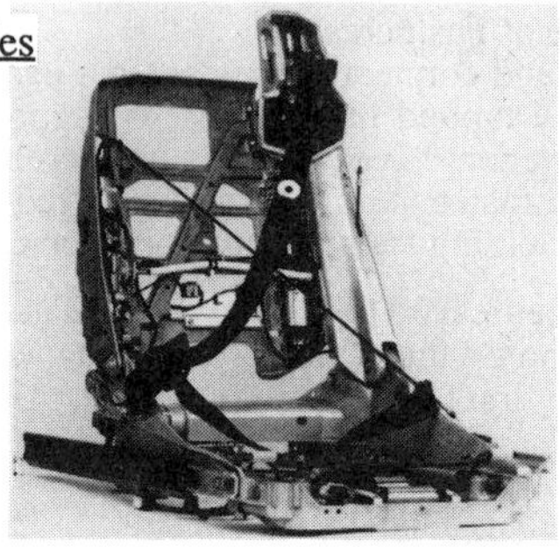

Fig. 13. Mercedes-Benz seat frame.

Instrument Panel

Automotive engineers at Audi recently developed a major structural cross-car beam assembly (Fig.14). Not only does the part weigh slightly over 4 kg, but it requires a high degree of both strength and ductility in order to protect occupants during a crash situation. After reviewing many materials, AM50 alloy was chosen to meet these requirements. In the development process, engineers soon discovered that magnesium could indeed be cast into very large and complex shapes with average wall thicknesses of 2.5mm. Other major benefits include opportunities in component consolidation and dimensional tolerances. From end to end, the panel is 1,450mm with a tolerance of +/- 1.0mm (12).

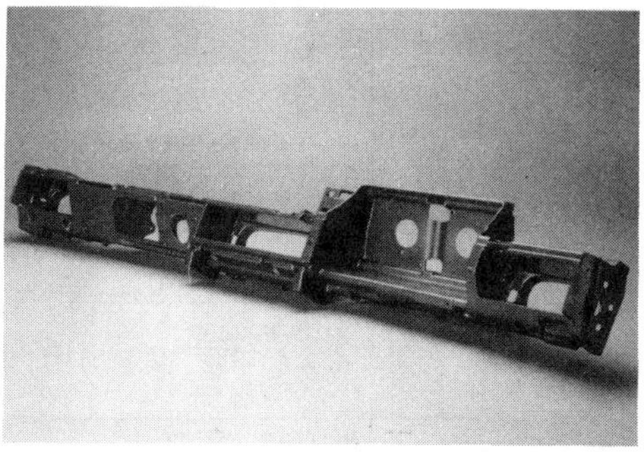

Fig. 14. Audi instrument panel.

4-Wheel Drive Transfer Case

Ford Motor has utilized magnesium on transfer case technology for over 4 years. The first transfer case was introduced in 1986 by Borg Warner. The most recent addition to the family of transfer cases is presented by the Dana Corporation (Fig.15). This case is currently used on the Aerostar Van program where it contributes to vehicle weight savings of over 25 percent.

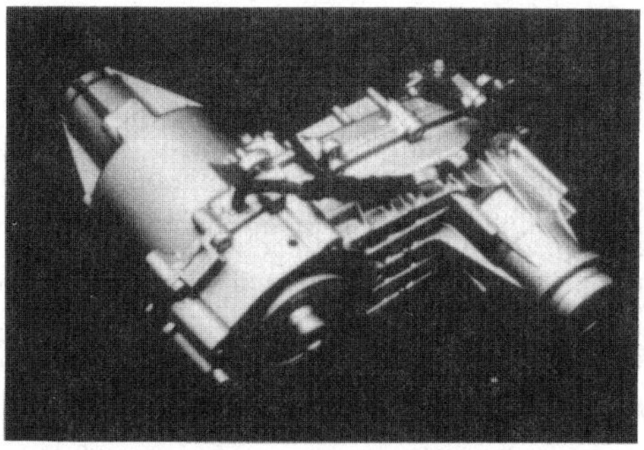

Fig. 15. 4-Wheel drive transfer case.

Transmission Clutch Housing

For 8 years now, AZ91D alloy has been used by Ford Motor Company for manual transmission clutch housings on its Ranger and Bronco series trucks (Fig. 16). Many issues regarding atmospheric and galvanic corrosion have been resolved. Salt water corrosion performance has proved out very well compared to many comparable aluminum castings (13). Galvanic corrosion, on the other hand, has been limited to a minimal amount. This is due primarily to the simple addition of coated fasteners and intermediate shims and gaskets between mating parts of dissimilar metals.

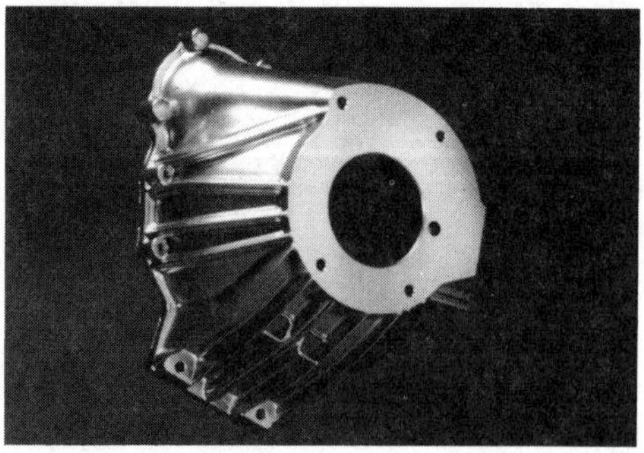

Fig. 16. Transmission clutch housing.

Computer Housing

The NeXT computer housing (Fig. 17) is made entirely from magnesium die castings. AZ91D magnesium alloy was chosen over a high heat thermoplastic, ABS, in order to provide long-lasting strength and durability to the assembly. Magnesium provides additional benefits in that walls are generally cast thinner than an ABS plastic without sink marks or warpage. Being a metal, magnesium naturally provides for EMI/RFI shielding without requiring special paints or coatings.

Fig. 17. Computer CPU housing.

Conclusion

Magnesium has been used in a wide variety of applications, yet has not received the research, marketing and product acceptance recognition it justly deserves.

To realize the unique characteristics and benefits of magnesium, many product and process improvements have been made in magnesium alloys over the last few years. These improvements have, thus far, significantly changed the way magnesium is perceived by many engineers and product planners. The result has been a number of new and different applications and a renewed consideration of magnesium as a structural material. The growth potential for the lightweight metal appears unlimited.

Certainly, not all the work has been done. The industry is just beginning. In order to take full advantage of magnesium, more work must be done in the areas of research and development to enhance its acceptance by industry and enable new products to be developed which will have superior performance and cost-effectiveness.

References

1. Mezoff, J. G., "Structural Magnesium - Prologue and Challenge", Keynote Address for Watertown Arsenal, June 16, 1987.
2. Ibid.
3. House, S. E., and Waltrip, J. S., "Safe Handling of Magnesium Alloys", SAE Technical Paper 900786, 1990.
4. Interview with David Hawke, Materials Specialist, Norsk Hydro Magnesium, February 19, 1990.
5. Hanawalt, J. D., et al., "Corrosion Studies of Magnesium and its Alloys", Trans. Am. Inst., Mining Met. Eng., 147, pp. 273-299, 1942.
6. Albright, D. L., et al., "Properties of Die Cast Magnesium Alloys of Varying Aluminum Content", SAE Technical Paper 900792, 1990.
7. Hustoft, O. M., and Estergaard, E. E., "Gravity Metering for Magnesium Cold Chamber Die Casting", SDCE International Congress, No. G-T87-002, 1987.
8. Ruden, T. J., "Magnesium vs. Plastics: A Comparison for Structural Applications", SDCE International Congress, 1987a.
9. Cole, B. K., "Hot Chamber Magnesium Die Casting in High-Tech Computer Applications", SDCE International Congress, 1987.
10. "Norsk Hydro Magnesium Design Manual", 1986.
11. Cameron, T. B., "Use of Threaded Fasteners in Die Cast Magnesium", SAE Technical Paper 860289, 1986.
12. Pennington, J., "Magnesium Casting Protects Auto Occupants", Modern Metals, Feb. 1990, p. 113.
13. Hawke, D. L., "Field Corrosion Performance of Magnesium Powertrain Components in Light Trucks", SAE Technical Paper 890206, 1989.

Development and production of magnesium wheels

E. Aasen
Fundo a.s., Norway

O. Holta
Norsk Hydro a.s., Magnesium Division, Norway

Abstract

Due to large potentials for saving unsprung weight, several car manufacturers have shown interest in magnesium wheels.

Fundo a.s which is a subsidiary of Norsk Hydro and an aluminium wheel supplier, has therefore started development and a preliminary small scale pilot production of magnesium wheels.

Since car wheels represent a safety component and are normally exposed to severe mechanical loading and corrosive environment, stringent requirements are set for design and production.

In order to ensure that these requirements are met, and simultaneously fully utilize the possibilities of the magnesium alloy so that the wheels will be as light as possible, Fundo has taken its approach of Computer Aided Engineering (CAE) into use also for magnesium wheels.

However, some of the special material data (mechanical properties) which the CAE approach requires, were initially not available for the AZ91 magnesium alloy which Fundo today sees as most relevant for wheel production.

A test program was therefore initiated in order to establish the required material data. Testing was performed on standard test specimens as well as on complete wheels, and comprise static strength, fatigue strength, and corrosion resistance.

In particular, the effect from variations in the production process was considered to be of great importance.

So far, the results have shown that grain size and casting defects have considerable influence on static as well as fatigue strength.

A magnesium wheel with geometry identical to an aluminium wheel would have a weight 2/3 that of the aluminium wheel. However, due to magnesium's somewhat lower mechanical properties, some material has to be added. In practice, a magnesium wheel will normally end up with a weight which is 25-30% lower than one in aluminium.

The AZ91 high purity alloy has base metal corrosion properties similar to aluminium alloys, but certain measures have to be taken to avoid galvanic corrosion due to contact with steel and iron.

A conclusion is that magnesium seems to be well suited for wheels even if this is a safety component which is exposed to severe mechanical loading and a corrosive environment. And with the qualification for wheels, there should also be a market for magnesium in a number of other structural applications where low weight is of importance.

1. Introduction

Even if magnesium wheels have been on the racing car market for many years, their use in series production cars has, for several reasons, been practically zero.

However, in the last few years there has beeen a trend to increase wheel and tyre dimensions, and this has resulted in increased unsprung weight which is undesirable for comfort and road handling characteristics. Especially for car models with large wheels, there is thus a need for lighter wheels.

Since magnesium gives potentials for weight saving, and the new high purity alloys give far better corrosion resistance than previous magnesium alloys (Ref. 1), several car manufacturers now show interest in magnesium wheels.

Fundo a.s which is a subsidiary of Norsk Hydro and a supplier of aluminium wheels to the car industry for several years, has therefore, together with the Magnesium Division of Norsk Hydro, started development and a preliminary small scale pilot production of magnesium wheels.

2. Requirements for design and testing

Wheels are normally exposed to severe mechanical loading and corrosive environment whether made from steel, aluminium or magnesium. They are considered to be a safety component since a structural failure of a wheel while driving on the road may have catastrophic consequences.

Stringent requirements have therefore been established for design and testing of wheels. There exist no separate requirements for magnesium wheels, and those for aluminium are assumed to ensure sufficient safety. These are partly covered by national and international standards, and partly by each car manufacturer's own requirements. Even if there exist numerous different standards and requirements, almost all contain the following main requirements:

- Fatigue strength
- Impact strength
- Corrosion resistance

Fatigue strength is normally tested in a dynamic cornering fatigue machine, fig. 1. This simulates the dynamic loading which a wheel is exposed to under cornering. By a given applied rotating bending moment, the wheel shall have a lifetime exceeding a specified number of cycles. An example of lifetime requirement is indicated in fig 8.

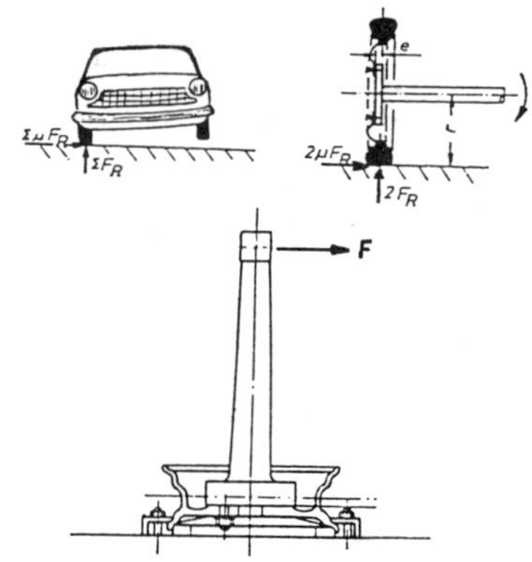

Fig 1. Cornering fatigue testing

Impact strength is normally tested in a machine where a weight falls on to a wheel with tyre from a given drop height, see fig 2. This simulates the case when a car skids sideways into a kerb. The wheel shall withstand the test without losing air, and without any other serious failures.

Corrosion properties are normally tested in salt spray chambers, and the philosophy is that the wheel must have a satisfactory corrosion resistance without any protective layer such as paint or others.

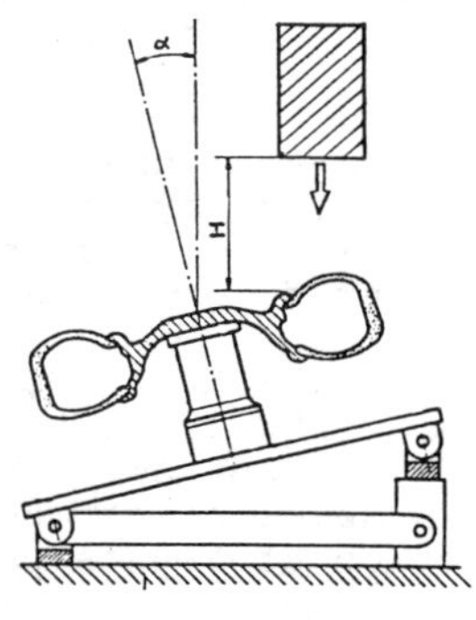

Fig 2. Impact testing

In order to meet the requirements mentioned above, it is important to do so with proper dimensioning of the wheel, and with sufficient mechanical properties of the alloy. Especially important parameters are fatigue strength and toughness (fracture elongation). However, for impact testing, yield strength and ultimate strength may also have an influence.

To meet the requirements of corrosion resistance, the corrosion properties of the alloy are, of course, important, but design measures to avoid contact corrosion must also be considered.

3. Design and dimensioning

As mentioned in the previous section, proper dimensioning of wheels is important to ensure that the requirements for mechanical strength are met. However, this shall also ensure the weight to be as low as possible.

This is normally no simple task since the styling dictates the outside surface of the wheel, and a minimum clearance to brakes dictates the inside boundary.

In order to be able to design really optimum wheels, Fundo has for several years utilized computer aided engineering (CAE) for aluminium wheels. This methodology is now also used for magnesium wheels, and ensures that the properties of magnesium are fully utilized.

Fundo's CAE comprises finite element method (FEM) analyses which are integrated with Computer Aided Design and Manufacturing (CAD/ CAM), Ref. 2.

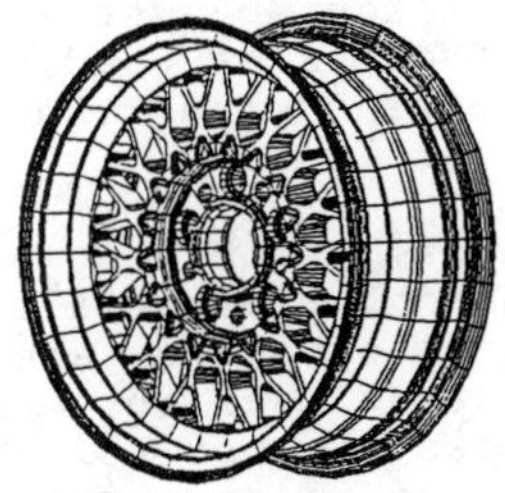

Fig 3. 3-D CAD Model

Fig 3 shows a 3-dimensional CAD-model of a wheel generated by the solid-modeler in the CAE system I-DEAS.

Since I-DEAS is also a FEM pre- and postprocessor, a finite element mesh can, relatively simply, be generated.

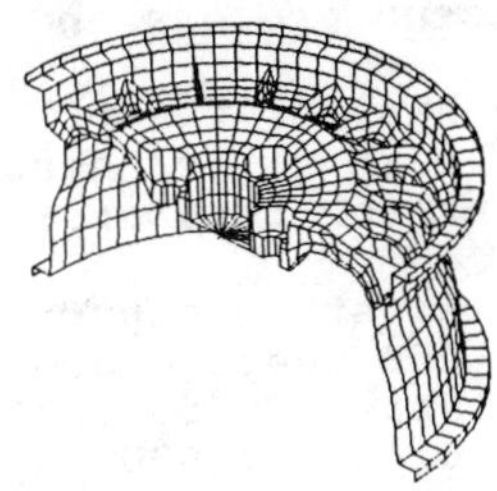

Fig 4. Finite element model

The mesh can consist of either shell or solid elements, dependent on the wheel design.

Fig. 4 shows an example of a finite element model. Due to symmetry in geometry and loading, only one half of the wheel is modelled.

Calculated deformation due to the bending moment in a cornering fatigue machine is shown in fig 5, and the stress level in the highest stressed area (stiffening ribs) of a wheel is shown in fig 6.

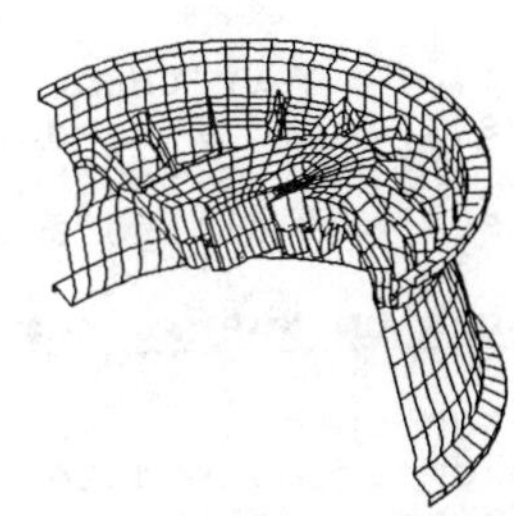

Fig 5. Calculated displacements

Since the peak stresses often are caused by a stress concentration in the transition between a stiff and a flexible area of the wheel, a smoother transition as shown in fig. 7, will often result in lower peak stresses and, simultaneously, a lower weight.

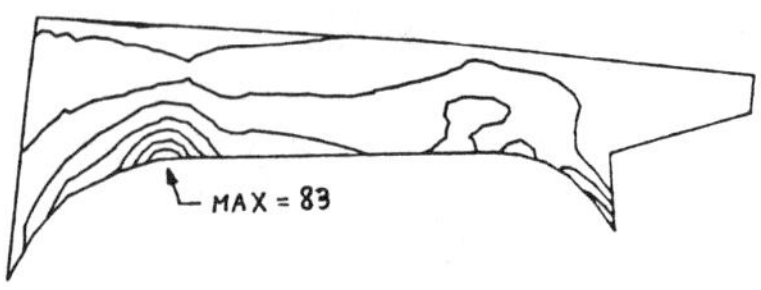

Fig 6. Stresses, initial geometry

The geometry shown in fig 7 normally also results in improved impact strength. High stresses will occur over a larger area, plastic energy will be taken up over a correspondingly larger area, and no local overstressing with resulting cracks will occur.

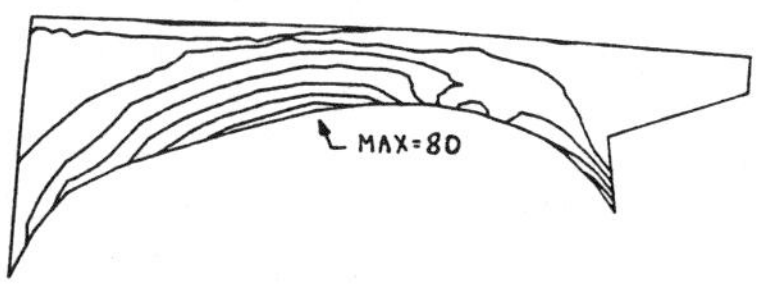

Fig 7. Stresses, optimized geometry

The success of the use of CAE is, however, dependent on available material data, especially for fatigue, and experience in its use on real wheels. For aluminium casting alloys, such data are to a large extent available, but for the magnesium alloys which Fundo today sees as most relevant for wheel production, only limited data has up to now been available.

A program was therefore initiated in order to establish the required material data. A further description of this program is given in section 6.

4. Chosen process and alloy for pilot production

Fundo has, for the pilot production, chosen the low pressure die (LPD) casting technique and the AZ91 high purity alloy. The reasons for this are:

- Existing experience with LPD casting of aluminium wheels.
- LPD casting process for magnesium already developed by Norsk Hydro, Ref. 3.
- Dies for LPD casting less expensive than for high pressure casting.
- The AZ91 alloy in LPD casting can be heat treated and has satisfactory mechanical properties.
- The AZ91 high purity alloy has good corrosion properties.

However, even if the chosen process and alloy are considered as the best suited for magnesium wheels today, it is difficult to say if this will be the case also for the future. Alternative processes and alloys are, therefore, continuously studied.

5. Magnesium wheels compared to aluminium wheels

Since the specific weight of magnesium is 2/3 that of aluminium, 1,8 kg/dm³ compared to 2,7 kg/dm³ , a magnesium wheel which is cast in the same die as an aluminium wheel will have a weight 2/3 that of the aluminium wheel.

Such a magnesium wheel will, however, due to somewhat lower mechanical strength, have lower fatigue life than the aluminium wheel with identical geometry.

Fig 8 shows the results of dynamic cornering fatigue testing of aluminium and magnesium wheels in dimension 5½ x 14 and with identical geometry. The aluminium wheels were cast in the AlSi11 alloy and weight 6,7 kg, whereas the magnesium wheels were in the AZ91-T4 alloy and weight 4,7 kg. M_b is the applied bending moment used in the cornering fatigue machine, see fig 1. N is number of cycles to failure.

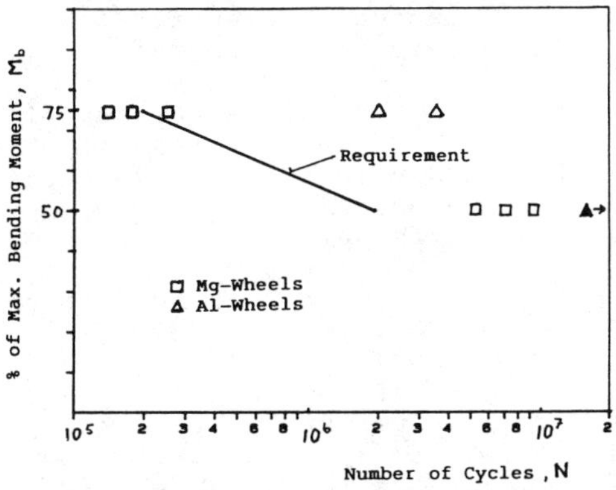

Fig 8. Results from fatigue testing of Al- and Mg-wheels.

As can be seen, the lifetime of the magnesium wheel is considerably lower than that of the aluminium wheel. A strengthening of the magnesium wheel is therefore necessary in order to obtain lifetimes comparable with those of the aluminium wheel. A necessary reduction of stresses in critical areas is about 20 % using the data in fig. 8.

However, stress reduction, and consequently adding of material, is necessary only in local areas which are exposed to high stresses. Even in a fully optimized aluminium wheel, the general stress level is relatively low. The reason for this is that the styling and requirements for good castability requires a minimum amount of material.

The necessary addition of material in a magnesium wheel will be dependent upon the styling, but an average of about 10%is assumed. The magnesium wheel will then end up with a weight of 70-75% of that of an aluminium wheel in similar styling and dimension, i.e. a weight saving of 25-30%.

Looking at the results from fatigue testing of several magnesium wheels, relativly large variations in fatigue lifetime can be observed. Since the worst case (lower boundary of an S-N curve) always has to be considered, it is important to get information about the cause of the variation so that measures can be taken to reduce them.

6. Mechanical properties of the AZ91 alloy in castings

In order to get information about the variances as mentioned above, and in general obtain information about mechanical properties required for CAE, a parametric study based on the AZ91 alloy was initiated.

The study, which is still going on, comprises the following test program:

1. Measurement of static properties as yield strength ($R_{p0.2}$), ultimate tensile strength (R_m) and elongation (A_5) on test specimens taken from wheels. Parameters which are varied are, among others:

    - Alloying elements (use of Si or not)
    - Melt treatment (use of grain refiner or not, type of grain refiner, temperatures in melting and casting furnaces).
    - Different heat treatments.

    Up to now a total of 10 different variations (test series) have been investigated.

2. Fatigue testing on test specimens selected according to results from static testing.

3. Fatigue testing of wheels in order to verify results obtained by test specimens.

Test specimens are further examined and characterized with respect to chemical composition, grain size and porosity.

Simultaneously, the measured data for the AZ91 alloy were compared to data for other alloys.

An example of results from the static testing is shown in fig 9.

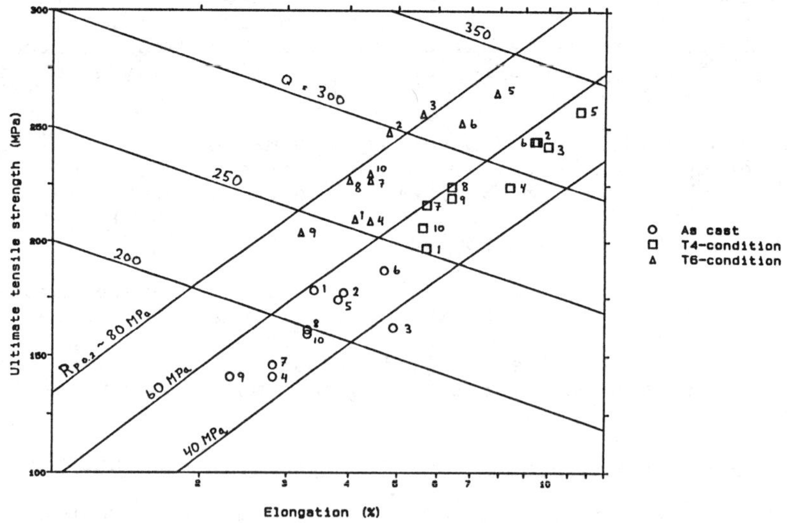

Fig 9. Ultimate tensile strength vs. elongation for samples taken from wheels cast in AZ91 alloys.

The figure shows ultimate tensile strength vs elongation for a number of samples. The numbers beside each measuring point refer to test series numbers. A quality index, Q, as proposed by Drouzy (Ref.4) is also included. This index is defined according to the equation

$$Q = R_m + K_1 \cdot \log A$$

where R_m is the ultimate tensile strength and A the elongation measured in %. The constant K_1 is, under assumption that Q on average should remain independent of ageing time, estimated to 73.

Curves for constant yield stress, $R_{po.2}$, are estimated by least squares fit to the equation

$$R_{po.2} = K_2 \cdot R_m - K_3 \cdot \log A + K_4$$

where K_2, K_3 and K_4 are constants.

A higher Q value indicates a higher quality of the casting which is assumed to be obtained by less porosity, smaller grain size or refined structure due to higher cooling rates.

Solution treatment (move from condition as cast to T4) also considerably improves the quality index of the AZ91 alloy, whereas artificial ageing increases the strength at the expense of ductility.

A comparison of the quality index Q, with measured grain size gives data as shown in fig. 10.

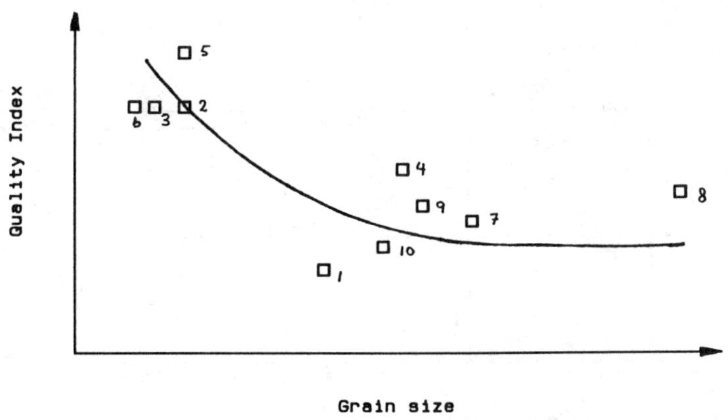

Fig. 10. Quality index, Q, as a function of measured grain size.

The variances in the quality index for a given grain size are caused by different levels of casting defects. The test specimens from series 1 have casting defects above the average, whereas specimens from series 4 and 8 have defects below the average.

The quality index is thus largely dependent upon grain size and casting defects.

The cooling rate is assumed to have a secondary effect on the quality index, but this effect is not included in fig. 10. The reason is that all the test specimens used in the figure are taken from the same location of the wheel, and have experienced same cooling rates.

However, further investigations are necessary to clarify if the reason for higher quality indexes are caused by smaller grain size and less casting defects which usually are an effect of higher cooling rates, or if there are other reasons as e.g a direct relationship between dendrite arm spacings (DAS) and the quality index.

Test specimens fatigue data from test series 1-3 are shown as S-N curves in fig. 11.

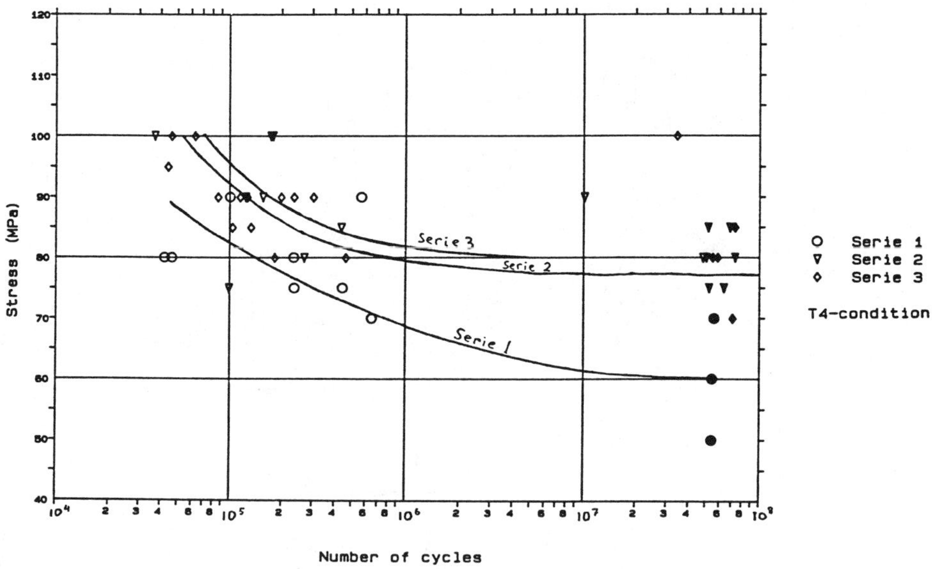

Fig. 11 Test specimen S-N curves for test series 1-3.

Plotting the stress values for infinite number of cycles (fatigue limit), against measured grain size, yields a curve as shown in fig 12.

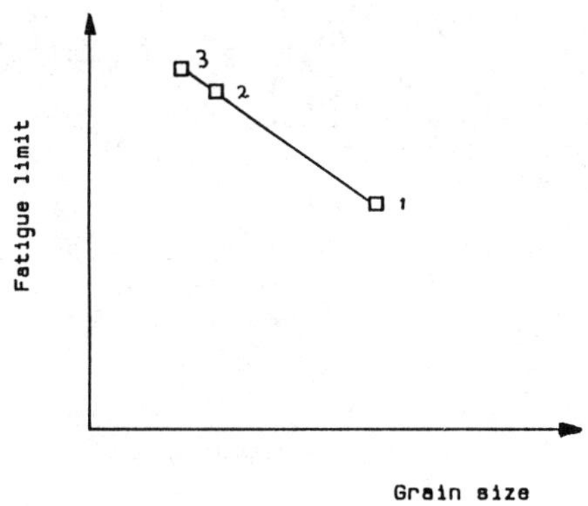

Fig 12. Fatigue limit as a function of grain size.

The fatigue testing has not yet been finished, but it can clearly be seen that, as for the quality-index, the grain size has considerable influence on fatigue strength. Similarly, casting defects will also have a large influence.

The influence from casting defects can be seen from fig. 11 where such defects are the reason for the variances in number of cycles to failure at a given stress level.

Similar variances have been experienced with cornering fatigue testing of complete wheels, see fig 8. Therefore, it is important to accurately control the production process (melt treatment and casting) and have a satisfactory quality control so that the risk of such failures may be eliminated.

Work which is going on within Norsk Hydro is also aimed at improving the production process. This relates to melt treatment as well as to the casting itself.

7. Corrosion protection

High purity magnesium alloys such as the AZ91HP, have been shown to have base metal corrosion properties which are similar to the alloys used for aluminium wheels, see eg. Ref 1.

However, galvanic corrosion may be a problem with magnesium in contact with steel and iron.

The areas in the vicinity of the wheel bolts or wheel nuts may be a problem area in this context as long as aluminium nuts are not used. Aluminium bushings which cover the wheel nut or bolt, have therefore, in special cases, been inserted, see fig. 13.

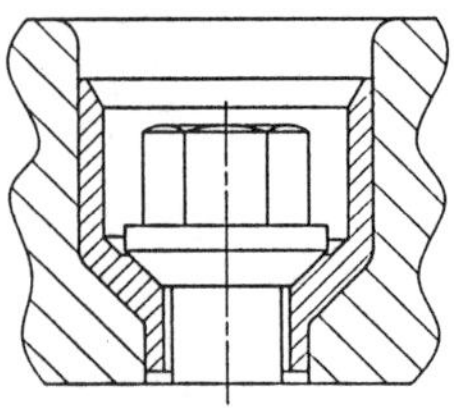

Fig. 13 Aluminium bushing

A similar problem may theoretically occur in the area where the magnesium wheel is in contact with the wheel hub and the brake disc.

To eliminate contact corrosion 100% in this area, one solution can be to isolate the wheel from the steel parts by e.g. plasma-spraying aluminium to the surfaces in contact with steel and iron.

However, since this area is not on the styling side of the wheel, minor corrosion is of no importance as long as it does not influence the functional properties of the wheel.

A proper design of the area where the wheel is in contact with the brake disc (see fig 14) can also avoid severe contact corrosion.

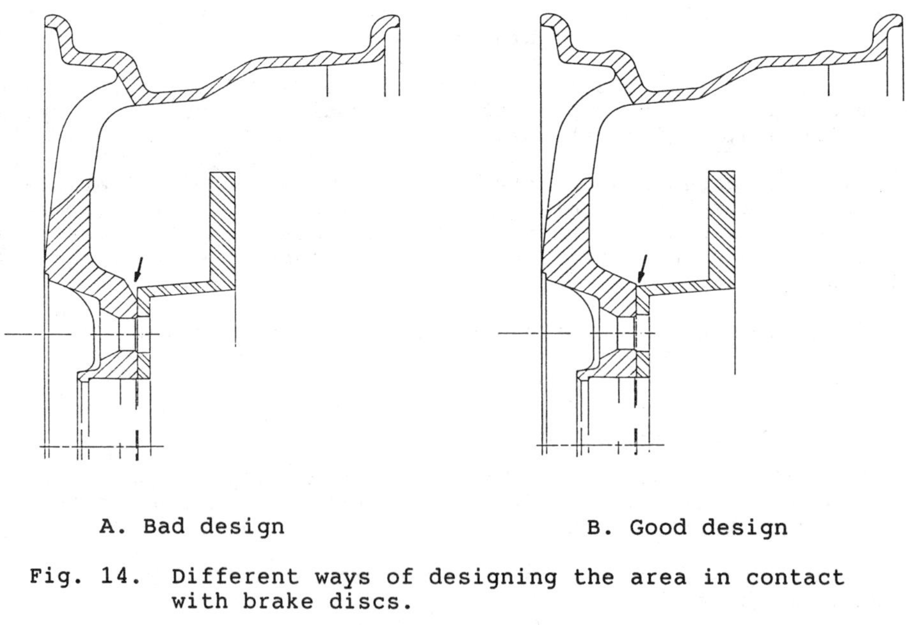

A. Bad design B. Good design

Fig. 14. Different ways of designing the area in contact with brake discs.

However, there is a need to obtain more experience of the corrosion environment under real driving conditions, on winter roads with much salt. Will the conditions be as bad as in a saltspray test chamber, or will centrifugal forces and heat from the brakes avoid salt and water to create corrosive conditions?

8. Conclusions

Even though all the studies on magnesium wheels related to mechanical and corrosion properties have not been completed yet, the following conclusions can be drawn:

- Magnesium wheels bring large potentials for weight savings, 25–30% compared to aluminium wheels.
- Mechanical properties are satisfactory, but good control of the production process is important to avoid too large a variation in the properties. Work which is going on within Norsk Hydro is also aimed at improving the production process. This relates to melt treatment as well as to the casting itself.
- Corrosion properties of high purity alloys are similar to those of aluminium as long as certain measures are taken to avoid galvanic corrosion.

Thus, magnesium is well applicable for wheels which are a safety component and may be one of the components on a car which is exposed to most severe mechanical loading and a corrosive environment. Further, with this qualification, there should also be a market for magnesium in a number of other structural applications where low weight is of importance.

9. References

1. F. Käumle, N.C. Tömmeraas, and J.A. Bolstad, "The Second Generation Magnesium Road Wheel", SAE Technical Paper Series, No. 850420, 1985.

2. E. Aasen, "Development and Production of Aluminium Wheels for the Car Industry. How I-DEAS is used for Integrated CAE and Communication with other CAD/CAM Systems". 1988 I-DEAS™/ CAEDS® Users' Conference, October 10-11, Stevenage, United Kingdom.

3. H. Westengen and O. Holta, "Low Pressure Permanent Mold Casting of Magnesium - Recent Developments", SAE Technical Paper Series, No. 880509, 1988.

4. M. Drouzy, S. Jacob and M. Richard, "Interpretation of Tensile Results by Means of Quality Index and Probable Yield Stress", AFT Int. Cast Metals Journal, vol. 5, no. 2, 1980, pp. 43-50.

Authors' Index